Landscape Evolution: Denudation, Climate and Tectonics over Different Time and Space Scales

The Geological Society of London
Books Editorial Committee

Chief Editor

BOB PANKHURST (UK)

Society Books Editors

JOHN GREGORY (UK)
JIM GRIFFITHS (UK)
JOHN HOWE (UK)
PHIL LEAT (UK)
NICK ROBINS (UK)
JONATHAN TURNER (UK)

Society Books Advisors

MIKE BROWN (USA)
ERIC BUFFETAUT (FRANCE)
JONATHAN CRAIG (ITALY)
RETO GIERÉ (GERMANY)
TOM MCCANN (GERMANY)
DOUG STEAD (CANADA)
RANDELL STEPHENSON (NETHERLANDS)

Geological Society books refereeing procedures

The Society makes every effort to ensure that the scientific and production quality of its books matches that of its journals. Since 1997, all book proposals have been refereed by specialist reviewers as well as by the Society's Books Editorial Committee. If the referees identify weaknesses in the proposal, these must be addressed before the proposal is accepted.

Once the book is accepted, the Society Book Editors ensure that the volume editors follow strict guidelines on refereeing and quality control. We insist that individual papers can only be accepted after satisfactory review by two independent referees. The questions on the review forms are similar to those for *Journal of the Geological Society*. The referees' forms and comments must be available to the Society's Book Editors on request.

Although many of the books result from meetings, the editors are expected to commission papers that were not presented at the meeting to ensure that the book provides a balanced coverage of the subject. Being accepted for presentation at the meeting does not guarantee inclusion in the book.

More information about submitting a proposal and producing a book for the Society can be found on its web site: www.geolsoc.org.uk.

It is recommended that reference to all or part of this book should be made in one of the following ways:

GALLAGHER K., JONES, S. J. & WAINWRIGHT, J. 2008. *Landscape Evolution: Denudation, Climate and Tectonics over Different Time and Space Scales*. Geological Society, London, Special Publications, **296**.

FROSTICK, L., MURPHY, B. & MIDDLETON, R. 2008. Exploring the links between sediment character, bed material erosion and landscape: implications from a laboratory study of gravels and sand–gravel mixtures *In*: GALLAGHER, K., JONES, S. J. & WAINWRIGHT, J. (eds) *Landscape Evolution: Denudation, Climate and Tectonics over Different Time and Space Scales*. Geological Society, London, Special Publications, **296**, 117–127.

GEOLOGICAL SOCIETY SPECIAL PUBLICATION NO. 296

Landscape Evolution: Denudation, Climate and Tectonics over Different Time and Space Scales

EDITED BY

K. GALLAGHER
Université de Rennes, France

S. J. JONES
University of Durham, UK

and

J. WAINWRIGHT
University of Sheffield, UK

2008
Published by
The Geological Society
London

THE GEOLOGICAL SOCIETY

The Geological Society of London (GSL) was founded in 1807. It is the oldest national geological society in the world and the largest in Europe. It was incorporated under Royal Charter in 1825 and is Registered Charity 210161.

The Society is the UK national learned and professional society for geology with a worldwide Fellowship (FGS) of over 9000. The Society has the power to confer Chartered status on suitably qualified Fellows, and about 2000 of the Fellowship carry the title (CGeol). Chartered Geologists may also obtain the equivalent European title, European Geologist (EurGeol). One fifth of the Society's fellowship resides outside the UK. To find out more about the Society, log on to www.geolsoc.org.uk.

The Geological Society Publishing House (Bath, UK) produces the Society's international journals and books, and acts as European distributor for selected publications of the American Association of Petroleum Geologists (AAPG), the Indonesian Petroleum Association (IPA), the Geological Society of America (GSA), the Society for Sedimentary Geology (SEPM) and the Geologists' Association (GA). Joint marketing agreements ensure that GSL Fellows may purchase these societies' publications at a discount. The Society's online bookshop (accessible from www.geolsoc.org.uk) offers secure book purchasing with your credit or debit card.

To find out about joining the Society and benefiting from substantial discounts on publications of GSL and other societies worldwide, consult www.geolsoc.org.uk, or contact the Fellowship Department at: The Geological Society, Burlington House, Piccadilly, London W1J 0BG, UK: Tel. +44 (0)20 7434 9944; Fax +44 (0)20 7439 8975; E-mail: enquiries@geolsoc.org.uk

For information about the Society's meetings, consult *Events* on www.geolsoc.org.uk. To find out more about the Society's Corporate Affiliates Scheme, write to enquiries@geolsoc.org.uk

Published by The Geological Society from:
The Geological Society Publishing House, Unit 7, Brassmill Enterprise Centre, Brassmill Lane, Bath BA1 3JN, UK

(*Orders*: Tel. +44 (0)1225 445046, Fax +44 (0)1225 442836)
Online bookshop: www.geolsoc.org.uk/bookshop

The publishers make no representation, express or implied, with regard to the accuracy of the information contained in this book and cannot accept any legal responsibility for any errors or omissions that may be made.

© The Geological Society of London 2008. All rights reserved. No reproduction, copy or transmission of this publication may be made without written permission. No paragraph of this publication may be reproduced, copied or transmitted save with the provisions of the Copyright Licensing Agency, 90 Tottenham Court Road, London W1P 9HE, UK. Users registered with the Copyright Clearance Center, 27 Congress Street, Salem, MA 01970, USA: the item-fee code for this publication is 0305-8719/08/$15.00.

British Library Cataloguing in Publication Data
A catalogue record for this book is available from the British Library.
ISBN 978-1-86239-250-2

Typeset by Techset Composition Ltd, Salisbury, UK
Printed by Antony Rowe Ltd, Wiltshire, UK

Distributors

North America
For trade and institutional orders:
The Geological Society, c/o AIDC, 82 Winter Sport Lane, Williston, VT 05495, USA
Orders: Tel. +1 800-972-9892
            Fax +1 802-864-7626
            E-mail: gsl.orders@aidcvt.com

For individual and corporate orders:
AAPG Bookstore, PO Box 979, Tulsa, OK 74101-0979, USA
Orders: Tel. +1 918-584-2555
            Fax +1 918-560-2652
            E-mail: bookstore@aapg.org
            Website http://bookstore.aapg.org

India
Affiliated East-West Press Private Ltd, Marketing Division, G-1/16 Ansari Road, Darya Ganj, New Delhi 110 002, India
Orders: Tel. +91 11 2327-9113/2326-4180
            Fax +91 11 2326-0538
            E-mail: affiliat@vsnl.com

Contents

Acknowledgements	vii
GALLAGHER, K., JONES, S. J. & WAINWRIGHT, J. The Earth's dynamic surface: an overview	1
ALLEN, P. A. Time scales of tectonic landscapes and their sediment routing systems	7
BRACKEN, L. J. & WAINWRIGHT, J. Equilibrium in the balance? Implications for landscape evolution from dryland environments	29
FLEURANT, C., TUCKER, G. E. & VILES, H. A. Modelling cockpit karst landforms	47
BARDOU, E. & JABOYEDOFF, M. Debris flows as a factor of hillslope evolution controlled by a continuous or a pulse process?	63
BRIANT, R. M., COOPE, G. R., GIBBARD, P. L., PREECE, R. C. & BOREHAM, S. Limits to resolving catastrophic events in the Quaternary fluvial record: a case study from the Nene valley, Northamptonshire, UK	79
VITA-FINZI, C. Fluvial solar signals	105
FROSTICK, L., MURPHY, B. & MIDDLETON, R. Exploring the links between sediment character, bed material erosion and landscape: implications from a laboratory study of gravels and sand–gravel mixtures	117
JONES, S. J. & FROSTICK, L. E. Inferring bedload transport from stratigraphic successions: examples from Cenozoic and Pleistocene rivers, south central Pyrenees, Spain	129
CALVET, M. & GUNNELL, Y. Planar landforms as markers of denudation chronology: an inversion of East Pyrenean tectonics based on landscape and sedimentary basin analysis	147
KING, R. B. South African pediments and interfluves	167
MITCHELL, N. C. Summary of progress in geomorphologic modelling of continental slope canyons	183
Index	195

Acknowledgements

We would like to thank the authors who made the effort to contribute to this volume. We would also like to thank those who attended the meeting, but are not explicitly represented here. Also, we would like to acknowledge support from the Geological Society of London, BP, Total, British Geomorphological Research Group and the British Sedimentology Research Group.

K. GALLAGHER, Université de Rennes
STUART J. JONES, Durham University
JOHN WAINWRIGHT, University of Sheffield

The Earth's dynamic surface: an overview

KERRY GALLAGHER[1], STUART JONES[2] & JOHN WAINWRIGHT[3]

[1]Géosciences Rennes, Université de Rennes 1, Campus de Beaulieu, Rennes, 35042, France
[2]Department of Earth Sciences, South Road, Durham University, Durham DH1 3LE, UK
[3]Sheffield Centre for International Drylands Research, Department of Geography,
University of Sheffield, Sheffield S10 2TN, UK

Catastrophe and continuity in landscape evolution

Debate about the relative roles of catastrophic v. continuous processes of landform evolution is as old as the discipline of Earth Science itself. Over the last 10 years or so, research in the Earth Sciences has focussed strongly on the Earth's surface and particularly in terms of quantifying rates of processes. This research parallels developments in geomorphology and sedimentology in the quantification of surface processes since the 1950s and 1960s. These surface processes are the manifestation of the large-scale interaction of climate and tectonics operating over a wide range of spatial and temporal scales. Thus, recent research had required integration of the historically distinct subjects of geomorphology, sedimentology, climatology and tectonics. Partly as a cause and partly as a consequence of this integration, there have been many recent developments in quantitative modelling and both laboratory and field-based analytical tools. Together, these have provided new insights into absolute and relative rates of denudation, and the factors that control the many dynamic processes involved.

One of the outstanding issues concerns the balance between tectonics, climate and denudation, and in particular the limiting effects of one on the others and the nature of dynamic feedback mechanisms. The fact that processes can be considered catastrophic or continuous, depending on the timescale of observation or interest, can hinder the predictability of models, depending on how they are formulated. Certain conditions may lead to a steady-state situation in which denudation balances tectonic uplift, leading to a more or less constant topography. Steady-state topography means that detailed study of present day landforms can provide important insights into the nature of surface processes back in time. Such assumptions underpin debates in geomorphology relating to the process-form linkage and the understanding of characteristic forms in the landscape. Alternatively, the recognition of non-steady-state situations and a clearer understanding of why these situations occur provide the key for resolving the climate-tectonics-landscape evolution feedback loop. The transition between the two states will reflect the process response time, and therefore the transitory state may provide a clearer picture of the time lag of topographic response to changes in the rates of climate change and tectonic forcing. However, the response time is not necessarily constant and may have changed considerably at key points in the past, such as the evolution of plants on land in the Palaeozoic and the acceleration of human activity within the landscape in the Holocene.

In terms of denudation (physical erosion and chemical and mechanical weathering), there are clearly catastrophic processes, such as landsliding, which operate discretely and on short timescales and more continuous processes, such as chemical weathering, which can be considerably more protracted. The distinction between discrete catastrophe and continuous modification depends also on the time and spatial scales of interest. These considerations also impact directly on the questions of if and how steady-state topography can be achieved, how the processes controlling this state can be quantified and resolved, what causes departures from a steady-state condition and how topography reflects the coupling between denudation, climate and tectonics.

Some of the key current research areas in the world are tectonically active regions, such as New Zealand (southern Alps), Taiwan and Olympic Mountains (USA). However, the link between tectonics and denudation is complicated in these convergent zones (e.g. Willett *et al.* 2001), as there is a significant horizontal component to the deformation and, additionally, climatic variations often produce marked asymmetry in denudation, which itself then feedbacks into the isostatic component of vertical motion.

In practice, this research field necessarily involves a broad range of disciplines including field geologists, geomorphologists, structural geologists, geochemists, climatologists and geophysical modellers. These researchers address the observational constraints on

---

From: GALLAGHER, K., JONES, S. J. & WAINWRIGHT, J. (eds) *Landscape Evolution: Denudation, Climate and Tectonics Over Different Time and Space Scales*. Geological Society, London, Special Publications, **296**, 1–5.
DOI: 10.1144/SP296.1   0305-8719/08/$15.00 © The Geological Society of London 2008.

spatial variations and controls on erosion and weathering, the contribution of geochemical and geophysical data in quantifying rates of erosion and surface deformation, and the insights or otherwise of process-oriented numerical models, linking tectonics, climate and denudation.

The aim of the 2004 William Smith conference was to encapsulate the current state of some of the research relevant to this area and promote discussion on the outstanding issues and future research directions (such as technical and analytical developments and the robust integration of modelling and observations). In particular,

- how the geological record preserves the nature and variability of erosion processes over a wide range of time and spatial scales (from years to millions of years);
- how this record can be interrogated through observation and laboratory analysis; and
- how physical models can be integrated with these data to provide a deeper understanding of the interactions between surface processes, climate and tectonics.

This publication contains a selection of the papers presented at the conference. In the first paper, **Allen** presents an overview of the timescales relevant to understanding the links between tectonic forcing and landscape response, including the sediment routing systems, and provides a conceptual comment on this subject in a recent essay in *Nature* (Allen 2005), solicited from the meeting. He makes the point that the division of catastrophe v. continuity is somewhat artificial, and suggests it is better to consider the overall system in terms of response times to forcing or perturbations (implicitly acting with variable periodicities), producing steady (buffered) or transient (reactive) conditions. These conditions depend on the relative timescales and coupling of the internal (e.g. tectonics) and external (topography, climate) systems.

These concepts are considered in the context of normal fault systems, such as the western USA, where the tectonic displacement field is reasonably well understood and the depositional systems recording the erosional response are well characterized in terms of fault length, displacement and relief along the fault block. This approach leads to a characteristic timescale for steady-state footwall relief of about 10^6 years. Analysis of bedrock incision in catchment areas implies response time of 10^5 years for high concavity regions to more than 10^6 years for larger low concavity catchments. Here response time is that required to achieve new equilibrium conditions. He also defines a relaxation time, effectively a time constant for the response (e.g. catchment denudation, sediment flux), which is typically an order of magnitude less than the response time determined from bedrock incision models.

Allen extends the analysis beyond the catchment to consider the fluvial response times (flood plains and alluvial) to periodic forcing and their influence on the upstream system. Typical response times are 10^4-10^6 years, being slower for larger gravel fractions in the sediment load and for larger alluvial systems.

This range in, and the complications between, different timescales for equilibrium and relaxation response times, the duration and periodicity of forcing, and the sediment-transport system, clearly have implications for the inference of tectonic signals from the sedimentary record. This aspect is highlighted as a major challenge for the future.

The concept of equilibrium is central to many debates regarding landform evolution, but appears to have become blurred in terms of definition and somewhat muddied in terms of usage. **Bracken & Wainwright** review the origins and definitions of the terminology and highlight the fact that an appropriate definition depends on context, for example the difference in process regimes and responses between temperate and arid regions. A problem is how to measure or demonstrate geomorphological equilibrium, and previous attempts have included monitoring channel form, grade and correlations between system properties. These appear to suffer from a degree of circular reasoning that makes them generally unsatisfactory, particularly when arguing that form is a proxy for process equilibrium. An unavoidable issue in dryland environments is the fact that much of the activity is concentrated into rare large floods, and they discuss the case for equilibrium and non-equilibrium in the terms of the factors controlling the observed variations in morphology in these environments. They go on to consider the scale dependence (in both space and time) of equilibrium and highlight the common problems associated with choosing these when attempting to define equilibrium conditions. Feedback between different processes is also critical to the attainment of equilibrium, but again the significance depends on the scale, and becomes more difficult over longer timescales and large lengthscales, as different processes can interact with variable complexity. Finally, they discuss the impact of non-linearity, thresholds and chaotic behaviour on the inference of equilibrium. Overall, equilibrium seems too difficult to demonstrate unequivocally, and many ideas of this state are difficult or impossible to test in practice.

Fleurant *et al.* employ the CHILD (Channel-Hillslope Integrated Landscape Development) model in a novel context to investigate the formation of karst landscapes. In particular they concentrate on the development of cockpit karst, using modelling as a means of evaluating different hypotheses found in the literature. The different elements of a landscape-evolution model driven by dissolution processes are discussed and assembled from an extensive review of the literature. Simulations have been carried out over 10 Ma time periods, reflecting the timescale of known cockpit karstification in Jamaica. GIS analysis of measured karst features is used to evaluate the performance of the model. Results suggest that conceptual models that use isotropic dissolution to explain these features are unable to reproduce observed patterns in the landscape. A more complex model version that uses anisotropic dissolution is able to reproduce the observed features closely. This result supports interpretations of cockpit karst as epikarst phenomena, and provides a broader context for the understanding of the longer term soil-landscape evolution in Jamaica.

The theme of continuous versus catastrophic process is addressed directly in the contribution of **Bardou & Jaboyedoff**. They investigate the evolution of Alpine hillslopes as a response to débris flow activity, which they relate to classic concepts of frequency and magnitude in geomorphology. Data from a wide range of catchments in the Swiss Alps were used, covering a period of at least 25 years. A range of analytical methods suggests a break in behaviour of events, with a threshold event size of about 10^5 m^3. As with other similar studies, the events show power-law scaling patterns. Stochastic modelling of the behaviour of this type of environment suggests that these different patterns can be explained by initial conditions as well as continuous processes such as weathering and catastrophic extreme events. The overall behaviour of such catchments needs to be understood as a complex set of interacting thresholds. Two case studies are then presented that illustrate the way in which these thresholds might evolve. The authors conclude with a brief assessment of how different types of catastrophic or continuous process in the landscape affect the management of risk in these environments at the present time.

Briant et al. address the issue of resolving of even shorter term (days, weeks) events such as flooding from the Quaternary geological record. Such events may dominate the record but the main problem is the lack of extremely high resolution absolute geochronological tools appropriate to this timescale (up to a few tens of thousands of years), coupled with lack of resolution in relative timing from sedimentology and palaeontology. The importance of this question comes from the desire to link longer term fluvial regimes to climate, which also clearly influences flood frequency. Additionally, the deposits produced by rare but large-scale floods may look similar to more frequent high-flow hydrological regimes. The authors examine these issues with a careful case study from central England, concluding that it is possible to link average fluvial activity to climate variations over timescales similar to the marine isotopic substage scale (about 10 ka, Shackleton 1969), but much finer resolution is currently not practical.

Vita-Finzi demonstrates that fluvial history and behaviour can be linked to solar activity. The fluvial history of the Mediterranean basin has long been recognized as having latitudionally diachronous, locally bipartite, episodes of fluvial aggradation during the period AD 500–1900. The theoretical consideration of Vita-Finzi, with support from general climate models, suggests that a decrease in solar radiation would lead to equatorward displacement of the subtropical jet streams and associated mid-latitude depressions. Vita-Finzi identifies a gradual decline in the Sun's activity from c. 7000 BP followed by a resurgence after AD 500. This episode would account for the observed fluvial aggradation in the Mediterranean basin. Recognition of solar signals in the fluvial record has important implications for not only modelling of fluvial systems but also environmental analysis and flood forecasting.

Frostick et al. report on a series of flume experiments designed to visualize the process of entrainment for gravels and sand–gravel mixtures in order to identify differences that may account for enhanced rates of bedload transport in bimodal sediments common in upland river systems. They use new image analysis techniques of the experimental runs to reveal important differences in entrainment processes. Observations suggest that the presence of sand increases the rates of gravel entrainment and leads to a distinctive patchiness in break-up which will encourage bedform development. In mixtures where sand is removed prior to gravel entrainment the bed becomes destabilized and allows larger areas to be come entrained. These observations show the importance of bed material character in controlling river form and processes. They conclude by emphasizing the need for the inclusion of grain-size measures in the flux/power relationships if such models are to capture some of the complexities of the controls on sediment transport in the natural environment. It is therefore essential that more consideration is given to the range of grain sizes available for transport from reach to drainage basin scales, with less emphasis on the mean grain size in landscape models.

Extending the theme of sediment transport to a field setting, **Jones & Frostick** address the question of determining the behaviour of ancient rivers, as clearly these are an important aspect of landscape development models. They deal with this in terms of palaeohydraulic conditions, parameterized in terms of stream power, bed load transport rates and efficiency. To infer these quantitatively, they need to determine palaeo-slope, -depth, -velocity and sediment thickness in Cenozoic/Quaternary gravel bed deposits in the Pyrenees in northern Spain. This determination requires identification and characterization of accretion directions (frontal or lateral) and size of accretionary foresets, measurement of sedimentological depth indicators and clast grain sizes. Such an approach is necessarily empirical and so they also address the uncertainties and bias in their approach systematically for the key parameters. The approach they develop tends to provide lower limits on the efficiency of bedload transport. They conclude that the main factors determining whether a gravel-bed river incises or aggrades are the sediment supply and the efficiency and rate of transport mechanisms to remove it.

Moving from the fluvial scale to the more regional considerations of landscape development, **Calvet & Gunnell** reassess the potential for inferring topographic evolution from erosion surfaces, although strongly making the point that a variety of other evidence needs to be considered in such a study. Thus, they also consider stratigraphy, geometrical relationships and palaeontological evidence. It is probably fair to say that erosion surfaces and geomorphological inferences based on them have received a torrid time going back at least 40 years (Chorley 1965, Bishop 1980, Summerfield 1991). These authors argue that some of these criticisms arise out of oversimplifying the formation of erosion surfaces, including the implicit assumption that they represent peneplanation at sea-level. They remap potential erosion surfaces based on the topographic dip from a digital elevation model, and quality-control these measurements with geological maps and field observations. They establish a relative chronology based on the other methods mentioned earlier and suggest that these surfaces reflect a regional control on the landscape as a consequence of well-defined base levels around the Pyrenees in the Miocene. They infer two generations of landforms: the residual summit surface and a lower pediment, and dismiss possible explanations in terms of exhumed unconformities or a Quaternary periglacial influence, advocating a post-orogenic, middle to late Cenozoic origin for these surfaces. They add more constraints from fauna preserved in karst infills and from the nature of clastic sequences in intermontane basins and conclude that the surfaces probably formed at elevations of 1 km or so about the contemporary sea-level, but still 1.5 km below their present-day elevation. Their preferred interpretation reflects a compromise between low-elevation peneplanation followed by major post-orogenic uplift, and planation at high elevation with little or no subsequent uplift.

A related morphometric approach is taken by **King** in his analysis of landforms in South Africa. An extensive dataset was compiled using data derived from a range of map and remotely sensed sources. These data are used to evaluate the major controls on semi-arid landforms in an area of relatively little tectonic activity. Slope shape seems to be most significantly controlled by rainfall amount and local relief, with rock type exerting a less strong control. At intermediate rainfall amounts, vegetation cover seems be a significant factor in affecting the shape of the landscape. These landscapes seem to be relatively recent (i.e. post-dating Late Pliocene uplift) and the result of pedimentation produced by diffuse overland-flow processes.

Finally, **Mitchell** presents an overview of recent work on erosion in submarine channels, drawing analogies with the morphologies produced by subaerial erosion in river channels. The morphologies can be characterized by longitudinal profiles, displaying variable concavity and gradient-area graphs in which steeper gradients are associated with smaller areas, and tributaries join the main channels with no major change in bathymetry (analogous to Playfair's law in geomorphology). Although there are many similar features to those observed in river channels, it is difficult to make direct observations of erosion in the submarine environment. One clear difference is that submarine channels are more directly influenced by slope failure and subsequent sediment flow erosion. This tends to occur in discrete local channels at different times. However simple but relevant models for associated channel erosion are quantitatively similar to stream power erosion laws in fluvial bedrock channels. Oceanographic currents can also lead to soft sediment erosion and locally to incision of tens of metres. Additionally, sediment can accumulate between channels preferentially (on interfluves), while it is flushed from within the channels themselves. This process can lead to enhanced channel relief. Currently there are no methods to quantify erosion analogous to those available to subaerial geomorphologists such as cosmogenic exposure dating or thermochronology, and this is a direction for future research. Mitchell discusses simple erosion laws, noting the similarities and differences to the sub-aerial situation. The former include the basic methods of erosion, abrasion, plucking and quarrying by the flow of relatively high density material and the apparent behaviour of knickpoints, while the latter include the sensitivity of flow strength to the nature of the solid load and the controls on shear failure.

Clearly, there remain many outstanding research questions in this field. In terms of conceptual advances, for example, the relative contributions of climate and tectonics in terms of cause of and effect, the nature of feedbacks, both positive and negative, of controlling factors, their scale dependence in both space and time and the role of thresholds v. continuous processes, are still regarded as open questions. The interrogation of the geological record is always problematic, given its incompleteness and ambiguity. Consequently, although the resolving power of absolute dating tools improves continually, the resolution of short-time scale phenomena in long timescale records remains elusive. Furthermore, understanding the links in, and controls of, a geomorphological system in terms of the processes of *in situ* chemical and physical breakdown, through transport and sediment routing to depositional basins, requires more integrated studies over a wide range of time and spatial scales.

Currently, many landscape-evolution models are empirically based, often incorporating lumped parameters of little instrinsic physical relevance, and so are difficult to determine independently of the model itself. The next generation of landscape-evolution models needs to move beyond the empirical, providing a sounder basis for understanding past processes in landscape evolution over geological timescales and, perhaps more topically, forecasting landscape development on human timescales.

References

ALLEN, P. 2005. Striking a chord. *Nature*, **434**, 961.

BISHOP, P. 1980. Popper's principle of falsifiability and the irrefutability of the Davisian cycle. *Professional Geographer*, **32**, 310–315.

CHORLEY, R. J. 1965. A re-evaluation of the geomorphic system of W. M. Davis. *In*: CHORLEY, R. J. & HAGGETT, P. (eds) *Frontiers in Geographical Teaching*. Methuen, London, 21–38.

SHACKLETON, N. J. 1969. The last interglacial in the marine and terrestrial records. *Proceedings of the Royal Society of London. Series B, Biological Sciences*, **174**, 135–154.

SUMMERFIELD, M. A. 1991. *Global Geomorphology, an Introduction to the Study of Landforms*. Longman, Harlow, Chapter 18.

WILLETT, S., SLINGERLAND, R. & HOVIUS, N. 2001. Uplift, shortening, and steady-state topography in active mountain belts *American Journal of Science*, **301**, 455–485.

Time scales of tectonic landscapes and their sediment routing systems

PHILIP A. ALLEN

Department of Earth Science and Engineering, Imperial College London, South Kensington Campus, London SW7 2AZ, UK (e-mail: philip.allen@imperial.ac.uk)

Abstract: In regions undergoing active tectonics, the coupling between the tectonic displacement field, the overlying landscape and the redistribution of mass at the Earth's surface in the form of sediment routing systems, is particularly marked and variable. Coupling between deformation and surface processes takes place at a range of scales, from the whole orogen to individual extensional fault blocks or contractional anticlines. At the large scale, the attainment of a steady-state between the overlying topography and the prevailing tectonic conditions in active contractional orogens requires an efficient erosional system, with a time scale dependent on the vigour of the erosional system, generally in the range 10^6-10^7 years. The catchment–fan systems associated with extensional fault blocks and basins of the western USA are valuable natural examples to study the coupling between tectonic deformation, landscape and sediment routing systems. Even relatively simple coupled systems such as an extensional fault block and its associated basin margin fans have a range of time scales in response to a tectonic perturbation. These response times originate from the development of uniform (steady-state) relief during the accumulation of displacement on a normal fault ($c. 10^6$ years), the upstream propagation of a bedrock knickpoint in transverse catchments following a change in tectonic uplift rate ($c. 10^6$ years), or the relaxation times of the integrated catchment–fan system in response to changes in climatic and tectonic boundary conditions (10^5-10^6 years). The presence of extensive bedrock or alluvial piedmonts increases response times significantly. The sediment efflux of a mountain catchment is a boundary condition for far-field fluvial transport, but the fluvial system is much more than a simple transmitter of the sediment supply signal to a neighbouring depocentre. Fluvial systems appear to act as buffers to incoming sediment supply signals, with a diffusive time scale ($c. 10^5-10^6$ years) dependent on the length of the system and the extent of its floodplains, stream channels and proximal gravel fans. The vocabulary for explaining landscapes would benefit from a greater recognition of the importance of the repeat time and magnitude of perturbations in relation to the response and relaxation times of the landscape and its sediment routing systems. Landscapes are best differentiated as 'buffered' or 'reactive' depending on the ratio of the response time to the repeat time of the perturbation. Furthermore, landscapes may be regarded as 'steady' or 'transient' depending on the ratio of the response time to the time elapsed since the most recent change in boundary conditions. The response of tectonically and climatically perturbed landscapes has profound implications for the interpretation of stratigraphic architecture.

Introduction: coupled tectonic–surface processes systems

One of the outstanding scientific questions in the natural sciences is the linkage between landscape form and the wide range of processes that shape it. My particular concern in this overview is the evolution of landscape over a tectonically active crustal template ('tectonic landscapes'), where the landscape is constantly adjusting to a dynamic tectonic displacement field. Such landscapes are characterized by complex sediment routing systems, which efficiently redistribute mass at the Earth's surface. The interaction between the deforming crustal template and its landscape and sediment routing systems operates at a range of characteristic time scales.

Landscape is the critical interface between an internal system driven essentially by tectonic fluxes, which interacts with an external system dominated by the effects of topography and climate on weathering and erosion. The very challenging feature of these two systems is that they operate at a range of temporal and spatial scales. Landscape is perturbed by variations in both internal and external mechanisms; the results of these perturbations might potentially be recognized by changes in the morphometric properties of this critical interface, or by the mass fluxes of rock, particulate sediment or solutes through and over it. Instead of the somewhat static visualizations of the past, which viewed tectonic movements as effectively instantaneous events causing uplift of the land surface, upon which geomorphic agents worked over aeons of time, it is now clear that, in regions of active tectonics, the rates of operation of tectonic and geomorphic processes are similar. Consequently, the possibilities for complex coupling are enormous.

From: GALLAGHER, K., JONES, S. J. & WAINWRIGHT, J. (eds) *Landscape Evolution: Denudation, Climate and Tectonics Over Different Time and Space Scales.* Geological Society, London, Special Publications, **296**, 7–28.
DOI: 10.1144/SP296.2 0305-8719/08/$15.00 © The Geological Society of London 2008.

Fundamental to the understanding of landscapes is the idea of steady-state (Willett & Brandon 2002), which implies that properties of the landscape, such as longitudinal river profiles, dynamically adjust towards equilibrium with prevailing boundary conditions. If these boundary conditions are perturbed, it follows that the landscape, or its component geomorphic subsystems, respond over a characteristic time scale. Attainment of or approach to adjustment to new ambient conditions does not require that the landscape becomes static, only that statistically (Ellis et al. 1999) its dynamic subsystems (such as hillslopes) and pointwise events (such as landslides) operate to achieve a steady-state in terms of the mean values of chosen parameters, such as relief, long profile and sediment efflux, or in terms of a uniform scaling between, for example, catchment slope and drainage area. Consequently, the concept of steady-state is complex, and a landscape may, for example, be in steady state with regard to one chosen parameter while it is in a transient state with regard to another.

Early attempts to incorporate Earth surface processes into analytical and conceptual models of mountain building and basin development treated erosion as a function of a range of environmental and topographic variables, notably mean elevation or relief (Ahnert 1970, 1984; Pinet & Souriau 1988; Summerfield & Hulton 1994) and precipitation (Fournier 1960). Erosion and sediment transport in early coupled tectonic–surface process numerical models were treated essentially as a diffusional problem (Flemings & Jordan 1989; Sinclair et al. 1991; Paola et al. 1992), whereby the sediment flux was scaled on the local topographic slope, and bed aggradation or erosion scaled on the curvature. Subsequent models have made use of the different, but coupled, processes operating on hillslopes and in bedrock rivers (Humphrey & Heller 1995), the efflux of the hillslope–bedrock channel system setting a boundary condition for far-field sediment dispersal. The landscape evolution models in use over the last decade (e.g. Kooi & Beaumont 1994; Johnson & Beaumont 1995; Tucker & Slingerland 1996; Braun & Sambridge 1997; Densmore et al. 1998; Tucker et al. 2001), although differing in detail, generally calculate bedrock incision and sediment transport by a stream power rule. These models have been successful at generating realistic landscapes, with catchment systems etched into uplifting orogenic, passive margin escarpment, fold–thrust belt anticlines and tilted fault block templates (review in Beaumont et al. 2000), and have led to insights into the use of topography as a tectonic recorder and into the coupling and feedbacks between various surface processes. However, it has proven difficult to directly test landscape evolution model output with field observations.

A drawback with the further development of numerical coupled tectonic–surface process models is that there is commonly a disconnection between the output of such models and the observational data against which they must be tested. As an example, measurements of fault slip rate, exhumation rate, sediment transport rate and sediment accumulation rate are all made at different temporal and spatial resolution. Model output, likewise, is specific to certain time scales and cannot be readily scaled up or down without a re-examination of system behaviour. There is therefore a need, at least initially, to select tectonic–geomorphic systems where there is high potential to constrain with measurements the nature of the tectonic displacement field as well as the surface processes, ideally with a closed sediment budget to facilitate mass balance. This goal is best achieved in the fault blocks and basins of extensional terrains such as the Basin and Range province of western USA and will be focussed upon in this review. Here, it is possible to estimate the spatial and time scales over which extensional faults grow, die or link by stress feedback (Cowie 1998) by observing multiple fault systems in different stages of evolution, and to document the evolution of catchments in uplifting footwalls and depositional systems in hangingwalls (Densmore et al. 1998; Cowie et al. 2000; Dawers & Underhill 2000; Densmore et al. 2004). In settings such as this, geophysical imaging and mapping of fault geometries potentially allow detailed reconstruction of the three-dimensional displacement fields that drive sediment transport and mass redistribution. Thermochronological analysis (fission track and (U-Th)/He dating) allows large-scale patterns of denudation to be evaluated over geological time scales. Techniques using concentrations of cosmogenic nuclides in bulk stream samples allow catchment-wide denudation rates to be estimated (Brown et al. 1995; Granger et al. 1996; Schaller et al. 2001) and distinct fan depositional segments to be placed within a chronological framework (Dühnforth et al. 2005).

A full coupling between tectonics and erosion implies that the tectonic displacement field strongly influences patterns of surface erosion, but also that spatial and temporal variations in erosion influence rates and patterns of deformation. Erosional unroofing, ice sheet melting, evaporative drawdown and depositional blanketing may all feed back strongly into deformation through the effect of unloading and loading on the lithosphere's state of stress and temperature. A simple example is the effect of localized river incision on the regional uplift of mountain peaks (Molnar & England 1990; Burbank 1992; Montgomery 1994). Erosion at the surface of a critically tapered orogenic wedge has been shown to have major impact on the exhumation of high P/T rocks, rates of frontal propagation v. out-of-sequence deformation, topographic profiles and gross depositional environments in adjacent foreland basins (Sinclair & Allen 1992; Willett 1992, 1999; Willett et al. 1993, 2006; Beaumont et al. 1996; Whipple & Meade 2004). Erosional unloading under conditions of localized high monsoonal rains has also been postulated to be responsible for activity of new fault zones, as in the Nepalese Himalaya (Wobus et al. 2005), and flexural rebound following mountain glacier and glacial lake unloading is thought to have triggered abrupt changes in slip rates on extensional faults in the Wasatch area, eastern Basin and Range province (Hetzel & Hampel 2005).

The precise response of the lithosphere to various spatial patterns of loading and unloading depends on the rheology of the underlying crust, but the crustal rheology may be coupled to surface processes. A weak plastic crust allows a strong local coupling of erosion and deformation (Koons et al. 2002), whereas a strong elastic crust behaves impassively to the very short wavelength negative loads caused by erosion (Leonard 2002). Simpson (2004a, b) suggested that the cutting of canyons by rivers through doubly plunging anticlines at their highest structural and topographic position, as is classically seen in the Zagros (Oberlander 1985) and Apennine (Alvarez 1999) belts, is a strong indicator of the local (10–20 km) feedback between erosional unloading and crustal deformation. The prerequisite is that the crust is already deforming plastically under regional deviatoric compressive stresses.

Landscapes in tectonically active environments can usefully be thought of as being composed of a number of interacting components: upper crustal structures such as faults or folds, hillslopes, debris-flow or fluvial channels, and alluvial or debris-flow fans. In this paper, I briefly consider the responses of each of these elements to tectonic perturbations, and then review the behaviour of linked catchment–fan systems in Basin and Range-type settings and their relationship to larger-scale, downstream river systems. A constant theme is that the different components of landscapes and their sediment routing systems respond to tectonic perturbations with different dynamics and response times.

may persist over very long time scales. The rate of lowering of a mountain belt at mean elevation H following the cessation of active uplift (Ahnert 1970; Pelletier 2004) is

$$\frac{\partial H}{\partial t} = -\frac{1}{\tau_d} H \left(\frac{\rho_c}{\rho_m - \rho_c} \right) \quad (1)$$

where τ_d is the characteristic time scale for the decay of mountain topography, and the density terms in brackets (for crust and mantle) incorporate the effects of Airy isostasy. If the rate of erosion (m a^{-1}) scales linearly on the mean elevation (m) by a coefficient equal to 0.61×10^{-7} (Pinet & Souriau 1988), τ_d becomes $\sim$70 Ma, which is substantially smaller than the age of some ancient orogens. Baldwin et al. (2003) and Pelletier (2004) suggested that the presence of extensive piedmonts, whether composed of bedrock or alluvium, could extend the time scale of decay of mountain belt topography.

Of greater interest in the present context is the response time of tectonically active orogens such as the Himalayas and Southern Alps. At the large scale, mountain belt landscape is perturbed by underlying tectonic fluxes of rock driven by continental convergence. There is therefore a balance between the rate of erosional unroofing and the flux towards the surface through tectonic advection (Fig. 1). Willett (1999) expressed this balance by a dimensionless erosion number N_e equal to kL/U, where U is the tectonic uplift rate of rocks, L is the half-width of the uplifting piece of crust, and k is a coefficient proportional to the bedrock incision efficiency and precipitation, with units of time^{-1}. If N_e tends to infinity, erosion ruthlessly planes off the topography despite high rates of tectonic uplift. If N_e is small ($<$1), erosion weakly offsets tectonic uplift rates, causing lower rates of exhumation. The erosion number therefore strongly controls the time taken for a mountain belt to achieve equilibrium with prevailing tectonic fluxes. Based on numerical experiments (Willett 1999), the time scale (t or dimensionless t^*) for the attainment of a steady-state topography ranges from $t^* = tV_p/h < 2$ to $t^* \approx 12$ for N_e of 10–2, respectively. For typical values of the convergence velocity V_p (0.005–0.01 m a^{-1}) and crustal thickness h (30–35 km), this time scale is of order $10^6–10^7$ years.

Tectonics and erosion at the orogenic scale

The fact that old Palaeozoic orogenic belts, such as the Appalachians and Urals, still exist as elevated topography today suggests that mountain belts

Growth of relief in the footwalls of extensional faults

The accumulation of displacement on faults during the progressive growth of normal fault arrays provides an initial framework for assessing the

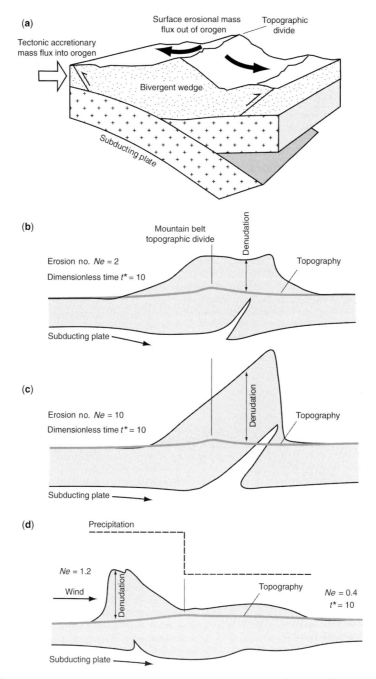

Fig. 1. Coupling between tectonics and erosion at the scale of a bivergent orogenic wedge. The time scale of the erosional response at the surface to variations in the tectonic advection of rock is dependent on the erosion number Ne (Willett 1999). (**a**) Schematic diagram illustrating bivergent wedge. (**b**) Envelope of deformed crust (grey) and topographic profile, showing spatial pattern of denudation at $Ne = 2$ at dimensionless time $t^* = 10$. (**c**) Envelope of deformed crust (grey) and topographic profile, showing spatial pattern of denudation at $Ne = 10$ at dimensionless time $t^* = 10$. Grey crust above the topographic profile shows high amount of erosional efflux and strong exhumation at high values of Ne. (**d**) Envelope of deformed crust (grey) and topographic profile, showing spatial pattern of denudation at dimensionless time $t^* = 10$ where Ne varies from windward (1.2) to leeward (0.4) as a result of an asymmetrical precipitation pattern. Denudation is focused on the windward side of the orogen.

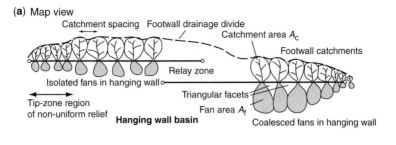

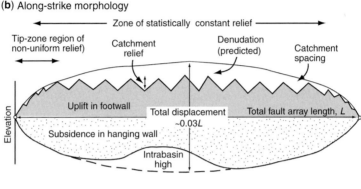

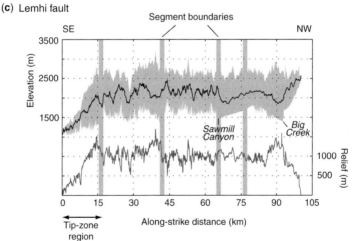

Fig. 2. Linked extensional fault segments and their catchment–fan systems, based on Densmore et al. (2004). (a) Footwall catchments and hanging wall fans along two fault segments connected by a relay zone. (b) Anticipated footwall topography and relief, and hangingwall subsidence along the same fault segments, showing a tip-zone region of non-uniform relief. (c) Distribution of footwall relief and topography along the Lemhi Fault of Idaho, showing that the tip-zone region is c. 15 km in length.

evolution of landscape and sediment routing systems over a deforming template (Fig. 2a, b). This framework is based on the observation that there is a clear proportionality between fault length and displacement (Cowie & Scholz 1992; Dawers et al. 1993; Schlische et al. 1996). Normal fault arrays grow by a combination of cumulative displacement on the fault, its lateral propagation at the fault tip, and linkage with adjacent segments. Such processes of growth have predictable outcomes for both subsidence and depositional environments in hangingwall basins (Gupta et al. 1998; Cowie et al. 2000) and for topographic evolution of uplifting footwalls. The pattern of fault displacement provides an envelope of tectonically generated topography upon which surface processes act (Densmore et al. 2004, 2007a).

Densmore et al. (2004) analysed the footwall topography on six presently active faults in Idaho and Montana using both USGS 30 m horizontal resolution and higher resolution 10 m digital elevation data (Fig. 2c). Fault systems are in various stages of development. Faults are initially single segment with footwall topography dominated by inherited relief. However, with time, fault segments link and the footwall topography becomes dominated by the effects of tectonic displacement. Near the fault tip, a region of increasing relief develops, from zero at the fault tip to a maximum of 10–15 km toward the centre of the fault segment. Beyond this distance, or tip-zone distance, the relief is uniform (Densmore et al. 2004, 2007a), despite the presumably increasing fault displacement as the centre of the fault is approached (Densmore et al. 2005). The presence of uniform relief suggests that the landscape must be limited by rock strength (Schmidt & Montgomery 1995; Densmore et al. 1998) and that it must be in a statistical steady-state (Ellis et al. 1999).

If we assume that fault growth is dominated by lateral tip propagation, and divide the tip zone length by the fault tip propagation rate, we effectively make a space for time (ergodic) substitution, and derive a range-scale characteristic response time for the attainment of steady-state relief. If the fault tip propagation rate is estimated from the displacement-length scaling relation, we obtain a time scale of the order of 10^6 years for the generation of steady-state footwall relief. If, on the other hand, the fault grows by discrete linkage events separated by periods when the fault is pinned, the time since the initiation of faulting will be roughly uniform along each segment (McLeod et al. 2000). This makes the ergodic substitution and the estimation of the time scale for the attainment of uniform relief less straightforward.

The time scale over which fault arrays initiate, interact and link to form a mature system with major border faults presents us with another time scale relevant to landscape and sediment routing systems. High-resolution seismic stratigraphic data offer good possibilities for timing fault activity in extensional basins (McLeod et al. 2000; Dawers & Underhill 2000). Faults may grow rapidly in the early rift phase, reaching their final length in 1–3 Ma (Morley 1999; Meyer et al. 2002; Gawthorpe et al. 2003). The phase of linkage based on seismic stratigraphic studies in the North Sea lasted 3–4 Ma (McLeod et al. 2000) and c. 5 Ma (Dawers & Underhill 2000), but was delayed about 10 Ma from the initiation of rifting. We can therefore expect landscapes and sediment routing systems to be perturbed by strong and complex variations in slip rate along fault arrays over time scales of 1–10 Ma.

Catchment evolution: analytical approaches

The texture or pattern of relief in mountain landscapes is primarily set by the presence of bedrock river channels flanked by hillslopes (Burbank et al. 1996). The processes forming bedrock channels and hillslopes act on the tectonic displacement field generated by fault growth. Bedrock channels dispose of the products of hillslope erosion by transporting them to neighbouring depocentres such as basin-margin fans, playas or fluvial systems.

A large literature has developed on the dynamics of bedrock rivers (e.g. Howard & Kerby 1983; Seidl & Dietrich 1992; Howard et al. 1994; Stock & Montgomery 1999 as a selection). The stream power rule, involving the calculation of a rate of bedrock river incision as a function of stream power or different formulations of shear stress or excess shear stress, has been extensively used in modelling studies. The advantage is the availability of an analytical solution to the problem of bedrock river incision, but the disadvantage is that its coefficients and exponents are generally poorly understood, commonly reflecting the lumping together of a range of complexly interacting physical processes. Nevertheless, despite this handicap, the stream power rule has been very widely applied, its parameter values partially constrained and its sensitivity to variations in its main parameters comprehensively explored (e.g. Sklar & Dietrich 1998; Whipple & Tucker 1999; Snyder et al. 2000, 2003; Tucker & Whipple 2002).

The stream power rule for detachment-limited situations (where the vertical rate of lowering of the stream bed is limited by the rate at which bed particles can be detached) states that the rate of bedrock river incision is a power law function of drainage area A and stream gradient S

$$\frac{\partial y}{\partial t} = KA^m S^n \qquad (2a)$$

or where there is an ongoing tectonic uplift rate U

$$\frac{\partial y}{\partial t} = U - KA^m S^n \qquad (2b)$$

where K is a dimensional coefficient of erosion, m and n are positive empirical area and slope coefficients (Howard & Kerby 1983; Whipple & Tucker 1999), and $m/n = \theta$ is commonly known as the concavity (Montgomery 2001; Dietrich et al. 2003). There is a large literature discussing the physical significance and range of values taken by the unknowns in the stream power formulation. The values of these parameters can be estimated, given certain assumptions, from logarithmic plots of slope v. drainage area (Snyder et al. 2000).

The topic of interest here is the rate of lowering of the longitudinal river profile in the face of a tectonic uplift of rocks. Whipple (2001) presented two scenarios in which bedrock rivers were perturbed tectonically: (a) a spatially uniform block-type tectonic uplift of rocks, and (b) an instantaneous fall of base level at the outlet. This allowed two response times to be calculated for the attainment of a new equilibrium under the changed tectonic (or climatic) boundary conditions (Whipple & Tucker 1999; Whipple 2001). The response time is defined as the time required for an erosional knickpoint to reach the upstream end of the channel network, which for convenience I refer to as an analytical response time. Since the celerity (velocity of the knickpoint 'wave') decreases as the upstream limit is approached due to diminishing water discharges, the numerical value of the analytical response time is strongly affected by the slow, final, asymptotic attainment of new equilibrium conditions. The response time for a block-type (spatially uniform) uplift τ_a is given by

$$\tau_a = \beta K^{-1/n}(U^{1/n} - 1)(fU^{1/n} - 1)(fU - 1)^{-1} \quad (3)$$

where fU is the ratio of final to initial rock uplift rate (equal to 2 for a doubling of the tectonic uplift rate), and β is a grouping of constants defined by

$$\beta = \left\{ k_a^{-m/n}\left(1 - \frac{hm}{n}\right)^{-1} \right.$$

$$\times \left(L^{1-hm/n} - x_c^{1-hm/n}\right), \frac{hm}{n} \neq 1 \quad (4)$$

$$\left\{ k_a^{-m/n} \ln\left(\frac{L}{x_c}\right), \frac{hm}{n} = 1 \right.$$

where L is the total catchment length, x_c is the position of the channel head, and k_a and h are empirical coefficients in Hack's Law relating catchment area A to downstream distance x ($A = k_a x^h$) (Hack 1957).

Using values of concavity typical of steep bedrock rivers, these analytical response times are of order 10^5 years irrespective of system size, but where concavities are low, as found in alluvial and badlands type rivers, analytical response times rise to significantly greater than 10^6 years for moderate to large system sizes (Fig. 3). The implication is that statistical steady-state is unlikely to be achieved where catchments are perturbed rapidly by changing tectonic or climatic boundary conditions. As a test, Densmore et al. (2007a) calculated analytical response times for catchments where bedrock rivers are perturbed by a non-uniform tectonic displacement field caused by the elastic deformation related to cumulative seismic displacements on a range-bounding fault (discussion in Ellis et al. 1999). They were particularly interested in the contrast between the tip-zone of growing faults and the central zone of uniform relief discussed above. In order to test whether catchments along the Basin and Range fault segments are in equilibrium with their tectonic displacement field, Densmore et al. (2007b) introduced a normalized response time t^* given by

$$t^* = \frac{\tau_a}{\tau_{onset}} \quad (5)$$

where τ_{onset} is the time since the initiation of rapid fault slip at the catchment position, taken as uniform at 6 Ma for the case of fault linkage, and derived from ergodic principles for the case of a propagating fault tip (see above). Catchments in the tip-zone had very high values of t^* compared with the central sector of uniform relief (Fig. 4). This suggests that tip-zone catchments have transient landscapes, whereas those in the central sector are in equilibrium with the present-day tectonic displacement field. This change from transient to equilibrium landscapes is accompanied by a different scaling between slope and area and the replacement of episodic stream-flows by debris-flows as the dominant sediment transport mechanism (Densmore et al. 2007a).

Whatever the chosen scenario for the tectonic displacement field, the migration of knickpoints in the longitudinal river transmits information about the new base level upstream to the drainage divide. If τ_a is the analytical response time, as described above, we can also define a characteristic relaxation time τ_r by fitting signals such as mean catchment denudation rate, mean fan aggradation rate, sediment discharge or elevation on a longitudinal profile with an expression of the form $1 - \exp(-t/\tau_r)$. τ_r is generally half an order of magnitude smaller than τ_a. A relaxation time of this type for a catchment–fan system perturbed by a change of tectonic boundary conditions can be estimated from simulations using a numerical landscape evolution model (LEM) (Allen & Densmore 2000). These authors allowed a statistically steady-state landscape to develop numerically on an inclined plane. A change of slip rate on a range-bounding fault was then applied

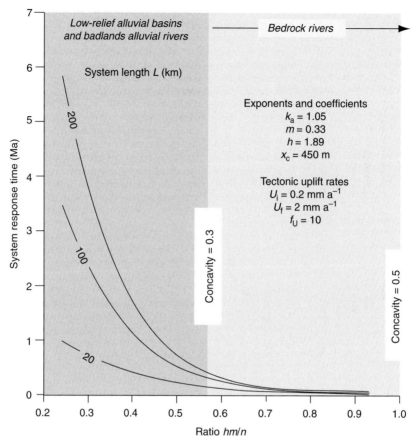

Fig. 3. The analytical response time calculated using the stream power rule (Whipple 2001) v. hm/n for different system lengths. System response times are <1 Ma for system lengths <200 km for the high concavities typical of bedrock rivers.

instantaneously, and the time evolution of catchment denudation rate and fan aggradation rate were tracked following the step change. Both responded with a relaxation time of c. 50 ka.

Hillslope erosion

In the previous discussion an assumption was made that hillslopes respond instantaneously to a local base level fall caused by incision of the river into bedrock during knickpoint migration. Consequently, if this assumption holds, the hillslope subsystem can be ignored from the consideration of the response time of a catchment to a tectonic perturbation, so that hillslope erosion is not the rate-limiting process. This is despite the fact that the vast bulk of sediment efflux of the catchment is derived by hillslope erosion. This assumption is worth briefly evaluating.

Hillslopes have traditionally been regarded as diffusive (Carson & Kirkby 1972; Anderson 1994; Fernandes & Dietrich 1997), which means that the mass flux over the hillslope is treated as related to the local topographic gradient by a transport coefficient k, and the erosion or deposition is related to the local topographic curvature by the diffusivity κ, where $\kappa = k/\rho_b$, and ρ_b is the bulk density of the mobile regolith. The form of a hillslope profile is therefore dependent on the diffusivity κ, but the amplitude of the profile is set by the rate of channel incision and the length of the hillslope from the ridge crest to the valley bottom L. The time constant for a diffusive system such as this is given by

$$\tau = L^2/\kappa \qquad (6)$$

from which it can be seen that, for the slow processes of rainsplash and creep, where κ most commonly falls in the range $0.05-0.01\,\text{m}^2\,\text{a}^{-1}$

(Anderson & Humphrey 1990; Rosenbloom & Anderson 1994; Burbank & Anderson 2001), hillslope time constants on small hillslopes where $L = 50$ m range from 50 to 250 ka. However, the hillslopes in the steep transverse catchments of the Basin and Range are dominated by landsliding rather than soil creep, giving a non-linear dependence of the mass flux on the topographic gradient (Roering et al. 1999). Effective diffusivities in humid, tectonically active regions dominated by landsliding, such as the western flank of the Southern Alps, New Zealand, range between 15 and 50 $m^2 a^{-1}$ (Koons 1989; Hovius et al. 1997). Numerical modelling (Densmore et al. 1998) and field studies (Pearce & Watson 1986) suggest that the response time of landslide-dominated hillslopes is rapid compared with the time required for bedrock rivers to adjust to base level change by knickpoint migration. Consequently, it is justifiable to treat the rate of bedrock incision and velocity of knickpoint migration as the key variables determining the response time of a mountain catchment to a base level change.

Catchment–fan systems

The magnitude of sediment discharge, its calibre and the outlet spacing along a mountain range front are primary parameters controlling the gross geometries and depositional facies of fans along the proximal edge of a sedimentary basin (Fraser & DeCelles 1992; Paola et al. 1992; Heller & Paola 1992; Gordon & Heller 1993; Whipple & Trayler 1996; Allen & Hovius 1998; Allen & Densmore 2000). Transverse mountainous catchments and their adjoining fans are ideal for studying integrated sediment routing systems (Fig. 5). First, in many cases the catchment–fan is a closed system amenable to a balance of mass or volume. Second, the progradation distance of fans is a small and easily identifiable length scale, which can be recognized in modern settings in terms of surface slope and grain size, and in the stratigraphic record in terms of sedimentary facies.

The catchment–fan is the archetypal sediment routing system. Far from a basin-margin fan being a passive dumping ground for sediment liberated from a mountain catchment, the entire system is integrated (Densmore et al. 2007b). That this is

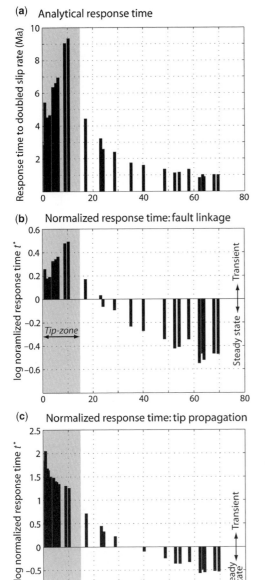

Fig. 4. Response times of catchments along the southern half of the footwall of the Lemhi Fault following a doubling of the slip rate (modified from Densmore et al. 2007a). The pre-perturbation slip rate is assumed to vary from 0 at the fault tip to 0.5 mm a^{-1} at the centre. The grey area is the tip-zone, characterized by increasing footwall relief. (a) Analytical response time calculated using equation (3) (Whipple 2001). (b) Normalized response time t^* (logarithmic scale) for fault growth by linkage, with the onset of fault slip uniform at 6 Ma. (c) Normalized response time t^* (logarithmic scale) for fault growth by tip propagation at

10 mm a^{-1}, so that the time since the onset of faulting varies from 0 at the fault tip to 7.5 Ma at the centre. Note that catchments in the tip zone are transient, whereas catchments in the central part of the footwall are likely to be in steady-state.

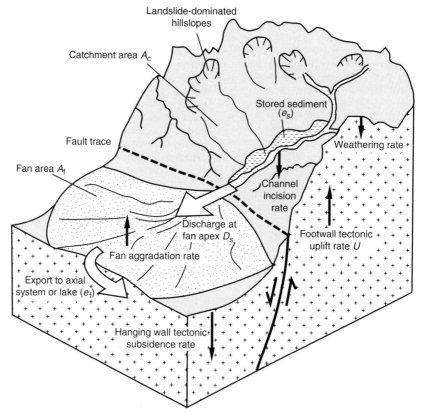

Fig. 5. Schematic illustration of a catchment–fan system, derived from Allen & Hovius (1998), upon which a mass balance can be carried out.

the case should immediately be suspected from the basic observation, made nearly half a century ago, that the surface slope of active depositional fan segments is identical to and continuous with that of the lower part of the upstream transverse valley (Bull 1964). Furthermore, the radial profiles of older fan segments are commonly colinear with the remnants of strath or fill terraces in the lower catchment valley. The catchment and the fan form one linked longitudinal profile, shaped by processes affecting the entire catchment–fan system with the fan toe as ultimate base level. The catchment and the fan are not only coupled (Humphrey & Heller 1995), they form part of one integrated system with a particular response time and behaviour when perturbed.

Densmore et al. (2007b) constructed a mass balance numerical model involving a transport-limited (where the rate of vertical lowering of the stream profile is limited by the rate at which sediment can be transported away) catchment, a spatially variable rock uplift pattern and an erosional law combining diffusive and concentrative terms to capture both hillslope diffusion and channel incision (Smith & Bretherton 1972; Simpson & Schlunegger 2003). In the numerical model, sediment derived from the catchment fills in hangingwall accommodation up to a certain linear surface slope. Consequently, fan slope is determined by precipitation rate as a proxy for the climatic impact on erosivity, and slip rate as a measure of the geometry of incremental accommodation in the hangingwall and of incremental tectonic uplift in the footwall. An instantaneous doubling of the slip rate (from 1 to 2 mm a^{-1}) results in an increase in fan slope, an increase in sediment flux at the catchment mouth, and initial backstepping of the fan toe, followed by a progradation of the fan toe. Interestingly then, a single perturbation, in this case tectonic, causes a number of effects on different aspects of the sediment routing system, each with its own response time. Whereas the fan slope and sediment efflux approach new equilibrium values with a time constant of

c. 600 ka, the duration of the backstepping-prograding activity of the fan toe, from the time of the slip rate perturbation until the fan toe reaches its previous extent, is significantly longer (2.5 Ma). In other words, one should avoid talking of response times too generally, since different parameters and indices respond at different time scales to the same perturbation. Catchment sediment efflux and fan slope decay to lower equilibrium values with a time constant of 500 ka when the slip rate is halved (from 1 to 0.5 mm a^{-1}), a value very similar to the case of doubling the slip rate.

The response time of a catchment–fan system must depend on the system diffusivity, which in the case of a diffusive-concentrative law is the square-backetted term (units of m^2 a^{-1}) in the expression for the rate of surface elevation change with time

$$\frac{\partial y}{\partial t} = U(x, t) + \frac{\partial}{\partial x}\left(\left[\kappa + c_{\mathrm{t}}(\alpha x)^m\right]\frac{\partial y}{\partial x}\right) \quad (7)$$

where surface elevation change is calculated along the single horizontal coordinate x, $U(x, t)$ is the tectonic uplift rate of rocks, and the system diffusivity is made up of two components, a linear hillslope diffusivity κ and a nonlinear 'channelized' or 'fluvial' diffusivity that depends on the precipitation rate α and downstream distance x. The response time constant therefore decreases following an increase in precipitation rate, and since equation (7) is essentially a diffusion equation, the time constant varies approximately as the inverse square of the precipitation. However, somewhat surprisingly, the time constant is independent of variations in the slip rate. Despite strongly affecting surface slope, hangingwall accommodation, base level change and tectonic fluxes in the footwall, the slip rate (within the geologically reasonable range of 0.1 and 2 mm a^{-1}) has an insignificant effect on the response time. The response times for the Basin and Range catchment–fan systems studied by Densmore et al. (2007b) ranged from 0.5 to 2 Ma, suggesting a somewhat sluggish response to tectonic and climatic changes.

The numerical model results discussed immediately above (Densmore et al. 2007b) refer to simple longitudinal profiles of catchment and fan. Basin-margin fans are, however, three-dimensional bodies that may occur as isolated cones or as coalesced bajadas (Fig. 6). The relationship between catchment area A_c and fan area A_f has long been noted (Bull 1962, 1964) and reported as a power law relation from geomorphic studies of modern fans (Hooke 1968; Bull 1977; Lecce 1990):

$$A_{\mathrm{f}} = cA_{\mathrm{c}}^n \quad (8)$$

where the coefficient c and exponent n have values (n is close to unity) attributed to factors such as variations in bedrock lithology, climate, rate of uplift of rock and rate and spatial distribution of subsidence (references and typical values of c and n are cited in Table 1 of Allen & Hovius 1998). A mass balance between the rock eroded in the catchment and the sediment deposited on the adjoining fan, taking into account the efficiencies of storage in the catchment and sediment loss from the downstream edge of the fan, gives a dimensionless sediment dispersal parameter ϕ:

$$\phi = \frac{A_{\mathrm{f}}}{A_{\mathrm{c}}} = \frac{V(1 - \lambda_{\mathrm{r}})}{\dot{y}(1 - \lambda_{\mathrm{s}})}e_{\mathrm{s}}e_{\mathrm{f}} \quad (9)$$

where $\dot{y}$ is the spatially averaged deposition rate on the fan (LT^{-1}), V is a measure of the catchment-averaged denudation rate due to landsliding (LT^{-1}), λ_{r} and λ_{s} are the rock and fan sediment porosities and e_{s} and e_{f} are the storage efficiency in the catchment (Hooke & Rohrer 1977) and the fan transport efficiency respectively (see Allen & Hovius 1998 for definitions). When catchment and fan area data are plotted logarithmically (Fig. 7), the best-fit trend is commonly linear with a slope of unity, showing that $n \approx 1$ in equation (8) and $\phi = $ constant. Deviations from a linear trend indicate that other factors, such as catchment storage efficiency (which might be expected to vary with catchment size), sediment export from the fan (which might be expected to vary with climate and sediment transport dynamics), transient behaviour between catchment efflux and fan deposition, or variations between isolated and coalesced fan geometries, may be important (Fig. 6). Fundamentally, however, the ratio between catchment area and fan area is believed to be an indicator of the importance of fault slip in providing both an uplift flux of rock in the footwall and accommodation in the hangingwall of Basin and Range-type catchment–fan systems (Whipple & Trayler 1996; Allen & Hovius 1998).

There are a number of implications of this argument. If ϕ does indeed reflect a variation in the spatial dispersal of sediment driven by tectonics, is it possible to identify systematic variations along a fault segment experiencing a lateral variation in slip rate (Fig. 6), or between different fault segments experiencing different slip rates? Assuming that systematic variation in ϕ requires that catchment denudation and fan aggradation closely reflect the tectonic displacement field causing footwall uplift and hanging wall subsidence in a dynamic steady-state, how long does it take for these conditions to be achieved, and can the transition from disequilibrium to steady-state be recognized in extensional provinces at the present day?

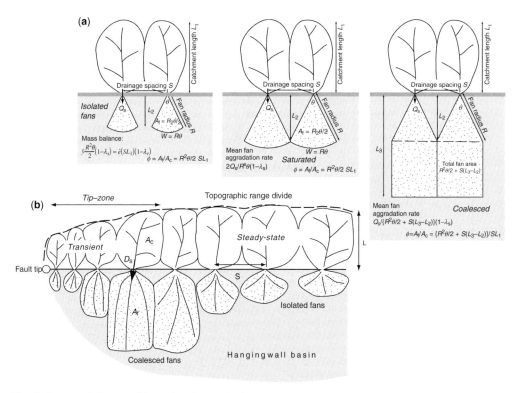

Fig. 6. Conceptual diagram of fan and catchment development. (**a**) The geometrical possibilities for the growth of fans in hangingwall basins, from isolated to saturated (just touching) to coalesced. (**b**) Concept of variation in the ratio of fan area to catchment area ϕ along a growing fault segment. Catchment areas provide variable discharges of sediment along the fault strike D_s, which is distributed over the fan surface. ϕ can be expected to vary depending on fault slip rate, geometry of fan surface and equilibrium of catchment–fan systems to ambient slip rates.

The occurrence of alluvial sedimentation (Carretier & Lucazeau 2005) and extensive piedmonts (Pelletier 2004) at the range-front affects the response of mountainous catchments to changes in tectonic and climatic boundary conditions. The presence of an alluvial apron increases catchment response times to a tectonic perturbation relative to catchments with fixed boundary conditions at the outlet, a result supported by sand box experiments (Babault et al. 2005). For example, using a tectonically uplifting block 20 km wide with fringing alluvial plains 15 km wide, the mean denudation rate of the block following an initial tectonic uplift of 2 mm a^{-1} (precipitation at 1 m a^{-1}) shows a slower response for a range front with a flanking alluvial piedmont compared to one without (Carretier & Lucazeau 2005). Once the drainage network is established, the mean denudation rate follows an exponential trend, from which relaxation times can be calculated. Whether the river incision model is detachment- or transport-limited, and whether the alluvial apron is characterized by channelized flow or sheet flow, relaxation times are 0.2–0.5 Ma for the range fronts without alluvial piedmonts and 0.7–1 Ma for those with range front basins. Carretier & Lucazeau (2005) also found a relaxation time of c. 1 Ma after an instantaneous increase in the tectonic uplift rate from 2 to 2.5 mm a^{-1}, with precipitation held constant at 1 m a^{-1}. Although the details of the modelling are different, these values are very similar to the time constants of Densmore et al. (2007b) and give confidence that for systems of this size, relaxation times are of order 1 Ma, with a significant amplification ($\times 1.5$ to $\times 10$) caused by the presence of a dynamic sedimentary piedmont.

Response times of river systems

The sediment efflux of mountain catchments sets a dynamic boundary condition for far-field transport by rivers. However, the fluvial system is not simply a passive transmitter of signals from the erosional source region. Sediment transported through a river system may spend long periods of

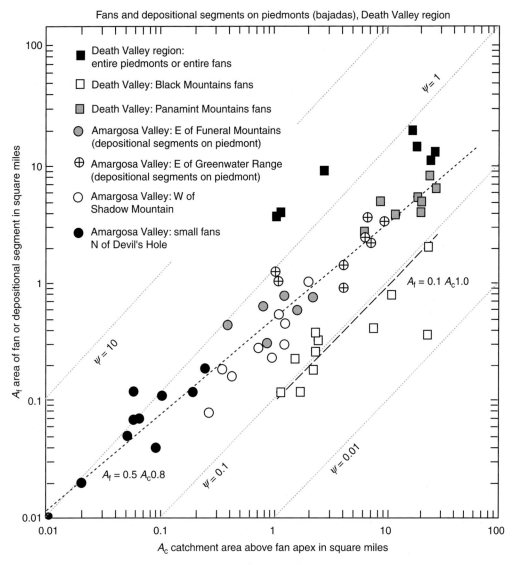

Fig. 7. Relationship between fan area A_f and catchment area A_c using data from 60 fans in the Death Valley region compiled by Denny (1965). Regressions are shown for Black Mountain fans, where $\phi = 0.1$, $c = 0.1$ and $n = 1$, and for all fans plotted, where ϕ varies between 0.3 and 1. See text for discussion.

time in storage within large floodplains and lakes, whereas other sediment may be transported much more rapidly through the channel system to a depositional site. Consequently, river systems may act as buffers to sediment efflux signals of mountain catchments. Since the coarser grain sizes in the fluvial system are prone to bedload transport within channels, their route to the final depositional site is likely to be faster than finer grain sizes that spend long periods of time trapped in floodplains. As a result, the buffering of fluvial systems acts selectively in terms of grain size in damping the upstream efflux signal.

One approach to investigating the way in which river systems buffer sediment output to the ocean is to make volumetric estimates of the sediment deposited in the ocean over geological time scales (Métivier et al. 1999; Zhang et al. 2001; Clift et al. 2004). Some authors claim that present-day rates are representative of the last 2 Ma, suggesting that present-day discharges are strongly buffered

(Métivier & Gaudemer 1999), whereas others believe, based on a recalculation of sediment volumes in basins fed by the major rivers of south and southeast Asia, that present-day sediment discharges into the ocean are several times the average rate for the Quaternary, and conclude that the large Asian fluvial systems are not strongly buffered (Clift et al. 2004). The observational basis for buffering in large alluvial systems is therefore currently controversial.

In contrast to the bedrock rivers considered in previous sections, a fundamental characteristic of alluvial channels is that they have sufficient sediment to equal or exceed the transport capacity. Models of the long-range transport of sediment by rivers (see Parker 1978a, b; Paola et al. 1992; Dade & Friend 1998) make use of a common set of assumptions (Paola 2000; Allen & Allen 2005, p. 254): (1) long-term river behaviour approximated by steady uniform flow down an inclined plane; (2) use of a formulation for flow resistance; (3) conservation of water discharge; and (4) use of the sediment continuity equation, which collectively argue for fluvial systems being regarded as diffusive with

$$\frac{\partial y}{\partial t} = \frac{\partial}{\partial x}\left(\kappa_a \frac{\partial y}{\partial x}\right) - U(x) \qquad (10)$$

where $U(x)$ is a tectonic subsidence or uplift rate term, y is the vertical coordinate, and the effective diffusivity or transport coefficient κ_a of the alluvial system is dependent principally on the discharge of water, a friction (flow resistance) factor, a dimensionless sediment transport rate and the dimensionless shear stress. The effective diffusivity κ_a (or parameters determining it) is observed to change abruptly between a steep, proximal gravelly zone and a lower gradient sandy zone in modern mixed gravel–sand systems (Paola et al. 1992; Paola & Seal 1995; Dade & Friend 1998; Parker et al. 1998; Sambrook-Smith & Ferguson 1995; Marr et al. 2000). This diffusivity is assumed to be constant within the gravel and sand regimes, but changes by a factor of about 10 at the boundary. Typical values for κ_a used by Marr et al. (2000) are 0.01 km^2 a^{-1} in the gravel regime and 0.1 km^2 a^{-1} in the sand regime. The separation of the gravel and sand regimes in an alluvial system by a discontinuity in effective diffusivity suggests that each zone may have its own response time, given by $\tau = L^2/\kappa_a$. For a basin length of 100 km the response time for the gravel regime is 10^6 year, whereas we have a response time of 10^5 years for the sand regime.

The response of the mixed-gravel fluvial system to a periodic forcing σ has been investigated using numerical experiments by Marr et al. (2000), who discriminated slow forcing when $\tau > \sigma$ and rapid forcing when $\tau < \sigma$. When the forcing is very rapid compared with the response time, the time for equilibrium to be re-established is almost constant for a fixed percentage of gravel in the sediment load, suggesting that the response time is controlled by the workings of the river system rather than by the frequency of the forcing. The value of this intrinsic response time is determined by the gravel fraction: fast ($<10^5$ years) with low gravel fractions and slow (10^5-10^6 years) for high gravel fractions.

It is clear that the stratigraphy of an alluvial system records a complex response to the various forcing mechanisms. The response of the system depends on the periodicity of the forcing compared with the basin response time σ/τ. Proximal and distal unconformities, variations in proximal and distal accumulation and movements of the gravel front and fluvial toe may be in phase or out of phase with external forcing.

The response times of alluvial systems have also been approximated from measurements of the sediment discharge leaving the catchment. If the system is assumed to be diffusive in character, and the discharge of water varies systematically with floodplain width W, the effective diffusivity κ_e of the channel-floodplain system is

$$\kappa_e = \frac{D_s}{W\left\langle\frac{\partial y}{\partial x}\right\rangle} \qquad (11)$$

where D_s is the sediment discharge (L^3 T^{-1}) and $<>$ denotes the spatial average of the slope. If L is the downstream length of the river-floodplain system and H_{max} the maximum relief between its upstream and downstream ends, the response time becomes

$$\tau = \frac{L^2}{\kappa_e} = \frac{L^2 W \langle \partial y/\partial x \rangle}{D_s} = \frac{LWH_{max}}{D_s} \qquad (12)$$

The large Asian river systems have typical values of $L \approx 10^6$ m, $W \approx 10^5$ m, $H_{max} \approx 1-2 \times 10^2$ m (slopes of $10^{-3}-10^{-4}$) and sediment discharges D_s of 10^{7-8} m^3 a^{-1} (Métivier & Gaudemer 1999). The characteristic response time is therefore in the region of 10^5-10^6 years. Castelltort and Van Den Driessche (2003) carried out a similar analysis on 93 of the world's major rivers, and found that response times varied between 10^4 years and more than 10^6 years. The response time depends on the scale of the channel–floodplain system. Large alluvial systems with extensive floodplains should therefore strongly buffer any variations in sediment supply with frequencies of less than 10^{5-6} years. This has strong implications for the detection of high-frequency driving mechanisms in the stratigraphy of sedimentary basins.

Towards a vocabulary for tectonically perturbed landscapes

Our vocabulary for landscapes in tectonically active regions should perhaps take greater account of the relationship between the response time scales of landscapes compared with the nature of its tectonic perturbations. Beaumont et al. (2000), for example, discriminated 'slow forcing', where the time scale of the tectonic forcing is very long compared with the response time, 'intermediate forcing', where they are comparable, and 'rapid forcing', where the time scale of the tectonic forcing is short compared with the response time. In intermediate forcing, the response is commonly damped and phase-shifted, so that a lag time is apparent. This lag time, for example between tectonic uplift and sediment discharge, may be very long (10^6–10^7 years) for large orogenic systems (Kooi & Beaumont 1996), but shorter for trains of folds in fold–thrust belts (order 10^5 years; Tucker & Slingerland 1996), which immediately suggests that some caution should be used in inferring the timing of tectonic events on the basis of the age of the stratigraphic response. Beaumont et al. (2000) also identify an 'impulsive forcing' where the response follows an approximately exponential decay curve, allowing a relaxation time scale to be calculated.

The forcing function, or perturbation, may be of different forms (Fig. 8). It may take the form of a function with a characteristic time scale for its duration t and amplitude A and it may repeat with a period σ. It may also be a discrete, spike-like function with negligible duration but amplitude A and period σ. The forcing may be a step change, in which case we let the time since a change in boundary conditions be T. We further assume that the landscape or sediment routing system has a response time to or relaxation time following the impulsive forcing or perturbation τ, and differentiate between an analytical response time, equal to the time required for the full attainment of new equilibrium conditions τ_a, and a relaxation time ('decay time') required to achieve a fraction $(1 - e^{-t/\tau})$ of the total response τ_r.

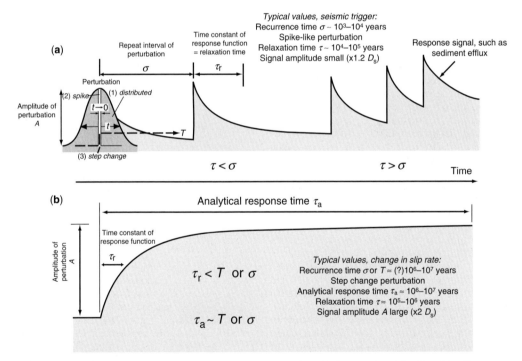

Fig. 8. The concept of a perturbation and its response function. (**a**) Perturbation may be (1) distributed, in this case in the form of a sinusoid with time scale t, (2) a spike of negligible time duration or (3) a step change in boundary conditions. The perturbation has a characteristic repeat interval σ, and the response function (such as sediment efflux) has a response time, here shown as a relaxation time τ. Two contrasting situations are shown, where $\tau < \sigma$ and $\tau > \sigma$. (**b**) A step change in boundary conditions provokes a long-term response, with an analytical response time for the full attainment of equilibrium at the asymptote τ_a, and a relaxation time attained when the response function has achieved $1/e$ of the total change in signal τ_r.

In a first scenario, the response time of the landscape or sediment routing system is very long compared with the repeat time (period) of the perturbation, which is equivalent to the 'rapid forcing' case of Beaumont et al. (2000). In this case, the landscape never fully adjusts to a single tectonic perturbation, so the tectonic events are strongly buffered by the sluggish reaction of the landscape. An example is the buffering of individual slip events on a range-bounding fault, which might involve a slip of c. 1–2 m repeated every 10^3–10^4 years, whereas the relaxation time for the longitudinal profile of the transverse rivers is expected to be in the region of 5×10^4 years or more, and the time scale for attainment of uniform relief c. 10^6 years. Considering a tectonically driven sediment efflux signal of a catchment, with amplitude A and duration t, it is likely that this signal will be modified by transfer through a large channel–floodplain system before entering the ocean. This is equivalent to the 'intermediate forcing' of Beaumont et al. (2000). If the alluvial channel–floodplain system has a characteristic time scale τ, the signal will reduce to an amplitude At/τ at the entry point into the ocean. If τ is large (c. 10^6 years), as suggested for the large alluvial systems of southeast Asia (Castelltort & Van Den Driessche 2003) and t/τ is $\ll 1$, the sediment routing system is highly buffered. Buffered systems therefore occur where $\tau \gg \sigma$ and where $t \ll \tau$ (Fig. 9).

In a second scenario, the response time of the landscape is fast compared with the long repeat time of the perturbation, which is partly equivalent of the 'slow forcing' case of Beaumont et al. (2000). In this case the landscape responds relatively rapidly to the perturbation, and can be termed 'reactive'. An example of a reactive landscape is a humid mountainous landscape with steep, narrowly spaced rivers, as on the western flank of the Southern Alps of New Zealand. Such a landscape responds quickly to a change of base level by rapid landsliding of hillslopes and bedrock river incision. Similarly, if the alluvial time scale τ is relatively small, as may occur on fan delta systems, sediment efflux signals of mountain catchments will be transferred with little modification to the ocean, and the shallow marine (and perhaps deep marine) stratigraphic record may closely record the tectonic driver in the source region both in terms of magnitude and timing. Reactive systems therefore occur where $\tau < \sigma$ and where $t > \tau$ (Fig. 9).

The idea of buffered v. reactive landscapes and sediment routing systems is related but not identical to the concept of steady-state or equilibrium. A landscape can be said to be 'steady' or 'in steady-state' if, through the operation of all of its linked geomorphic subsystems, it is statistically in

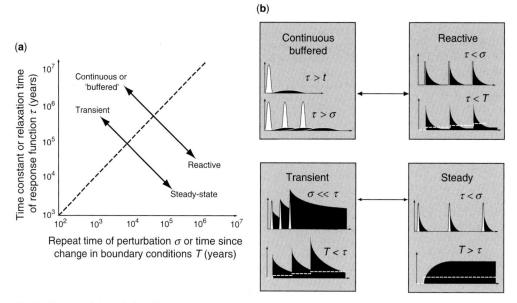

Fig. 9. Concept of a vocabulary for landscapes coupled to active tectonics. Landscapes are shown varying along two axes, from transient to steady, and from buffered to reactive. These theoretical types depend on a number of parameters: the time scale of the perturbation t; the time since a change in boundary conditions T; the response time and relaxation time of the response function τ; and the repeat time of the perturbation σ.

balance with the prevailing boundary conditions (e.g. Willett & Brandon 2002). Examples of steady landscapes are the catchments of the central parts of extensional fault segments. Here, the combination of steep slopes and adequate precipitation causes bedrock incision rates to keep pace with the tectonic fluxes caused by cumulative slip on the range-bounding fault. If secular changes in fault slip rate are infrequent (T is large), the catchment response time τ is relatively rapid (c. 5×10^4 years) and the footwall topography is steady. Aggradation on adjoining fans can also be considered steady under these conditions. At a larger scale, landscapes in steep mountain belts in a wet climate grow to equilibrium relatively quickly with slowly changing plate velocities. Any changes in the rate of tectonic advection of rocks to the orogenic surface are quickly adjusted to in an aggressive erosional environment. The wet flanks of the Southern Alps (New Zealand), Taiwan and central-eastern Himalayas are examples.

If, however, the response time τ is long compared with the time since a change in boundary conditions T, the landscape is out of equilibrium with the prevailing conditions and can be termed 'transient' (Fig. 9). An example is provided by the small catchments located within the tip-zone region of an extensional fault segment propagating at its tip. Despite their small size, these low gradient catchments respond slowly to their incorporation into the footwall block, and are in a transient stage with respect to their prevailing tectonic environment. Slip rate variations caused by glacial–interglacial changes in surface loads, such as ice and lake water masses, which operate at the 10^4 year time scale (Hetzel & Hampel 2005), are also likely to produce transient landscapes in uplifting footwalls. Alluvial basins and bedrock rivers subjected to eustatic changes in base level at the outlet are also commonly in a transient condition following high-frequency Quaternary glacial cycles. Downstream low gradients cause knickpoint celerities to be insufficient to establish steady conditions following high-frequency eustatic variation at the 10^4 year time scale. The result of partial adjustment to a series of high-frequency eustatic changes is a staircase of knickpoints each migrating at a different celerity (Bishop et al. 2004; Densmore et al. 2007b).

Outlook

There is a rich texture to the research area of tectonic landscapes and sediment routing systems. To further advance this field, there is an urgent need for more measurements that constrain ages, chronology and rates. In linked fault arrays and their catchment–fan systems, good observational data are required on fault slip rates, displacement-length profiles and denudation profiles, key geomorphic surfaces need to be dated and distinct fan segments characterized in terms of their geomorphic evolution and relation to climate change. Further data are required on medium- to long-term erosion and sediment transport rates. Key tools include thermochronology, especially (U – Th)/He, and cosmogenic nuclide dating. The general availability of digital elevation data, including very high resolution data from airborne laser swath mapping, allows the morphometry of erosional and depositional landscapes to be rapidly evaluated, including surface roughness and long profiles. Numerical modelling will continue to reveal important and sometimes counterintuitive results from integrated sediment routing systems. However, the most valuable results from numerical modelling will be derived from models that are well constrained by observational data.

The more we make steps in understanding landscapes and sediment routing systems, the more we discern their complex response to perturbations of all types. Had such information been available previously to guide stratigraphic models, it is arguable that the entire field of sequence stratigraphy or genetic stratigraphy would have developed differently. Stratigraphic architectures and gross depositional facies in siliciclastic depositional settings result from the interplay between accommodation generation and filling or partial filling with sediment. The extreme complexity of the sediment supply signal (what might be called the 'Q_s problem') makes forward modelling of stratigraphy a challenging exercise. Recognition of the Q_s problem should also help to introduce more realism in the simplistic inversion of driving mechanisms from the stratigraphic record. We might then be better able to read the epic narrative contained in sedimentary rocks.

Conclusions

There is no single response time of a landscape to a change in prevailing conditions, nor a single index of landscape sensitivity. In order to understand how landscapes and sediment routing systems interact with a tectonic displacement field, we need to investigate the varied response times of a linked set of geomorphic subsystems (Fig. 10). In this sense, landscapes and their associated sediment routing systems respond polychromatically, like the striking of a chord (Allen 2005). Coupled tectonic–surface process models therefore are geared to hearing the chords, discriminating the component frequencies and evaluating the strings and fret or finger positions used to create them.

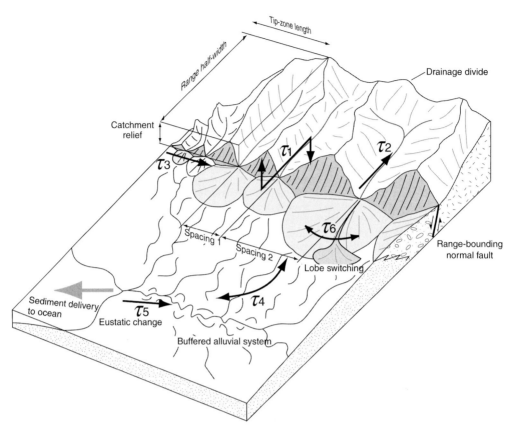

Fig. 10. Different response times in a sediment routing system characterized by an extensional fault block in the source region, and an alluvial system in the transport-depositional region. τ_1, relaxation time for the attainment of equilibrium conditions in catchment denudation rate and fan aggradation following a step change in the slip rate on the range-bounding fault; τ_2, response time and relaxation time for the development of new equilibrium profiles in bedrock rivers following a base-level fall at the outlet, block-type or non-uniform tectonic uplift of the footwall; τ_3, time scale for the attainment of uniform relief in the footwall estimated from the length of the tip-zone region in a laterally propagating fault segment; τ_4, response time of a buffered alluvial system to a sediment efflux signal from its source region; τ_5, response time for the upstream propagation of a knickpoint caused by a eustatic change; τ_6, time scale for fan resurfacing by lobe switching. For a similar concept see Carretier & Lucazeau (2005).

A survey of the landscapes and sediment routing systems in regions of growing extensional fault arrays, such as the Basin and Range province of western USA, exemplifies some of the issues that need to be considered in understanding such polychromatic systems. The landscape responds to a tectonic displacement field that includes the elastic deformation around a range-bounding fault undergoing cumulative coseismic slip, the lateral growth of the fault segment by tip propagation, the interference and linkage with other fault segments, and secular changes in long-term slip rate on the main border fault.

One landscape response is the development of a new equilibrium profile in bedrock rivers subjected to a sudden base level fall at the range front, determined by the celerity of knickpoint migration. Such knickpoint migration sets a new local base level for hillslope mass wasting and controls the sediment output to adjoining fans. Consequently, as long as there are no major changes in the quantity of stored sediment in the catchment over time, average fan aggradation rate responds with the same time scale as the eroding catchment regions. Another response is the development of young, transient catchments in the recently uplifted footwall close to the growing lateral fault tip. A plot of topographic relief along the strike of active faults shows that relief grows from zero at the fault tip to a statistically uniform value after a certain tip-zone distance, reflecting a characteristic time scale for the development of steady footwall topography. The relaxation time following a step change in slip rate or climate change is increased by the presence of an alluvial or bedrock piedmont adjacent to the range front.

Sediment routing systems are the distributors of mass over the Earth's surface. In the simplest systems, the products of catchment erosion are distributed over basin margin fans, with little export into downstream sediment transport systems. With a closed sediment budget, the ratio of fan to catchment area is primarily an indicator of the slip rate on border faults. The resurfacing of the fan by the migration and avulsion of active lobes introduces a further time scale in the sediment routing system. Commonly, however, a sediment pulse from a mountain catchment is transferred through an alluvial system to the ocean. Such alluvial systems have an effective diffusivity, and a characteristic time scale, determined by the ability of the floodplain to store sediment over long periods of time. Consequently, discrete, spike-like sediment efflux signals may be strongly damped before export to the ocean. If correct, this poses a challenge to stratigraphers attempting to invert causative mechanisms from the architecture of the sedimentary basin-fill.

I have benefited enormously over the last decade from discussions and collaboration with Alex Densmore (ETH-Zürich), Sanjeev Gupta (Imperial College London), Niels Hovius (Cambridge), Mike Ellis (Memphis), Hugh Sinclair (Edinburgh), Guy Simpson (ETH-Zürich) and Sébastien Castelltort (ETH-Zürich), all of whom are gratefully acknowledged. Sanjeev Gupta and Kerry Gallagher provided helpful reviews. This contribution stems from the William Smith lecture given at the Geological Society of London, 4 October 2004. I am grateful for the generous support of the Swiss National Science Foundation, projects 2100-067624.02, 200020-105225 and 200021-101695.

References

AHNERT, F. 1970. Functional relationships between denudation, relief, and uplift in large mid-latitude drainage basins. *American Journal of Science*, **208**, 243–263.

AHNERT, F. 1984. Local relief and the height limits of mountain ranges. *American Journal of Science*, **284**, 1035–1055.

ALLEN, P. A. 2005. Striking a chord. *Concepts Essay, Nature*, **434**, 21 April 2005, 961.

ALLEN, P. A. & ALLEN, J. R. 2005. *Basin Analysis: Principles and Applications*. Blackwell, Oxford.

ALLEN, P. A. & DENSMORE, A. L. 2000. Sediment flux from an uplifting fault block. *Basin Research*, **12**, 367–380.

ALLEN, P. A. & HOVIUS, N. 1998. Sediment supply from landslide-dominated catchments: implications for basin-margin fans. *Basin Research*, **10**, 19–35.

ALVAREZ, W. 1999. Drainage on evolving fold–thrust belts: a study of transverse canyons in the Apennines. *Basin Research*, **11**, 267–284.

ANDERSON, R. S. 1994. Evolution of the Santa Cruz Mountains, California, through tectonic growth and geomorphic decay. *Journal Geophysical Research*, **99**, 20161–20179.

ANDERSON, R. S. & HUMPHREY, N. F. 1990. Interaction of weathering and transport processes in the evolution of arid landscapes. *In*: CROSS, T. A. (ed.) *Quantitative Dynamic Stratigraphy*. Prentice Hall, Englewood Cliff, NJ, 349–361.

BABAULT, J., BONNET, S., CRAVE, A. & VAN DEN DREISSCHE, J. 2005. Influence of piedmont sedimentation on erosion dynamics of an uplifting landscape: an experimental approach. *Geology*, **33**, 301–304.

BALDWIN, J. A., WHIPPLE, K. X. & TUCKER, G. E. 2003. Implications of the shear stress river incision model for the time scale of post-orogenic decay of topography. *Journal Geophysical Research*, **108**, 2158.

BEAUMONT, C., ELLIS, S., HAMILTON, J. & FULLSACK, P. 1996. Mechanical model for subduction-collision: tectonics of Alpine-type compressional orogens. *Geology*, **24**, 675–678.

BEAUMONT, C., KOOI, H. & WILLETT, S. 2000. Coupled tectonic-surface process models with applications to rifted margins and collisional orogens. *In*: SUMMERFIELD, M. A. (ed.) *Geomorphology and Global Tectonics*. Wiley, Chichester, 29–55.

BISHOP, P., HOEY, T. B., JANSEN, J. D. & ARTZA, I. L. 2004. Knickpoint recession rate and catchment area: the case of uplifted rivers in eastern Scotland. *Earth Surface Processes and Landforms*, **30**, 767–778.

BRAUN, J. & SAMBRIDGE, M. 1997. Modelling landscape evolution on geological time scales: A new method based on irregular spatial discretization. *Basin Research*, **9**, 27–52.

BROWN, E. T., STALLARD, R. F., LARSEN, M. C., RAISBECK, G. M. & YIOU, F. 1995. Denudation rates determined from the accumulated of *in situ*-produced ^{10}Be in the Luquillo experimental forest, Puerto Rico. *Earth and Planetary Science Letters*, **129**, 193–202.

BULL, W. B. 1962. *Relations of Alluvial Fan Size and Slope to Drainage Basin Size and Lithology in Western Fresno County, California*. United States Geological Survey Professional Paper, **450-B**, 51–53.

BULL, W. B. 1964. *Geomorphology of Segmented Alluvial Fans in Western Fresno County, California*. United States Geological Survey Professional Paper, **353-E**, 89–129.

BULL, W. B. 1977. The alluvial fan environment. *Progress in Physical Geography*, **1**, 222–270.

BURBANK, D. W. 1992. Causes of recent Himalayan uplift deduced from depositional patterns in the Ganges Basin. *Nature*, **357**, 680–682.

BURBANK, D. W. & ANDERSON, R. S. 2001. *Tectonic Geomorphology*. Blackwell Science, Oxford.

BURBANK, D. W., LELAND, J. ET AL. 1996. Bedrock incison, rock uplift and threshold hillslopes in the northwestern Himalayas. *Nature*, **379**, 505–510.

CARRETIER, S. & LUCAZEAU, F. 2005. How does alluvial sedimentation at range fronts modify the erosional dynamics of mountain catchments. *Basin Research*, **17**, 361–381.

CARSON, M. A. & KIRKBY, M. J. 1972. *Hillslope Form and Process*. Cambridge University Press, Cambridge.

CASTELLTORT, S. & VAN DEN DRIESSCHE, J. 2003. How plausible are high-frequency sediment supply-driven cycles in the stratigraphic record? *Sedimentary Geology*, **157**, 3–13.

CLIFT, P. D., LAYNE, G. D. & BLUSZTAJN, J. 2004. *In*: CLIFT, P. D., WANG, P., HAYES, D. & KUHNT, W. (eds) *Continent–Ocean Interactions in the East Asian Marginal Seas*. American Geophysical Union, Washington, DC.

COWIE, P. A. 1998. A healing–reloading feedback control on the growth of seismogenic faults. *Journal Structural Geology*, **23**, 1901–1915.

COWIE, P. A. & SCHOLZ, C. H. 1992. Displacement–length scaling relationship for faults: data synthesis and discussion. *Journal of Structural Geology*, **14**, 1149–1156.

COWIE, P. A., GUPTA, S. & DAWERS, N. H. 2000. Implications of fault array evolution for synrift depocentre development: insights from a numerical fault growth model. *Basin Research*, **12**, 241–262.

DADE, W. B. & FRIEND, P. F. 1998. Grain size, sediment transport regime and channel slope in alluvial rivers. *Journal of Geology*, **106**, 661–675.

DAWERS, N. H. & UNDERHILL, J. R. 2000. The role of fault interaction and linkage in controlling syn-rift stratigraphic sequences: Statfjord East area, northern North Sea. *American Association of Petroleum Geologists Bulletin*, **84**, 45–64.

DAWERS, N. H., ANDERS, M. H. & SCHOLZ, C. H. 1993. Growth of normal faults: Displacement-length scaling. *Geology*, **21**, 1107–1110.

DENNY, C. S. 1965. *Alluvial Fans in the Death Valley Region, California and Nevada*. United States Geological Survey Professional Paper, **466**, Washington, DC.

DENSMORE, A. L., ELLIS, M. A. & ANDERSON, R. S. 1998. Landsliding and the evolution of normal fault-bounded mountain ranges. *Journal of Geophysical Research*, **103**, 15203–15219.

DENSMORE, A. L., DAWERS, N. H., GUPTA, S., GUIDON, R. & GOLDIN, T. 2004. Footwall topographic development during continental extension. *Journal of Geophysical Research*, **109**, F03001.

DENSMORE, A. L., DAWERS, N. H., GUPTA, S. & GUIDON, R. 2005. What sets topographic relief in extensional footwalls? *Geology*, **33**, 453–456.

DENSMORE, A. L., GUPTA, S., ALLEN, P. A. & DAWERS, N. H. 2007a. Transient landscapes at fault tips. *Journal of Geophysical Research – Earth Surface*, **112**, F03S08, doi: 10.1029/2006JF000560.

DENSMORE, A. L., ALLEN, P. A. & SIMPSON, G. 2007b. Development and response of a coupled catchment–fan system under changing tectonic and climatic forcing. *Journal of Geophysical Research – Earth Surface*, **112**, F01002, doi: 10.1029/2006JF000474.

DIETRICH, W. E., BELLUGI, D. G., SKLAR, L. S., STOCK, J. D., HEIMSATH, A. M. & ROERING, J. J. 2003. Geomorphic transport laws for predicting landscape form and dynamics. *In*: WILCOCK, P. R. & IVERSON, R. M. (eds) *Prediction in Geomorphology*. American Geophysical Union, Geophysical Monograph, **135**, 103–132.

DÜHNFORTH, M., DENSMORE, A. L., IVY-OCHS, S., ALLEN, P. A. & KUBIK, P. W. 2005. Timing and controls on debris-flow fan deposition and abandonment on alluvial fans in Owens Valley, California. *Geophysical Research Abstracts*, **7**, 03158, European Geosciences Union, 2005.

ELLIS, M. A., DENSMORE, A. L. & ANDERSON, R. S. 1999. Development of mountainous topography in the Basin Ranges, USA. *Basin Research*, **11**, 21–41.

FERNANDES, N. F. & DIETRICH, W. F. 1997. Hillslope evolution by diffusive processes: the time scale for equilibrium adjustments. *Water Resources Research*, **33**, 1307–1318.

FLEMINGS, P. B. & JORDAN, T. E. 1989. A synthetic stratigraphic model of foreland basin development. *Journal of Geophysical Research*, **94**, 3851–3866.

FOURNIER, F. 1960. *Climat et Erosion. La Relation entre l'Erosion du Sol par l'eau et les Précipitations Atmosphériques*. Presses Universitaires de France, Paris.

FRASER, G. S. & DECELLES, P. G. 1992. Geomorphic controls on sediment accumulation at margins of foreland basins. *Basin Research*, **4**, 233–252.

GAWTHORPE, R. L., JACKSON, C. A. L., YOUNG, M. J., SHARP, I. R., MOUSTAFA, A. R. & LEPPARD, C. W. 2003. Normal fault growth, displacement localization and the evolution of normal fault populations: the Hamman Faraun fault block, Suez Rift, Egypt. *Journal of Structural Geology*, **25**, 883–895.

GORDON, I. & HELLER, P. L. 1993. Evaluating major controls on basinal stratigraphy, Pine Valley, Nevada: implications for syntectonic deposition. *Bulletin of the Geological Society of America*, **105**, 47–55.

GRANGER, D. E., KIRCHNER, J. W. & FINKEL, R. 1996. Spatially averaged long-term erosion rates measured from *in-situ* produced cosmogenic nuclides in alluvial sediments. *Journal of Geology*, **104**, 249–257.

GUPTA, S., COWIE, P. A., DAWERS, N. H. & UNDERHILL, J. R. 1998. A mechanism to explain rift basin subsidence and stratigraphic patterns through fault array evolution. *Geology*, **26**, 595–598.

HACK, J. T. 1957. *Studies of Longitudinal Stream Profiles in Virginia and Maryland*. United States Geological Survey Professional Paper, **294-B**, 42–97.

HELLER, P. L. & PAOLA, C. 1992. The large-scale dynamics of grain size variation in alluvial basins, 2. Applications to syntectonic conglomerate. *Basin Research*, **4**, 91–102.

HETZEL, R. & HAMPEL, A. 2005. Slip rate variations on normal faults during glacial–interglacial changes in surface loads. *Nature*, **435**, 5 May 2005, 81–84.

HOOKE, R. L. 1968. Steady-state relationships on arid region alluvial fans in closed basins. *Americal Journal of Science*, **266**, 609–629.

HOOKE, R. L. & ROHRER, W. L. 1977. Relative erodibility of source area rock types, as determined by second order variations in alluvial fan size. *Geological Society of America Bulletin*, **88**, 1177–1182.

HOVIUS, N., STARK, C. P. & ALLEN, P. A. 1997. Sediment flux from a mountain belt derived by landslide mapping. *Geology*, **25**, 231–234.

HOWARD, A. D. & KERBY, G. 1983. Channel changes in badlands. *Bulletin of the Geological Society of America*, **94**, 739–752.

HOWARD, A. D., SEIDL, M. A. & DIETRICH, W. E. 1994. Modeling fluvial erosion on regional to continental scales. *Journal of Geophysical Research*, **99**, 13971–13986.

HUMPHREY, N. F. & HELLER, P. L. 1995. Natural oscillations in coupled geomorphic systems: an alternative origin for cyclic sedimentation. *Geology*, **23**, 499–502.

JOHNSON, D. D. & BEAUMONT, C. 1995. Preliminary results from a planform kinematic model of orogen evolution, surface processes and the development of clastic foreland basin stratigraphy. *In*: DOROBEK, S. L. & ROSS, G. M. (eds) *Stratigraphic Evolution of Foreland Basins*. Special Publication of the Society of Economic Paleontologists and Mineralogists, **52**, Tulsa, OK, 3–24.

KOOI, H. & BEAUMONT, C. 1994. Escarpment evolution on high elevation rifted margins: insights derived from a surface processes model that combines diffusion, advection and reaction. *Journal of Geophysical Research*, **99**, 12191–12209.

KOOI, H. & BEAUMONT, C. 1996. Large-scale geomorphology: classical concepts reconciled and integrated with contemporary ideas via a surface processes model. *Journal of Geophysical Research*, **101**, 3361–3386.

KOONS, P. O. 1989. The topographic evolution of collisional mountain belts: a numerical look at the Southern Alps, New Zealand. *American Journal of Science*, **289**, 1041–1069.

KOONS, P. O., ZEITLER, P. K., CHAMBERLAIN, C. P., CRAW, D. & MELTZER, A. S. 2002. Mechanical links between erosion and metamorphism in Nanga Parbat, Pakistan Himalaya. *American Journal of Science*, **302**, 749–773.

LECCE, S. A. 1990. The alluvial fan problem. *In*: RACHOCKI, A. H. & CHURCH, M. (eds) *Alluvial Fans: a Field Approach*. Wiley, New York, 3–24.

LEONARD, E. M. 2002. Geomorphic and tectonic forcing of late Cenozoic warping of the Colorado Piedmont. *Geology*, **30**, 595–598.

MARR, J. G., SWENSON, J. B., PAOLA, C. & VOLLER, V. R. 2000. A two-diffusion model of fluvial stratigraphy in closed depositional basins. *Basin Research*, **12**, 381–398.

MCLEOD, A. E., DAWERS, N. H. & UNDERHILL, J. R. 2000. The propagation and linkage of normal faults: insights from the Strathspey-Brent-Statfjord fault array, northern North Sea. *Basin Research*, **12**, 263–284.

METIVIER, F. & GAUDEMER, Y. 1999. Stability of outlet fluxes of large rivers in South and East Asia during the last 2 million years: implications for floodplain processes. *Basin Research*, **11**, 293–303.

MÉTIVIER, F., GAUDEMER, Y., TAPPONNIER, P. & KLEIN, M. 1999. Mass accumulation rates in Asia during the Cenozoic. *Geophysical Journal International*, **137**, 280–318.

MEYER, V, NICOL, A., CHILDS, C., WALSH, J. J. & WATTERSON, J. 2002. Progressivie localization of strain during the evolution of a normal fault population. *Journal of Structural Geology*, **24**, 1215–1231.

MOLNAR, P. & ENGLAND, P. 1990. Late Cenozoic uplift of mountain ranges and glocal climate change: Chicken or egg? *Nature*, **346**, 49–52.

MONTGOMERY, D. R. 1994. Valley incision and the uplift of mountain peaks. *Journal of Geophysical Research*, **99**, 13913–13921.

MONTGOMERY, D. R. 2001. Slope distributions, threshold hillslopes, and steady state topography. *American Journal of Science*, **301**, 432–454.

MORLEY, C. K. 1999. Patterns of displacement along large normal faults: implications for basin evolution and fault propagation, based on examples from East Africa. *Bulletin of the American Association of Petroleum Geologists*, **83**, 613–634.

OBERLANDER, T. M. 1985. Origin of drainage transverse to structures in orogens. *In*: MORISAWA, M. & HACK, J. T. (eds) *Tectonic Geomorphology*. The Binghampton Symposia in Geomorphology, International Series, **15**. Allen & Unwin, London, 155–182.

PAOLA, C. 2000. Quantitative models of sedimentary basin filling. *Sedimentology*, **47**(suppl. 1), 121–178.

PAOLA, C. & SEAL, R. 1995. Grain size patchiness as a cause of selective deposition and downstream fining. *Water Resources Research*, **31**, 1395–1407.

PAOLA, C., HELLER, P. L. & ANGEVINE, C. L. 1992. The large-scale dynamics of grain size variation in alluvial basins. 1, Theory. *Basin Research*, **4**, 73–90.

PARKER, G. 1978a. Self-formed straight rivers with equilibrium banks and mobile bed. Part 1: the sand-silt river. *Journal of Fluid Mechanics*, **89**, 109–125.

PARKER, G. 1978b. Self-formed straight rivers with equilibrium banks and mobile bed. Part 2: the gravel river. *Journal of Fluid Mechanics*, **89**, 127–146.

PARKER, G., PAOLA, C., WHIPPLE, K. X. & MOHRIG, D. C. 1998. Alluvial fans formed by channelized fluvial and sheet flow. 1: Theory. *Journal of Hydraulic Engineering*, **124**, 985–995.

PEARCE, A. J. & WATSON, A. J. 1986. Effects of earthquake-induced landslides on sediment budget and transport over a 50-yr period. *Geology*, **14**, 52–55.

PELLETIER, J. D. 2004. The influence of piedmont deposition on the time scale of mountain belt denudation. *Geophysical Research Letters*, **31**, L15502.

PINET, P. & SOURIAU, M. 1988. Continental erosion and large-scale relief. *Tectonics*, **7**, 563–582.

ROERING, J. J., KIRCHNER, J. W. & DIETRICH, W. E. 1999. Evidence for nonlinear, diffusive sediment transport on hillslopes and implications for landscape morphology. *Water Resources Research*, **35**, 853–870.

ROSENBLOOM, N. A. & ANDERSON, R. S. 1994. Hillslope and channel evolution in a marine terraced landscape, Santa Cruz, California. *Journal of Geophysical Research*, **99**, 14013–14029.

SAMBROOK-SMITH, G. H. & FERGUSON, R. I. 1995. The gravel–sand transition along river channels. *Journal of Sedimentary Research*, **65**, 423–430.

SCHALLER, M., VON BLANCKENBURG, F., HOVIUS, N. & KUBIK, P. W. 2001. Large-scale erosion rates from *in situ*-produced cosmogenic nuclides in European river sediments. *Earth and Planetary Science Letters*, **188**, 3–4.

SCHLISCHE, R. W., YOUNG, S. S., ACKERMANN, R. V. & GUPTA, A. 1996. Geometry and scaling relations of a population of very small rift-related normal faults. *Geology*, **24**, 683–686.

SCHMIDT, K. M. & MONTGOMERY, D. R. 1995. Limits to relief. *Science*, **270**, 617–620.

SEIDL, M. A. & DIETRICH, W. E. 1992. The problem of channel erosion into bedrock. *Catena Supplement*, **23**, 101–124.

SIMPSON, G. 2004a. Role of river incision in enhancing deformation. *Geology*, **32**, 341–344.

SIMPSON, G. 2004b. Dynamic interactions between erosion, deposition, and three-dimensional deformation in compressional fold belt settings. *Journal of Geophysical Research*, **109**, F03007.

SIMPSON, G. & SCHLUNEGGER, F. 2003. Topographic evolution and morphology of surfaces evolving in response to coupled fluvial and hillslope sediment transport. *Journal of Geophysical Research*, **108**, 2300, doi:10.1029/2002JB002162.

SINCLAIR, H. D. & ALLEN, P. A. 1992. Vertical versus horizontal motions in the Alpine orogenic wedge: stratigraphic response in the foreland basin. *Basin Research*, **4**, 215–232.

SINCLAIR, H. D., COAKLEY, B. J., ALLEN, P. A. & WATTS, A. B. 1991. Simulation of foreland basin stratigraphy using a diffusion model of mountain belt uplift and erosion: an example from the central Alps, Switzerland. *Tectonics*, **10**, 599–620.

SKLAR, L. & DIETRICH, W. E. 1998. River longitudinal profiles and bedrock incision models: Stream power and the influence of sediment supply. *In*: TINKLER, K. J. & WOHL, E. E. (eds) *Rivers over Rock: Fluvial Processes in Bedrock Channels*. AGU Geophysics Monograph Series, **107**. Washington, DC, 237–260.

SMITH, T. R. & BRETHERTON, F. P. 1972. Stability and the conservation of mass in drainage basin evolution. *Water Resources Research*, **8**, 1506–1529.

SNYDER, N. P., WHIPPLE, K. X., TUCKER, G. E. & MERRITTS, D. J. 2000. Landscape response to tectonic forcing: digital elevation model analysis of stream profiles in the Mendocino triple junction region, northern California. *Bulletin of the Geological Society of America*, **112**, 1250–1263.

SNYDER, N. P., WHIPPLE, K. X., TUCKER, G. E. & MERRITTS, D. J. 2003. Importance of a stochastic distribution of floods and erosion thresholds in the bedrock river incision problem. *Journal of Geophysical Research*, **108**, 2117.

STOCK, J. D. & MONTGOMERY, D. R. 1999. Geologic constraints on bedrock river incision using the stream power law. *Journal of Geophysical Research*, **104**, 4983–4993.

SUMMERFIELD, M. A. & HULTON, N. J. 1994. Natural controls on fluvial denudation rates in major world drainage basins. *Journal of Geophysical Research*, **99**, 13871–13883.

TUCKER, G. E. & SLINGERLAND, R. 1996. Predicting sediment flux from fold and thrust belts. *Basin Research*, **8**, 329–350.

TUCKER, G. E. & WHIPPLE, K. X. 2002. Topographic outcomes predicted by stream erosion models: sensitivity analysis and intermodel comparison. *Journal of Geophysical Research*, **107**, 2179.

TUCKER, G., LANCASTER, S., GASPARINI, N. & BRAS, R. 2001. The channel-hillslope integrated landscape development model (CHILD). *In*: HARMON, R. S. & DOE III, W. W. (eds) *Landscape Erosion and Evolution Modeling*. Kluwer Academic/Plenum, New York, 349–388.

WHIPPLE, K. X. 2001. Fluvial landscape response time: how plausible is steady state denudation? *American Journal of Science*, **301**, 313–325.

WHIPPLE, K. X. & MEADE, B. J. 2004. Controls on the strength of coupling among climate, erosion and deformation in two-sided, frictional orogenic wedges at steady state. *Journal of Geophysical Research*, **109**, F01011.

WHIPPLE, K. X. & TRAYLER, C. R. 1996. Tectonic control on fan size: the importance of spatially variable subsidence rates. *Basin Research*, **8**, 351–366.

WHIPPLE, K. X. & TUCKER, G. E. 1999. Dynamics of the stream power river incision model: Implications for height limits of mountain ranges, landscape response timescales, and research needs. *Journal of Geophysical Research*, **104**, 17661–17674.

WILLETT, S. D. 1992. Dynamic and kinematic growth and change of a Coulomb wedge. *In*: *Thrust Tectonics*, Chapman and Hall, London, 19–31.

WILLETT, S. D. 1999. Orogeny and orography: the effects of erosion on the structure of mountain belts. *Journal of Geophysical Research*, **104**, 28957–28981.

WILLETT, S. D. & BRANDON, M. T. 2002. On steady states in mountain belts. *Geology*, **30**, 175–178.

WILLETT, S. D., BEAUMONT, C. & FULLSACK, P. 1993. Mechanical model for the tectonics of doubly vergent compressional orogens. *Geology*, **21**, 371–374.

WILLETT, S. D., SCHLUNEGGER, F. & PICOTTI, V. 2006. Messinian climate change and erosional destruction of the central European Alps. *Geology*, **34**, 613–616.

WOBUS, C., HEIMSATH, A., WHIPPLE, K. X. & HODGES, K. 2005. Active out-of-sequence thrust faulting in the central Nepalese Himalaya. *Nature*, **434**, 1008–1011.

ZHANG, P., MOLNAR, P. & DOWNS, W. R. 2001. Increased sedimentation rates and grain sizes 2–4 Myr ago due to the influence of climate change on erosion rates. *Nature*, **410**, 891–897.

Equilibrium in the balance? Implications for landscape evolution from dryland environments

LOUISE J. BRACKEN[1] & JOHN WAINWRIGHT[2]

[1]*Department of Geography, University of Durham, Durham DH1 3LE, UK*

[2]*Sheffield Centre for International Drylands Research, Department of Geography, University of Sheffield, Winter Street, Sheffield S10 2TN, UK*

Abstract: Equilibrium is a central concept in geomorphology. Despite the widespread use of the term, there is a great deal of variability in the ways equilibrium is portrayed and informs practice. Thus, there is confusion concerning the precise meanings and usage of the concept. In this chapter we draw on examples from dryland environments to investigate the practical implications of applying and testing the concept of equilibrium. Issues that we cover include the importance of scale and spatial variability, time, the assumption of constant environmental feedbacks and non-linearities. The evaluation demonstrates that there are a range of problems inherent with using ideas of geomorphological equilibrium explicitly or implicitly to structure research in drylands. Many of these problems also apply to other environments.

The concept of equilibrium in geomorphology has been widely debated (e.g. Phillips 1992; Rhoads & Thorn 1993; Ahnert 1994; Thorn & Welford 1994). As we have previously discussed (Bracken & Wainwright 2006), there is thus much confusion about different understandings and different interpretations of the concept, which led us to re-evaluate how useful the concept is overall. While some authors argue that differences in terminology are unimportant (e.g. Phillips & Gomez 1994), we have argued that they actually add to the confusion, and thus undermine the foundations of the concept in underpinning geomorphological research. Often, these differences have arisen because of different underlying ideas about whether equilibrium should refer to the processes and/or form of the landscape. Generally (but not always: see Ahnert 1994), 'steady-state equilibrium' is used to refer to landscapes of constant form, while 'dynamic equilibrium' is more usually employed in relation to process-rate changes to produce constant forms. In contrast, 'disequilibrium' is often used to refer to landscapes that are *not yet* in equilibrium (however defined), while 'non-equilibrium' tends to be used for landscapes that do not attain equilibrium despite relatively long timescales of stability of external forcing factors (Table 1). A more detailed discussion of definitions and usage is provided by Bracken & Wainwright (2006).

Drawing on arguments from dynamical systems theory, Thorn & Welford (1994) suggested that this sort of confusion in terminology also occurs in the mathematics and physics literature, and thus simply importing further, different uses of the terminology from outside geomorphology is not necessarily helpful. Other disciplines, such as ecology, have also encountered similar problems (e.g. Turner *et al.* 1993; Perry 2002). While agreeing with this analysis, we suggested that, at a more general level, the complex, multivariate nature of geomorphic systems means that, in dynamical terms, it cannot be assumed *a priori* that they will necessarily exhibit equilibrial behaviour. Furthermore, the multivariate nature of the systems – whose process characteristics are often still poorly defined (e.g. the definition of stream–power relationships to erosional response; Dietrich *et al.* 2003) – means that a demonstration of *potential* equilibrium states is limited by problems of mathematical tractability (Bracken & Wainwright 2006). Any attempt to demonstrate the existence of equilibrium landscapes therefore depends on defining the study in relation to a precise and stated definition, and the collection of appropriate, high-quality, multivariate data in relation to well-understood representations of the process relationships. The spatial scale at which these data are collected will also affect the sorts of equilibrium that can be observed.

The choice of a specific definition is closely linked to evolving geomorphological methodologies, which are in turn related to disciplinary and environmental contexts of study. Much work has thus been focused on temperate environments and has often been interrelated with applied studies that needed to demonstrate that landscapes could be stable and thus managed (Graf 1984, 2001). One implication of this focus on temperate environments is the nature of change, which is

Table 1. *Summary of the key definitions of equilibrium and suggested differences between them*

Term	Proponent	Spatial scale[a]	Temporal scale (years)[a]	Form or process?
Dynamic equilibrium	Gilbert (1877) Ahnert (1994) Chorley & Kennedy (1971)	Hillslope–catchment	Relative to process rate	Constant form, processes maintain balance
Dynamic metastable equilibrium	Schumm (1973) Ahnert (1994) Schumm (1975) Tricart (1965) Pitty (1971)	Reach–catchment	Relative to process rate	Changing process regain balance Threshold jumps in form and process
Steady-state equilibrium	Mackin (1948) Rubey (1952) Hack (1960) Schumm & Lichty (1965) Richards (1982) Ahnert (1994)	Reach Catchment Catchment	10^1 10^1 Relative to process rate	Constant form, not necessarily static Balance of processes and form at system scale
Quasi-equilibrium	Petts & Foster (1985)			Oscillations about a mean, alternative term for steady-state equilibrium
Disequilibrium	Renwick (1992) Ahnert (1994) Thorn & Welford (1994)	Landform Landform	10^3–10^4 10^2	Changing form and process in an attempt to regain balance, but not yet in balance
Non-equilibrium	Renwick (1992) Ahnert (1994) Thorn and Welford (1994) Tooth & Nanson (2000)	Landform Landform Catchment	10^3–10^4 10^2 10^2	Absence of equilibrium, despite long timescales of stability

[a]These are implicit and inferred from reading the selected articles where possible.

typically dominated by more gradual changes, and by geomorphic responses that can be more frequently observed. In contrast, semi-arid and arid areas may be inactive for relatively long periods of time, and are typically dominated by more -xdramatic, infrequent events, so that process regimes are distinctly different. Temporal scales of investigation are thus also important in the analysis of geomorphological equilibrium. Thus, in this chapter, we will draw on a number of examples from dryland environments in order to investigate some of the practical implications of our earlier conceptual analysis, in relation to how to go about measuring equilibrium in geomorphic systems.

Measuring geomorphological equilibrium

Few people have actually suggested attempting to measure the existence of equilibrium in geomorphic systems. The measurement approaches suggested by Leopold (1973) and Richards (1982) are as follows: (i) analysing the temporal variation of channel form; (ii) examining the continuity of sediment transport; (iii) investigating the efficiency of channel form; and (iv) by correlating system properties. The analysis of variation in channel form is based on regime theory (Blench 1969), which proposes that channels adjust to average widths, depths and slopes that depend on the imposed discharge, sediment load and the ability of banks to erode. The steady-state equilibrium is considered to exist while these factors are unchanging and, assuming the relationship between form and process is constant, the morphology remains relatively constant over time (Richards 1982). Leopold (1973) proposed monumenting river cross sections to monitor channel change through its transient states. However, the regime theory that underlies this approach is an engineering concept that comes out of a modernist perspective that channels should behave in a way that is manageable and thus humanly modifiable. Increasingly, applied geomorphologists are finding that this concept is difficult to sustain (e.g. Brooks & Shields 1996; Downs & Gregory 2001).

The second method is based on the concept of grade (Mackin 1948). Mackin's definition of grade assumes spatial continuity of sediment transport so that the available slope is adjusted to provide just the velocity required to transport the load supplied (with available discharge and the prevailing channel characteristics). Richards (1982) therefore suggests monitoring sediment transport along a reach to establish whether it is occurring at a constant rate, which would suggest that equilibrium was being maintained. This concept is similarly problematic, however, and is also based on the assertion that process and form are unequivocally linked (Cooke & Reeves 1976; Lane & Richards 1997; Bracken & Wainwright 2006). It has been commonly accepted that this linkage can be demonstrated because of the existence of so-called characteristic forms of hillslope for specific processes. Yet, this idea ignores the underlying assumptions of the analysis that defines characteristic forms, which include that of the existence of equilibrium (Kirkby 1971; Carson & Kirkby 1972). Thus, an interpretation of equilibrium by pointing to empirical observations that a specific slope possesses a characteristic form is based on circular logic.

Furthermore, the approach suggests that negative feedbacks are not important in system response to maintain conditions. However, Simon (1992), for example, has shown that for two different fluvial systems, with very different characteristics and forcings, time-dependent reductions in specific energy at a point act to minimize the expenditure of energy over a reach during channel evolution as the fluvial system adjusts to a new equilibrium. His research suggests feedbacks exist within the channel system, but that the channels do not necessarily return to the initial equilibrium following interference in the system. Negative feedbacks within geomorphic systems are also supported by the work of Harvey in the Howgill Fells. Over the long term (decades or more) progressive changes in gully morphology have modified process interactions to lead ultimately to gully stabilization and hence slope forms considered to be at equilibrium (Harvey 1996, 2001). Yet this work also suggests changes in sediment generation and removal rates during the late Holocene due to changing climate (Harvey 1992). Hence environmental factors do not necessarily remain constant and thus changes may relate to external forcing as much as to internal feedbacks.

The third method develops the second approach further by focusing on monitoring the evolution of the cross-sectional form of the channel. Natural channels are considered in this analysis to be an approximation to the most efficient cross-sectional form, that is semi-circular. However, for any given cross section, several equilibrium slopes can maintain transport continuity because of the multivariate nature of the system, and hence number of degrees of freedom for the channel to adjust (Hey 1978; Richards 1982). It has therefore been assumed that, despite these degrees of freedom, channel cross-sections can be monitored to give an idea of equilibrium. Again, there are the same problems of process–form linkage inherent in this method, but there is also the issue that different spatio-temporal scales of equilibrium may mean that some elements of the system are in equilibrium (in a dynamical systems sense), while others may not (Thornes & Brunsden 1977). These issues of scale will be considered in more detail below.

The fourth method, the correlation of system properties, is based on the proposal that there is strong intercorrelation between system variables, and these may be used to diagnose equilibrium. For example, if width is plotted against discharge for a given stream and there is a lot of scatter, this is suggested to be diagnostic of a system in disequilibrium since not all points have a perfect correlation (Richards 1982). This approach suffers from the fallacy of affirming the consequent, as any correlation between system variables may come from the interactions within the system and thus tells us nothing about whether the system overall is at equilibrium.

Thus, it can be seen that all of the approaches that have been proposed to test the idea of geomorphic equilibrium are seriously flawed. Other reasons why Richards's (1982) criteria remain rarely tested are that the type of research necessary is unlikely to gain funding in the current atmosphere of research, and testing equilibrium is not a fashionable area for research and is not seen to further our understanding, not least because of embedded disciplinary concepts (Bracken & Wainwright 2006). Any suggested monitoring schemes are likely to be phenomenally expensive, which will also dissuade researchers from attempting any measurement. The techniques that would need to be used would be so invasive that they would significantly affect the system and not give representative results (see Lane et al. 1998).

The case for equilibrium in drylands

There has been a debate surrounding equilibrium in semi-arid areas because a single flood event cannot represent the range of morphologically significant discharges. Drylands are characterized by extreme flow variability with long periods of little or no flow, interspersed with occasional large flood events. The highly variable nature of precipitation in drylands, both temporally and spatially, typically produces many short-lived flow events, which evaporate and percolate to cause successive periods of valley aggradation when small floods deposit sediment, and incision and degradation when larger floods occur (Schumm 1961; Schick 1974; Bull & Kirkby 2002). These rapid and often dramatic changes in dryland channels can be contrasted with the idea that steady state occurs as channels are 'delicately adjusted' based on the assumptions derived from temperate channels that underpin Mackin's (1948, p. 463) definition. Pickup & Rieger (1979) proposed that channel forms in Australia result from a series of discharges rather than a single dominant discharge. The dimensions of the channel can only be explained by a single discharge of a given frequency when either the temporal fluctuation in discharge is low, or the particular channel dimension is insensitive to flow variations. In a similar vein, Wolman & Gerson (1978) argued that the effectiveness of an event reflects the morphological changes caused by both erosion and deposition, as well as the associated sediment transport that occurs during a flood event. In semi-arid channels, forms tend to represent the last major flood and recovery periods are relatively long (up to 100 years: Fig. 1) compared with the usually assumed steady-state timescales of geomorphic systems (Schumm & Lichty 1965: see discussion below) for smaller events to rework these major morphological changes (e.g. Graf 1979; Bull 1991). Richards (1982) noted that, if the recovery period exceeded the mean recurrence interval of the extreme flood responsible, equilibrium between the channel form and the representative channel-forming discharge of intermediate magnitude is not possible (see also Graf 1983, 1988; Warren 1985; Bull 1991; Cooke et al. 1993; Patton et al. 1993; Bourke & Pickup 1999).

Pickup & Reiger (1979) noted two types of non-equilibrium channels; those sensitive to variations in discharge, and those subject to extreme flow, while Bourke & Pickup (1999) proposed that the assemblage of forms was a product of a system where process and form are rarely in equilibrium. More recently Tooth & Nanson (2000) suggested that dryland river systems can exhibit both equilibrium and non-equilibrium conditions depending on factors such as catchment size, channel gradient, flood duration, unit stream power, channel confinement, sediment cohesion and bank strength. Tooth & Nanson (2000) applied Richards's (1982) four criteria to measure equilibrium to establish whether some Australian systems were in either equilibrium or non-equilibrium. Their study proposed that previous work had been undertaken on relatively small, steep and sparsely vegetated headwater reaches that were dominated by short-lived, high-magnitude floods. They concluded that, in middle reaches of intermediate to large, low gradient catchments, where long-duration floods generate moderate to low unit stream powers and boundary resistance is high, that there is evidence for (i) stable channels, (ii) insignificant sediment discontinuities, (iii) strong correlations between channel form and process and (iv) reaches adjusted to maximum sediment transport efficiency. Hence some reaches may be in a state of dynamic equilibrium in the sense of Gilbert or Ahnert (Fig. 2). Thus, the study of Tooth & Nanson (2000) also

Fig. 1. Illustration of the response of semi-arid catchment systems to events of long return periods. The Río Guadalfeo in southern Spain responded to major storm events on 18–19 October 1973, when up to 250 mm of rain fell in less than 10 h and on 6 November 1982, when 119 mm fell in less than 24 h. (**a**) A terrace cut-and-fill as a response to the 1973 event, dissecting the base of an alluvial fan from a tributary catchment; and (**b**) the two terrace levels on the left of the picture were formed during the two events.

emphasizes the importance of defining temporal and spatial scales of observation in the assessment of the existence of geomorphological equilibrium. However there are aspects of these studies that can be further questioned such as the assumption that steady-state equilibrium, on which Richards's (1982) criteria are founded, exists. If the form proxy for process equilibrium is questioned as discussed above and below, then it makes these techniques of measurement highly problematic. It also assumes a very simplistic approach, assuming that geomorphic systems are in simple equilibrium conditions, which contrasts with our understanding of dynamic systems, where multiple equilibria of different types are often found. Notable exceptions in the geomorphological literature are Thornes's

Fig. 2. An example from the Rambla de Torrealvilla in SE Spain to illustrate that different parts of a catchment can be in equilibrium whilst other areas may not. Here, side gullies continue to cut down, whist the main channel is relatively stable and possibly approaching equilibrium, although nickpoint retreat from the left of the image suggests that this stability is unlikely to be long-lived.

(1983) work on the dynamics of vegetation–erosion competition and Graf's (1983) and Thornes's (1982) applications of catastrophe theory to the understanding of dryland channels.

Issues arising from employing the geomorphological concept of equilibrium in drylands

Process and form revisited

The idea of constant form exists in Hack's (1960) work, but can also be traced to Gilbert's (1887, 1914) papers. Constant form hypotheses have been incorporated into several models of hillslope development, but ambiguity exists firstly as to whether constant form prevails throughout landscape development, or secondly, whether landforms converge on a constant form condition. The former is not very plausible since for this to happen down-wasting must be balanced by uplift, which is unlikely (Smith & Bretherton 1972). Such an adjustment cannot occur simply through the process of isostatic rebound following erosion, because the consequent amount of uplift is insufficient to balance the rate of removal (Anderson & Burbank 2001). The likelihood that other processes causing uplift can *exactly* balance the difference in rates is highly unlikely (Willett & Brandon 2002), notwithstanding some recent arguments that tectonic rates in some active orogens are controlled by feedback from erosion rates (Whipple & Meade 2004). The idea of convergence to a constant form is mathematically convenient since a constant solution is sought to equations for hillslope development (e.g. Carson & Kirkby 1972; Smith & Bretherton 1972). In terms of this concept, there is also the issue that there must be a period of differential downwearing to produce a varied landscape (Ahnert 1967). Hence constant form needs to be attained and then maintained. However, there is a possible contradiction with the first set of limitations relating to erosion–tectonics feedbacks. In particular this case implies that only relatively 'old' landscapes should be in steady state, but these landscapes are generally those that do not have strong tectonic inputs, and therefore it is still difficult to support a mechanism by which steady state could be achieved. Other problems also arise if a state of balance precludes the landscape evolving from one form to another. However, Hack (1960) suggested that, as long as forces operated at a gradual rate to maintain a balance in erosional processes, then the topography will remain in a state of balance, even though it is evolving. It is also important to remember that invariant forms need not be equilibrium forms (Kirkby 1974).

Even in temperate environments, case studies of fluvial forms reveal how widely actual rivers differ from the typical case generally identified (Miller & Gupta 1999). This difference raises questions concerning how channel morphology is related to the distribution of flows over time and to interactions between erosion and deposition. Kirkby (1999) suggests that, since many rivers have magnitude–frequency distributions very different from those of 'normal' rivers, and that these distributions may not be stationary over time, the notion of equilibrium form may not be relevant, let alone determined. For example, in dryland rivers, preservation of form is related to distance downstream and distance from the active channel, with the larger-scale forms preserved for longer periods (Bourke & Pickup 1999). The sediments deposited during high-magnitude events modulate the geomorphic response of low-magnitude floods and there is no evidence of whether this range of forms is stationary over time. Even extreme events will behave differently in similar environmental contexts based on the sequence of events and the past history of land-use and environmental change at the site. For example, Wainwright (1996a) found that extreme events in the Ouvèze basin of SE France caused relatively small impacts on the landscape because the landscape had been significantly eroded over a long period of time, while subsequent flood events in the Berre catchment of SW France had localized higher impacts where land-use change had only been relatively recent (Fig. 3; see also the review of antecedent conditions in land degradation and thus landscape evolution in Wainwright & Thornes 2003). In some respects, these issues are related to the sensitivity to initial conditions of chaotic systems, although the complexity of past trajectories of the complex landscape system makes it very difficult to evaluate such trajectories and their consequences (see also Vale 2003).

Many workers have agreed that ephemeral channels are in non-equilibrium because of the variable discharge regime, the episodic nature of landscape development and abrupt downstream transitions. The variability inherent in dryland environments produces an assemblage of fluvial forms that varies across a hierarchy of scales and suggests that form and process are rarely in equilibrium. This disparity is partly because it is possible to see the variability of forms preserved by the mix of large floods and small, localized flows. Yet in temperate alluvial rivers it is less easy to see the range of fluvial forms due to submergence by water and camouflage by vegetation, and the different environmental controls are brushed over in an attempt to conform to the concept of equilibrium. Hence is research suggesting that equilibrium exists where we can discount changes in environmental factors, or where they are less obvious, and does not exist where we can easily identify variations in controlling factors? Are we simplifying temperate processes and forms too much?

As noted by Richards (1990), the dominant critical rationalist approach in geomorphology implies that the link between process and form is unchallenged. A critical realist theoretical perspective would, on the other hand, imply a critical difference in the sense that the observable (landscape) form is generated by non-observable (geomorphic) processes. The link is not necessarily direct and attempts to demonstrate it can fall foul of the problem of affirming the consequent. There thus needs to be a qualitative assessment of the explanatory mechanisms outside the observations and the underlying models. An underlying concept of equilibrium as generally applied by geomorphologists is precisely this link between process and form. In a number of cases (e.g. Chorley & Kennedy 1971; Cooke & Reeves 1976; Pitty 1982; Schumm 1991) this link is seriously called into question by the equifinality of the same form produced by a range of different processes. However, if a form-based definition of equilibrium is taken, equifinality implies strongly equilibrial conditions, whereas a process-based definition would imply the opposite. This distinction would further explain how confusion arises because of a lack of a clearly defined concept of equilibrium. It also demonstrates how a contradiction can arise because of the differing implications of a study of form and process and the links between them.

The importance of scale and spatial variability

In Gilbert's (1909) definition of equilibrium, convex slopes were proposed as the fixed equilibrium geometry of slope form, assuming that both the soil depth and rate of weathering were constant. As demonstrated previously, if equilibrium conditions are assumed *a priori*, specific slope forms emerge from the action of different processes (Kirkby 1967; Carson & Kirkby 1972). Likewise, Hack (1960) assumed down-wasting occurred at the same rate throughout a catchment. Both workers can easily be criticized for these assumptions and, with the expansion of process monitoring and modelling, spatial and temporal variability is now understood to be a key issue in both understanding and predicting geomorphological

Fig. 3. Comparison of the major storm events in southern France. (**a, b**) From the event of 22 September 1992 in the Ouvèze, SE France, in which up to 300 mm of rain fell in just under 4 h, causing major modifications of the channel, but relatively little hillslope activity other than in badland areas and as debris flows on higher slopes (**b**) due to the extensive past history of slope erosion in the area (see Wainwright, 1996b for further details). (**c, d**) From the event of 12–13 November 1999 in the Berre catchment, SW France, in which up to 400 mm fell in 24 h, causing both major channel and hillslope erosion.

processes. This point also brings us again to the issue of scale.

Graf (1988) first suggested that geomorphological equilibrium states were scale-dependent and that choices of suitable temporal and spatial scales are necessary for determining which reaches of a particular river are in equilibrium or non-equilibrium. Different opinions concerning equilibrium may be reached depending on the spatial scale considered. Studies that have tended to investigate relatively small headwater basins in dryland environments have concluded that the systems are in non-equilibrium. However studies that have looked at larger fluvial systems have suggested that some

(c)

(d)

Fig. 3. (*Continued*).

sections may be in equilibrium (e.g. Tooth & Nanson 2000). Therefore the spatial scale may also influence conclusions drawn concerning the existence of equilibrium.

Other spatial considerations include whether different sections of a catchment can be in equilibrium whilst other sections are not? And if the river channel is in equilibrium do the slopes also have to be in equilibrium? The answer to these questions is likely to depend on both the definition of equilibrium used and the timescale over which equilibrium is considered (Table 1). For gravel streams, Pizzuto (1992) found that equilibrium timescales were unreasonably long, and suggested that channels were best described as systems with multiple timescales and multiple rates of response. Hence hydraulic geometry equations do not provide evidence for equilibrium landscapes, but reflect the constancy of network topology and the rapid response times of width and depth. Therefore network topology can be considered to be in equilibrium with the landscape, whilst some channel characteristics are not. Notwithstanding the possible existence of equilibrium in network topology, the state of adjacent hillslopes will be further complicated by the nature of slope-channel coupling, which Wainwright (2006) has argued can produce conditions of both equifinality and system divergence for otherwise similar configurations (see also Church 2002).

The question regarding whether some areas of a catchment can be in equilibrium whilst other areas are not can be further illustrated by considering flood generation in drylands. Our basic understanding of rainfall inputs and stream-channel behaviour in semi-arid areas is well established, but we only have limited information about how rainfall inputs are related to flood generation at different scales (Bracken et al. 2007). Approximately two-thirds of runoff events in the hyper-arid Negev and Sinai are generated in medium-sized catchments (up to 50 km^2) by high-intensity, short-duration storm events, which have extremely flashy hydrographs and have dramatic effects on geomorphology (Schick 1988). Longer-term floods were produced by low rainfall intensities which may last a few days, but peak discharges rarely reached those obtained by the intense cloud burst-type rain and the floods which tend to produce more dramatic changes to geomorphology (Schick 1988). In contrast Dick et al. (1997) focused on small basins (0.88 up to 1.03 km^2) and showed that rapid runoff was produced during short duration storms, although the largest flow event recorded was produced by a low-intensity storm of long duration, which occurred when antecedent moisture was high. Bracken et al. (2007) presented results from analysis of a 6 year data record in SE Spain from research conducted in two large basins (171 and 200 km^2). Floods were related to the total rainfall occurring in a spell of rain, rather than to rainfall intensity. Whilst most storms occurred over a period of less than 24 h and the number of rainfall events declined as the duration exceeded 8 h, analysis showed that floods were produced by storms lasting longer than 18 h. Only one flood event was produced by a very short (15 min) storm with high-intensity rainfall. Thus at different spatial scales the processes resulting in flood generation are likely to be different, with rainfall intensity more important at small spatial scales, and total rainfall more important at larger scales (Fig. 4). The response time to large rainfall/flood events will therefore also vary at different spatial scales according to the dominant processes operating. This therefore supports Pizzutto's (1992) suggestion of multiple timescales and multiple rates of response, and thus differential responses in different parts of a catchment can be expected.

Time

The notion of time is central to the geomorphological concept of equilibrium, despite the fact that most of the commonly used definitions are not explicit about the time frames involved (Table 1). This notion can lead, and has led, to problems in deciding whether variables are dependent or independent. The first and classic study of geomorphic time was by Schumm & Lichty (1965), who attempted to reconcile the dichotomy of views held in the 1960s in relation to the approaches of Davis and of Gilbert (with respect to the Davisian cycle and Gilbertian equilibrium and thus form- v. process-based methodologies). Schumm & Lichty (1965) introduced three timescales (cyclic, modern and steady) with implicit notions of embedded equilibrium (i.e. they call 'modern' time also 'graded' and the use of 'steady' to relate to the steady state of the present). However, this definition of different time periods is a qualitative one, can lead to confusion and adds issues of circularity into the concept of equilibrium.

Despite this attempt to be explicit about time scales, time has remained a key problem, especially in choosing the 'correct' time and spatial scale of a study, with kilometre-scale landscapes taking millions of years to develop and those on metre-scales taking millennia to decades (Church 1995). It has also been suggested that the inference of Schumm & Lichty (1965) that different scales of form and process are causally independent of each other is unsustainable since short-timescale and small-spatial-scale processes influence processes over longer timescales and spaces (Lane & Richards 1997). As Kennedy (1997) suggests, the argument of Lane & Richards (1997) is founded on the concepts of thresholds and complex response suggested by Schumm (1977), and therefore has significant implications in relation to possible multiple equilibrium behaviour.

Howard (1982) was the first to try to define the relevant timescales for using the equilibrium concept and proposed that the appropriate scale to describe equilibrium is related to the rate at which a particular system responds to change. This approach was an improvement on previous work, but defining the relaxation rate of any geomorphic system is usually impractical (Mayer 1992), contains implicit assumptions of the system dynamics

(a)

(b)

Fig. 4. Example of a flood event affecting a semi-arid catchment in a spatially variable way, Walnut Gulch, Arizona, July 1994. (**a**) Hillslope and headwater channel response is relatively rapid following the onset of rainfall; while (**b**) downstream response is much delayed and corresponds to a significant phase of sediment transport in the channel long after the hillslopes have stopped responding. Subsequent work on the scour and fill of channels in this catchment by Powell *et al.* (2005) demonstrated a complex response with significant implications for understanding potential equilibrium behaviour of these channels.

(and thus the lack of *a priori* definitions of equilibrium behaviour as discussed above), and means that the concept must lose some generality. Hence the equilibrium concept is convenient for the development of a short-term model caused by negative feedbacks, and is excellent for highlighting the (problematic) relationship between form and process. However, there are problems with change of timescales and shifting variables from independent to dependent. These issues are central to the problems in using equilibrium in a practical context.

At longer timescales, environmental factors change significantly. Over the Quaternary period, there have been numerous cycles from cold to warm conditions, and different global conditions existed previously through the Cenozoic period.

Wainwright (2006) demonstrated using a landform-evolution model that realistic climate changes over the last 740 ka produced significant transient responses in landscape behaviour, particularly when modulated in a complex way due to interactions with vegetation. Thus, most landforms have evolved under changing environmental factors. Geomorphological (and nonlinear dynamics, to a certain extent) approaches to equilibrium encourage us to consider these changes as perturbations to the landscape system. At shorter timescales, this approach may provide useful results, and at intermediate timescales, Wainwright (2006) suggested that analysis of phase spaces of the response of complex systems is a useful way forward in characterizing their behaviour.

It should be noted that parallel debates exist in the ecological literature regarding spatial and temporal scale. Turner et al. (1993, p. 224) note that 'much of the controversy surrounding concepts of landscape equilibrium can be eliminated with the explicit consideration of the spatial scales of disturbance [in] the landscape and the temporal scales of disturbance and recovery'. They suggest that researchers should at least try to think in terms of the correct 'scale ball-park'. Often, this explicitness requires detailed knowledge of the system behaviour, which is often not possible in complex geomorphological systems, nor in ecological systems (Levin 1992).

The assumption of constant environmental factors (or whatever happened to feedbacks?)

The concept of feedbacks primarily entered geomorphology with the focus on quantitative methods and then systems theory (e.g. Chorley & Kennedy 1971). In one sense they can be considered to be the mechanism by which equilibrium happens (or non-equilibrium in the case of positive feedbacks), and many current descriptions of geomorphic systems are based on the description of feedbacks (see especially Ahnert 1996). Conversely, it may be questioned whether they possess any explanatory power and are simply *ad hoc* statements that are used to complement the main hypothesis of how a process operates. Evaluation of the existence of feedbacks in geomorphic systems is therefore central to the evaluation of the usefulness of the equilibrium concept in geomorphology. In practical terms, the only way of testing their existence is by the evaluation of modelled systems with empirical data.

Over timescales appropriate to the development of the landforms, an approach that externalizes environmental variability as perturbations to the system potentially undermines the explanatory potential. For example, feedbacks relating to uplift and enhanced weathering are considered to have played a major role in the development of the Quaternary glaciations (e.g. Ruddiman & Raymo 1988; Raymo 1994). Considering these factors as external to the geomorphic system rather than as internal feedbacks is thus clearly problematic.

Hence whilst feedbacks are central to the concept of equilibrium and are key to returning systems to a balanced state, research has shown that changes in process rates do not necessarily occur in the direction suggested by the concept. Process rates are subject to change in response to system forcings, but are more easily measured and understood for short time periods. Over long time periods there are many changes in environmental conditions, so it is difficult to understand the range of forcings operating in geomorphological systems, and hard to unravel the complex interactions that control form and process. For instance Bracken & Kirkby (2005) demonstrated that in semi-arid areas runoff produced by short-lived storms with high, but not rare, rainfall intensities and total rainfall amounts, can cause significant hillslope sediment transport (Fig. 5). Measured sediment transport was related to runoff and a qualitative estimate of slope position but varied dramatically with lithology; marl sites produced most runoff and sediment transport, followed by the sites of mixed red and blue schist, then blue schist. Figure 6 shows how calculated sediment transport increased with level of erosion and tended to level off as rills and gullies start to dominate hillslope sediment transport. It is therefore possible to develop a good understanding of process rates at small scales over relatively short time periods. However, when data from different hillslopes are combined to investigate how sediment transport varies at the catchment scale (approximately 200 km^2), there is no clear relationship, although all rates do plot within an envelope curve (Fig. 7). This uncertainty is due to local variations in sediment transport caused by variations in sediment movement between rill and interrill areas, due to local influences of patchy vegetation, but also due to environmental forcings not investigated in the empirical data collection. Thus, the issue of unravelling complex process interactions and their consequences is non-trivial, even at relatively small spatial and temporal scales.

Nonlinearities

According to Howard's (1988) definition of equilibrium, the output of a system at any time is related to the input at the same time by a single, valued, temporarily invariant functional relationship. Nonlinearities exist whenever outputs of energy are not proportional to the inputs (Phillips 1992). Schumm (1979) proposed the existence of thresholds in geomorphic systems and hence these

Fig. 5. Major responses to a major storm event in the Rambla Nogalte (SE Spain). (**a**) Olive trees adjacent to the main river channel buried by hillslope-channel sediment transport during a large flood; and (**b**) removal of the surface layer of hillslope sediment.

systems must contain nonlinearities. His earlier papers on complex response in semi-arid channel systems (Schumm 1973, 1975), as well as the work of Thornes (1983) and Graf (1983) in relation to catastrophe theory, further emphasize the importance of nonlinearities. Therefore according to Howard's (1988) definition, equilibrium cannot exist because the relationship between system inputs and outputs is not temporarily invariant (Phillips 1992).

Phillips (1992) also considers dissipative systems, which are systems in which energy is dissipated in maintaining order in states removed from equilibrium, such as self-organizing processes which produce an orderly sequence of systems configurations, with each sequence induced by a fluctuation. Self-organization is known to occur across a range of scales in dryland processes, for example from rill formation (Favis-Mortlock 1998), to the development of banded vegetation (Thiery et al.

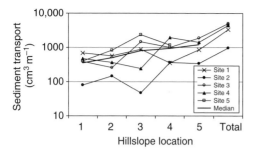

Fig. 6. Variations in calculated sediment transport from spray-painted lines. The heavy line shows the weak median trend of increasing sediment transport downslope.

1995; Valentin *et al.* 1999; Ludwig *et al.* 2005) the relationship between human action and erosion patterns (Wainwright 2007) and the frequency–size characteristics of landslides (Guzzetti *et al.* 2005). Phillips (1992) proposed that open geomorphic systems that exhibit thresholds are dissipative. This definition implies the existence of deterministic complexity in geomorphic systems, otherwise known as chaos. Chaos describes irregular, apparently random behaviour which arises deterministically due to non-linear couplings in systems. Chaos is distinguished from stochastic complexity by extreme sensitivity to initial conditions and by the increasing divergence between results and predictions with time (Phillips 1992). Despite recent advances in developing methods to assess the presence of chaos in geomorphic systems (e.g. Phillips 1999; Gomez & Phillips 1999; Sivakumar & Jayawardena 2002; Regonda *et al.* 2004), it often remains difficult to specify whether a specific system is chaotic, not least because such approaches require long data series, which are often not available.

Discussion and conclusions

Despite the apparently disparate nature of the issues raised in this paper, they point to a range of problems that are inherent with using ideas of geomorphological equilibrium either explicitly or implicitly to structure research programmes. As we have previously noted (Bracken & Wainwright 2006), many of these issues relate to untested ideas and hypotheses that have become entrenched in the collective geomorphological psyche. This chapter has further argued that these ideas are probably also untestable. One reason for this problem is the inherent temporal and spatial scales of different concepts of equilibrium, and the lack of specification of these scales as applied by different workers. Fundamentally, though, the inherent complexity of geomorphic systems means that even from first principles it is generally impossible to specify the longer-term behaviour of the system. The usefulness of approaches that are either implicitly or explicitly underpinned by equilibrium concepts is thus debateable. The convenience or comfort of employing such an approach must seriously be weighed against the potential problems, not least those relating to the production of misleading results or understandings.

In dryland environments, there are a number of specific problems in applying the concept of equilibrium. Spatial heterogeneity at hillslope to catchment scales controls the response to land-forming events. The impact of this heterogeneity is further accentuated by the large temporal variability in the size of these events. There is therefore the

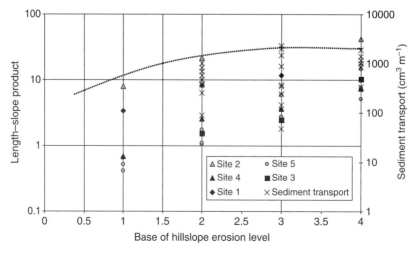

Fig. 7. The relationship between sediment transport and hillslope morphology. See text for further details.

consequence that spatial variation dominates landscape response, and hence studies of landscape response are inherently location specific. There is therefore a danger in extrapolating from a limited number of examples, and thus a tendency to overgeneralize, or conversely reject hypotheses because of a lack of recognition of the specificity of the location of specific investigations. Difficulties in relating measurements and observations across scales can further limit the potential of developing more general understanding from such data sources. In this context, the impact of external forcings can be difficult to interpret and interrelate, especially considering the highly anthropized landscapes that are typically investigated. Threshold behaviours seem widespread and further complicate these interpretations. Given these limitations, it is difficult to characterize the time-scale of observations needed to characterize a landscape before deciding it is in dis-equilibrium or non-equilibrium. An alternative is to use slow process rates as an indication, yet an environment can have slow process rates but still have large external forcings that can produce dramatic landscape change.

Are these problems then just specific to research in dryland environments? In a tropical setting, Gupta (2004) has demonstrated the complex response of the Mekong system in relation to tectonic processes and variability in monsoonal cycles. Even in temperate environments there has been an increasing recognition of the rôle of variability in system characterization and response. For example, ideas of variable source areas (Hewlett & Hibbert 1967; Beven 1983), connectivity of catchment properties (Todd et al. 2006) and 'old' and 'new' water (McDonnell et al. 1990; Weiler et al. 2003) are becoming more prevalent.

The implications for the study of landscape evolution are thus considerable. Given that there is no *a priori* requirement for landscapes to be in any form of equilibrium, and that there is no convincing demonstration for their existence, it is important not to introduce circular arguments into the interpretation of landform evolution. Further, given the necessity for the use of models in the understanding of landform evolution, not least because of the difficulty of extrapolating process understanding over time periods orders of magnitude longer than human experience, results should not be artificially constrained by imposing conditions that will produce outcomes that simply reinforce the initial, untested assumptions. As pointed out by Dietrich *et al.* (2003), such models are typically poorly tested, especially in comparison with the extensive literature that exists on the difficulties of model validation in other contexts (e.g. Oreskes et al. 1994; Anderson & Bates 2001; Mulligan & Wainwright 2003). Certain techniques applied to the understanding of long-term erosion rates almost certainly have similar assumptions built into them (cf. Von Blanckenburg 2005). In this case, field and laboratory measurements are thus not independent sources of data to test the behaviour of landform-evolution models.

A new generation of models are needed that exhibit self-organizing behaviour in the evolution of the landscape. These models will be less constrained by the pre-existing conceptual frameworks than existing models. To do so, there also needs to be a convergence in approach to the representation of fundamental processes. It is unlikely that significant progress can be made in understanding the principles of landform evolution until we understand why different formulations of sediment transport work better in different settings. The new models will also allow us to recognize the importance of complex response and complexity across a range of scales. The role that models play in informing observation must also be recognized. For example, despite the complexity demonstrated in the link between climate and uplift regimes by Willett & Brandon (2002), tests of this relationship using field data have been overly simplistic (e.g. Burbank *et al.* 2003; Reiners *et al.* 2003), notwithstanding the inherent limitations of the data employed in these cases (Hovius 2004). Complex responses across a range of spatial and temporal scales are perhaps most evident in dryland systems, but is increasingly recognized in other environments as noted above. There is a significant body of work that investigates sources of this complexity in terms of spatial variation and heterogeneity. For example, work on understanding spatial pattern (Bull *et al.* 2003), connectivity (Cammeraat 2002; Bracken & Croke, 2007; Müller *et al.* 2007) and scaling (Kirkby *et al.* 1996; Zhang *et al.* 2002) in catchment systems allows us to develop our understanding of complex responses to events that modify the landscape. However, this work has typically only been focussed on very short timescales. It is only when we can understand the evolution of these patterns and their interaction with other processes over long timescales that we can provide a complete understanding of landscape evolution.

References

AHNERT, F. 1967. The role of the equilibrium concept in the interpretation of landforms of the fluvial erosion and deposition. *In*: MACAR, P. (ed.) *L'Evolution des Versants*. Université de Liège, Liège, 23–41.

AHNERT, F. 1994. Equilibrium, scale and inheritance in geomorphology. *Geomorphology*, **11**, 125–140.

AHNERT, F. 1996. *Introduction to Geomorphology*. Arnold, London.

ANDERSON, D. W. & BURBANK, R. S. 2001. *Tectonic Geomorphology*. Blackwell Science, Oxford.

ANDERSON, M. G. & BATES, P. D. (eds) 2001. *Model Validation: Perspectives in Hydrological Science.* Wiley, Chichester.

BEVEN, K. 1983. Catchment geomorphology and the dynamics of runoff contributing areas. *Journal of Hydrology,* **65,** 139–158.

BLENCH, T. 1969. *Mobile Bed Fluviology.* University of Alberta Press, Edmonton.

BOURKE, M. C. & PICKUP, G. 1999. Fluvial form variability in arid central Australia. In: MILLAR, A. & GUPTA, A. (eds) *Varieties of Fluvial Form.* Wiley, Chichester, 249–271.

BRACKEN, L. J. & CROKE, J. 2007. The concept of hydrological connectivity and its contribution to understanding runoff-dominated geomorphic systems. *Hydrological Processes,* **21,** 1749–1763.

BRACKEN, L. J. & KIRKBY, M. J. 2005. Differences in hillslope runoff and sediment transport rates within two semi-arid catchments in southeast Spain. *Geomorphology,* **68,** 183–200.

BRACKEN, L. J. & WAINWRIGHT, J. 2006. Geomorphological equilibrium; myth and metaphor? *Transactions of the Institute of British Geographers,* **NS 31,** 167–178.

BRACKEN, L. J., COX, N. J. & SHANNON, J. 2007. The nature of rainfall inputs and their relationship to flood generation in semi-arid systems. *Journal of Hydrology,* doi: 10.1002/hyp.6641.

BROOKES, A. & SHIELDS, D. F. 1996. *River Channel Restoration: Guiding Principles for Sustainable Projects.* Wiley, Chichester.

BULL, L. J. & KIRKBY, M. J. 2002. Dryland river characteristics and concepts. In: BULL, L. J. & KIRKBY, M. J. (eds) *Dryland Rivers; Hydrology and Geomorphology of Semi-Arid Channels.* Wiley, Chichester, 3–16.

BULL, L. J., KIRKBY, M. J., SHANNON, J. & DUNSFORD, H. D. 2003. Predicting hydrologically similar surfaces (HYSS) in semi-arid environments. *Advances in Environmental Monitoring and Modelling,* **1,** 1–26.

BULL, W. B. 1991. *Geomorphic Responses to Climate Change.* Oxford University Press, Oxford.

BURBANK, D. W., BLYTHE, A. E., PUTKONEN, J., PRATT-SITAULA, B., GABET, E., OSKIN, M., BARROS, A. & OJHA, T. P. 2003. Decoupling of erosion and precipitation in the Himalayas. *Nature,* **426,** 652–655.

CAMMERAAT, L. H. 2002. A review of two strongly contrasting geomorphological systems within the context of scale. *Earth Surface Processes and Landforms,* **27,** 1201–1222.

CARSON, M. A. & KIRKBY, M. J. 1972. *Hillslope Form and Process.* Cambridge University Press, Cambridge.

CHORLEY, R. J. & KENNEDY, B. A. 1971. *Physical Geography, A Systems Approach.* Prentice Hall International, London.

CHURCH, M. 1995. Geomorphic response to river flow regulation – case-studies and time-scales. *Regulated Rivers – Research & Management,* **11,** 3–22.

CHURCH, M. 2002. Geomorphic thresholds in riverine landscapes. *Freshwater Biology,* **47,** 541–557.

COOKE, R. U. & REEVES, R. W. 1976. *Arroyos and Environmental Change in the American South-West.* Clarendon Press, Oxford.

COOKE, R. U., WARREN, A. & GOUDIE, A. S. 1993. *Desert Geomorphology,* 2nd edn. University College London Press, London.

DICK, G., ANDERSON, R. S. & SAMPSON, D. E. 1997. Controls on flash flood magnitude and hydrograph shape, Upper Blue Hills badlands, Utah. *Geology,* **25,** 45–48.

DIETRICH, W. E., BELLUGI, D. G., SKLAR, L. S., STOCK, J. D., HEIMSATH, A. M. & ROERING, J. J. 2003. Geomorphic transport laws for predicting landscape form and dynamics. In: WILCOCK, P. R. & IVERSON, R. M. (eds) *Prediction in Geomorphology.* Geophysical Monograph **135,** American Geophysical Union, Washington, DC, 1–30.

DOWNS, P. W. & GREGORY, K. J. 2001. *River Channel Management: Towards Sustainable Catchment Hydrosystems.* Arnold, London.

FAVIS-MORTLOCK, D. 1998. A self-organizing dynamic systems approach to the simulation of rill initiation and development on hillslopes. *Computers & Geosciences,* **24,** 353–372.

GILBERT, G. K. 1877. *Report on the Geology of the Henry Mountains.* United States Geographical and Geological Survey of the Rocky Mountains Region. United States Government Printing Office, Washington, DC.

GILBERT, G. K. 1909. The convexity of hilltops. *Journal of Geology,* **17,** 344–350.

GILBERT, G. K. 1914. The Transportation of Debris by Running Water. United States Geological Survey Professional Paper **86,** Washington, DC.

GOMEZ, B. & PHILLIPS, J. D. 1999. Deterministic uncertainty in bed load transport. *Journal of Hydraulic Engineering – ASCE,* **125,** 305–308.

GRAF, W. L. 1979. Mining and channel response. *Annals of the Association of American Geographers,* **69,** 262–275.

GRAF, W. L. 1983. Flood related channel change in an arid region river *Earth Surface Processes and Landforms,* **8,** 125–139.

GRAF, W. L. 1984. The geography of American field geomorphology. *Professional Geographer,* **36,** 78–82.

GRAF, W. L. 1988. *Fluvial Processes in Dryland Rivers.* Springer, Berlin.

GRAF, W. L. 2001. Damage control: restoring the physical integrity of America's rivers. *Annals of the Association of American Geographers,* **91,** 1–27.

GUPTA, A. 2004. The Mekong River: morphology, evolution and palaeoenvironment. *Journal of the Geological Society of India,* **64,** 525–533.

GUZZETTI, F., REICHENBACH, P., CARDINALI, M., GALLI, M. & ARDIZZONE, F. 2005. Probabilistic landslide hazard assessment at the basin scale. *Geomorphology,* **72,** 272–299.

HACK, J. T. 1960. Interpretation of erosional topography in humid temperate regions. *American Journal of Science,* **258,** 80–97.

HARVEY, A. M. 1992. Process interactions, temporal scales and the development of hillslope gully systems: Howgill Fells, northwest England. *Geomorphology* **5,** 323–344.

HARVEY, A. M. 1996. Holocence hillslope gully systems in the Howgill Fells, Cumbria. In: ANDERSON, M. G. & BROOKS, S. M. (eds) *Advances in Hillslope Processes,* Vol. 2. Wiley, Chichester, 247–270.

HARVEY, A. M. 2001. Coupling between hillslope and channels in upland fluvial systems; implications for landscape sensitivity, illustrated from the Howgill Fells, NW England. *Catena*, **42**, 225–250.

HEWLETT, J. D. & HIBBERT, A. R. 1967. Factors affecting the response of small watersheds to precipitation in humid regions. *In*: SOPPER, W. E. & LULL, H. W. (eds) *Forest Hydrology*. Pergamon Press, Oxford, 275–290.

HEY, R. D. 1978. Determinate hydraulic geometry of river channels. *Journal of the Hydraulics Division, Proceedings of the American Society of Civil Engineers*, **104**, 869–885.

HOVIUS, N. 2004. Time scales of erosion of the Taiwan orogen. William Smith Meeting, 'Earth's Dynamic Surface – Catastrophe and Continuity in Landscape Evolution'.

HOWARD, A. D. 1982. Equilibrium and time scales in geomorphology – application to sand-bed alluvial streams. *Earth Surface Processes and Landforms*, **7**, 303–325.

HOWARD, A. D. 1988. Equilibrium models in geomorphology. *In*: ANDERSON, M. G. (ed.) *Modelling Geomoprhological Systems*. Wiley, Chichester, 49–72.

KENNEDY, B. 1997. Classics in physical geography revisited – time, space and causality in geomorphology. *Progress in Physical Geography*, **21**, 419–423.

KIRKBY, M. J. 1967. Measurement and theory of Soil Creep. *Journal of Geology*, **75**, 359–378.

KIRKBY, M. J. 1971. Hillslope process–response models based on the continuity equation. *In*: BRUNSDEN, D. (ed.) *Slopes: Form and Process*. IBG Special Publication, London, **3**, 15–30.

KIRKBY, M. J. 1974. *Erosion and equilibrium*. Department of Geography. Leeds University, Working Paper, **57**.

KIRKBY, M. J. 1999. Towards an understanding of varieties of fluvial form. *In*: MILLER, A. J. & GUPTA, A. (eds) *Varieties of Fluvial Form*. Wiley, Chichester, 507–514.

KIRKBY, M. J., IMESON, A. C., BERGKAMP, G. & CAMMERAAT, L. H. 1996. Scaling up processes and models from the field plot to the watershed and regional areas. *Journal of Soil and Water Conservation*, **51**, 391–396.

LANE, S. N. & RICHARDS, K. S. 1997. Linking river channel form and process: time, space and causality revisited. *Earth Surface Landforms and Processes*, **22**, 249–260.

LANE, S. N., BIRON, P. M. ET AL. 1998. Three-dimensional measurement of river channel flow processes using acoustic Doppler velocimetry. *Earth Surface Processes and Landforms*, **23**, 1247–1267.

LEOPOLD, L. B. 1973. River channel change with time – an example. *Bulletin of the Geological Society of America*, **84**, 1845–1860.

LEVIN, S. A. 1992. The problem of pattern and scale in ecology. *Ecology*, **73**, 1943–1967.

LUDWIG, J. A., WILCOX, B. P., BRESHEARS, D. D., TONGWAY, D. J. & IMESON, A. C. 2005. Vegetation patches and runoff-erosion as interacting ecohydrological processes in semiarid landscapes. *Ecology*, **86**, 288–297.

MACKIN, J. H. 1948. Concept of the graded river. *Bulletin of the Geological Society of America*, **59**, 463–512.

MAYER, L. 1992. Some comments on equilibrium concepts and geomorphic systems. *Geomorphology*, **5**, 277–295.

MCDONNELL, J. J., BONELL, M., STEWART, M. K. & PEARCE, A. J. 1990. Deuterium variations in storm rainfall – implications for stream hydrograph separation. *Water Resources Research*, **26**, 455–458.

MILLER, A. J. & GUPTA, A. 1999. *Varieties of Fluvial Form*. Wiley, Chichester.

MULLER, E. N., WAINWRIGHT, J. & PARSONS, A. J. 2007. Impact of connectivity on the modeling of overland flow within semiarid shrubland environments. *Water Resources Research*, **43**, W09412, doi:10.1029/2006WR005006.

MULLIGAN, M. & WAINWRIGHT, J. 2003. Modelling and model building. *In*: WAINWRIGHT, J. & MULLIGAN, M. (eds) *Environmental Modelling: Finding Simplicity in Complexity* Wiley, Chichester, 1–73.

ORESKES, N., SHRADER-FRECHETTE, K. & BELLITZ, K. 1994. Verification, validation and confirmation of numerical models in the Earth Sciences. *Science*, **263**, 641–646.

PATTON, P. C., PICKUP, G. & PRICE, D. M. 1993. Holocene palaeofloods of the Ross River, central Australia, *Quaternary Research*, **40**, 201–212.

PERRY, G. L. W. 2002. Landscapes, space and equilibrium: shifting viewpoints. *Progress in Physical Geography*, **26**, 339–359.

PETTS, G. E. & FOSTER, I. D. L. 1985. *Rivers and Landscape*. Edward Arnold, London.

PHILLIPS, J. D. 1992. Nonlinear dynamical systems in geomorphology: revolution or evolution? *Geomorphology*, **5**, 219–229.

PHILLIPS, J. D. 1999. Divergence, convergence, and self-organization in landscapes. *Annals of the Association of American Geographers*, **89**, 466–488.

PHILLIPS, J. D. & GOMEZ, B. 1994. In defense of logical sloth. *Annals of the Association of American Geographers*, **84**, 697–701.

PICKUP, G. & REIGER, W. A. 1979. A conceptual model of the relationship between channel characteristics and discharge. *Earth Surface Processes and Landforms*, **4**, 37–42.

PITTY, A. F. 1982. *The Nature of Geomorphology*. Methuen, London.

PIZZUTO, J. E. 1992. The morphology of graded gravel rivers – a network perspective. *Geomorphology*, **5**, 457–474.

POWELL, D. M., BRAZIER, R. J., WAINWRIGHT, J., PARSONS, A. J. & KADUK, J. 2005. Streambed scour and fill in low-order dryland channels. *Water Resources Research*, **41**, art. no. W05019.

RAYMO, M. E. 1994. The initiation of Northern Hemisphere glaciation. *Annual Review of Earth and Planetary Sciences*, **22**, 353–383.

REGONDA, S. K., SIVAKUMAR, B. & JAIN, A. 2004. Temporal scaling in river flow: can it be chaotic? *Hydrological Sciences Journal*, **49**, 373–385.

REINERS, P. W., EHLERS, T. A., MITCHELL, S. G. & MONTGOMERY, D. R. 2003. Coupled spatial variations in precipitation and long-term erosion rates across the Washington Cascades. *Nature*, **426**, 645–647.

RENWICK, W. H. 1992. Equilibrium, disequilibrium and non eqilibrium landforms in the landscape. *Geomorphology*, **5**, 265–276.

RHOADS, B. L. & THORN, C. E. 1993. Geomorphology as science: the role of theory. *Geomorphology*, **6**, 287–307.

RICHARDS, K. S. 1982. *Rivers: Form and Process in Alluvial Channels*. Methuen, London.

RICHARDS, K. S. 1990. 'Real' geomorphology. *Earth Surface Processes and Landforms*, **15**, 195–197.

RUDDIMAN, W. F. & RAYMO, M. E. 1988. Northern hemisphere climatic regimes during the past 3 Ma: possible tectonic connections. *Philosophical Transactions of the Royal Society, London*, **318B**, 411–430.

SCHICK, A. P. 1974. Formation and obliteration of desert stream terraces – a conceptual analysis. *Zeitschrift für Geomorphologie Supplement Band*, **21**.

SCHICK, A. P. 1988. Hydrologic aspects of floods in extreme arid environments. *In*: BAKER, V. R., KOCHEL, R. C. & PATTON, P. C. (eds) *Flood Geomorphology*. Wiley, New York, 189–203.

SCHUMM, S. A. 1961. *Effect of sediment characteristics on erosion and deposition in ephemeral stream channels*. United States Geological Survey Professional Paper, Washington, DC, **352C**.

SCHUMM, S. A. 1973. Geomorphic thresholds and complex response of drainage systems. *In*: MORISAWA, M. (ed.) *Fluvial Geomorphology*, Proceedings of the 4th Annual Geomorphology Symposia Series, Binghamton. Allen & Unwin, London, 299–311.

SCHUMM, S. A. 1975. Episodic erosion: a modification of the geomorphic cycle. *In*: MELHORN, W. N. & FLEMAL, R. C. (eds) *Theories of Landform Development*, Allen & Unwin, London, 69–88.

SCHUMM, S. A. 1977. *The Fluvial System*. Wiley, New York.

SCHUMM, S. A. 1979. Geomorphic thresholds – concept and its applications. *Transactions of the Institute of British Geographers*, **4**, 485–515.

SCHUMM, S. A. 1991. *To Interpret the Earth: Ten Ways to be Wrong*. Cambridge University Press, Cambridge.

SCHUMM, S. A. & LICHTY, R. W. 1965. Time, space and causality in geomorphology. *American Journal of Science*, **263**, 110–119.

SIMON, A. 1992. Energy, time, and channel evolution in catastrophically disturbed fluvial systems. *Geomorphology*, **5**, 345–372.

SIVAKUMAR, B. & JAYAWARDENA, A. W. 2002. An investigation of the presence of low-dimensional chaotic behaviour in the sediment transport phenomenon. *Hydrological Sciences Journal*, **47**, 405–416.

SMITH, T. R. & BRETHERTON, F. P. 1972. Stability and the conservation of mass in drainage basin evolution. *Water Resources Research*, **8**, 1506–1529.

THIERY, J. M., D'HERBÈS, J. M. & VALENTIN, C. 1995. A model simulating the genesis of banded vegetation patterns in Niger. *Journal of Ecology*, **83**, 497–507.

THORN, C. E. & WELFORD, M. R. 1994. The equilibrium concept in geomorphology. *Annals of the Association of American Geographers*, **84**, 666–696.

THORNES, J. B. 1982. The ecologoy of erosion. *Geography*, **70**, 222–235.

THORNES, J. B. 1983. Evolutionary geomorphology. *Geography*, **68**, 225–235.

THORNES, J. B. 2004. Stability and instability in the management of Mediterranean desertification. *In*: WAINWRIGHT, J. & MULLIGAN, M. (eds) *Environmental Modelling: Finding Simplicity in Complexity*. Wiley, Chichester, 303–314.

THORNES, J. B. & BRUNSDEN, D. 1977. *Geomorphology and Time*. Methuen, London.

TODD, A. K., BUTTLE, J. M. & TAYLOR, C. H. 2006 Hydrologic dynamics and linkages in a wetland-dominated basin. *Journal of Hydrology*, **319**, 15–35.

TOOTH, S. & NANSON, G. C. 2000. Equilibrium and non-equilibrium conditions in dryland rivers. *Physical Geography*, **21**, 183–211.

TRICART, J. 1965. Les discontinuities dans le phénomenes d'erosion. *International Association of Scientific Hydrology*, **59**, 233–243.

TURNER, M. G., ROMME, W. H., GARDNER, R. H., O'NEILL, R. V. & KRATZ, T. K. 1993. A revised concept of landscape equilibrium: disturbance and stability on scaled landscapes. *Landscape Ecology*, **8**, 213–227.

VALE, T. R. 2003. Scales and explanations, balances and histories: musings of a physical geography teacher. *Physical Geography*, **24**, 248–270.

VALENTIN, C., D'HERBÈS, J. M. & POESEN, J. 1999. Soil and water components of banded vegetation patterns. *Catena*, **37**, 1–24.

VON BLANCKENBURG, F. 2005. The control mechanisms of erosion and weathering at basin scale from cosmogenic nuclides in river sediment. *Earth and Planetary Science Letters*, **237**, 462–479.

WAINWRIGHT, J. 1996*a*. Infiltration, runoff and erosion characteristics of agricultural land in extreme storm events, SE France. *Catena*, **26**, 27–47.

WAINWRIGHT, J. 1996*b*. Hillslope response to extreme storm events: the example of the Vaison-la-Romaine event. *In*: ANDERSON, M. G. & BROOKS, S. M. (eds) *Advances in Hillslope Processes*. Wiley, Chichester, 997–1026.

WAINWRIGHT, J. 2006. Degrees of separation: hillslope-channel coupling and the limits of palaeohydrological reconstruction. *Catena*, **66**, 93–106.

WAINWRIGHT, J. 2007. Can modelling enable us to understand the rôle of humans in landscape evolution? *Geoforum*, doi:10.1016/j.geoforum.2006.09.011.

WAINWRIGHT, J. & THORNES, J. B. 2003. *Environmental Issues in the Mediterranean: Processes and Perspectives from the Past and Present*. Routledge, London

WARREN, A. 1985. Arid geomorphology. *Progress in Physical Geography*, **97**, 434–441.

WEILER, M., MCGLYNN, B. L., MCGUIRE, K. J. & MCDONNELL, J. J. 2003. How does rainfall become runoff? A combined tracer and runoff transfer function approach. *Water Resources Research*, **39**, art. no. 1315.

WHIPPLE, K. X. & MEADE, B. J. 2004. Controls on the strength of coupling among climate, erosion, and deformation in two-sided, frictional orogenic wedges at steady state. *Journal of Geophysical Research*, 109 F01011 doi:10.1029/2003JF000019.

WILLETT, S. D. & BRANDON, M. T. 2002. On steady states in mountain belts. *Geology*, **30**, 175–178.

WOLMAN, M. G. & GERSON, R. 1978. Relative times and effectiveness of climate watershed geomorphology. *Earth Surface Processes and Landforms*, **3**, 189–208.

ZHANG, X., DRAKE, N. A. & WAINWRIGHT, J. 2002. Scaling land-surface parameters for global scale soil-erosion estimation. *Water Resources Research*, **38**, 1180.

Modelling cockpit karst landforms

C. FLEURANT[1], G. E. TUCKER[2] & H. A. VILES[3]

[1]*Département Paysage, Institut National d'Horticulture, 49045 Angers, France*
(e-mail: cyril.fleurant@inh.fr)

[2]*Department of Geological Sciences, Cooperative Institute for Research in Environmental Sciences, University of Colorado, Boulder, CO, USA*

[3]*Centre for the Environment, University of Oxford, South Park, Oxford OX1 3QY, UK*

Abstract: The purpose of this article is to present a model of the formation processes of cockpit karst landscapes. The CHILD software was used to simulate landscape evolution including dissolution processes of carbonate rocks. After examining briefly how the CHILD model operates, two applications of this model involving dissolution of carbonate rocks are presented. The simulated landscapes are compared with real landscapes of the Cockpit Country, Jamaica, using morphometric criteria. The first application is based on the hypothesis that dissolution of carbonate rocks is isotropic over time and space. In this case, dissolution is constant throughout the whole area studied and for each time step. The simulated landscapes based on this hypothesis have morphometric features which are quite different from those of real landscapes. The second application considers that dissolution of carbonate rocks is anisotropic over time and space. In this case, it is necessary to take into account subsurface and underground processes, by coupling surface runoff and water infiltration into the fractured carbonates.

Over large temporal and spatial scales, morphometric evolution of landscapes is governed by a number of complex processes and factors: climatic conditions, erosion and sedimentation. A better understanding of landscape evolution requires numerical tools that make it possible to explore such complex processes. Over the last 10 years, particular focus has been put on the development of such models (Willgoose *et al.* 1991; Beaumond *et al.* 1992; Braun & Sambridge 1997; Kaufman & Braun 2001; Tucker *et al.* 2001*a, b*). The main models are SIBERIA (Willgoose *et al.* 1991), GOLEM (Tucker & Slingerland 1994), CASCADE (Braun & Sambridge 1997), CAESAR (Coulthard *et al.* 1998) and CHILD (Tucker *et al.* 1999), as reviewed in Coulthard (2001). Very few of them explore the dissolution processes of carbonates and their role in the production of karst morphologies. Consequently, the aim of the present study is to model the morphometric evolution of cockpit karst landscapes, including the various geomorphic processes involved.

There has been considerable research into quantification of underground karstic processes (Day 1979*a*; Gunn 1981; Groves & Howard 1994; Clemens *et al.* 1997; Siemers & Dreybrodt 1998; Kaufman & Braun 1999, 2000; Bauer *et al.* 2005), but very few studies have explored dissolution by surface waters (Ahnert & Williams, 1997; Kaufman & Braun 2001). Starting from a three-dimensional network and using simple erosion laws, Ahnert & Williams (1997) recreated tower karst landscapes, whilst Kaufman & Braun (2001) demonstrated how CO_2-enriched surface runoff shapes landscapes typical of large karstic valleys through dissolution processes.

Karst landscapes offer a wide variety of morphometric styles depending on numerous variables (White 1984) such as temperature, partial pressure of carbon dioxide and precipitation. Changing the value of one of the variables can modify the landscape and result in doline karst, cockpit karst or tower karst.

Cockpit karst landscapes are usually described as a succession of summits and depressions with irregular size and position. Several morphometric techniques have been developed to describe the geometry of cockpit karst landscapes (Day 1979*b*; Lyew-Ayee 2004).

However, within cockpit karst the depressions are often said to be star-shaped (Sweeting 1972). Grund (1914 in Sweeting 1972) put forward a hypothesis concerning their formation, i.e. that cockpit karsts with star-shaped depressions could be more extensively developed doline karsts. Cockpit karst landscapes are usually thought to occur only in humid tropical areas. Only regions with total precipitation exceeding 1500 mm a year are likely to develop this type of karst (Sweeting 1972). However, rain is not enough; very hard carbonates and a pre-existing fracture network are also necessary. In fact, the geological features and structure are a major component in the formation of cockpits karst terrains (Lyew-Ayee 2004).

The objective of the present paper was to describe the integration of carbonate dissolution processes in the CHILD landscape evolution model (Tucker et al. 2001a, b). Interest in integrating dissolution into the model is governed by the need to understand the evolution of cockpit karst landscapes. In the present paper, we describe the functioning principles of the CHILD model and consider the contribution of the new processes used to model karst dissolution.

Two applications of this karst dissolution model are described: isotropic time–space dissolution and anisotropic time–space dissolution. Both applications will be tested on the slope scale, that is at the scale of one single cockpit.

The morphometric features of the simulated cockpits will be compared with the geometrical properties of real cockpits, thanks to data collected in the typesite cockpit karst area in Jamaica and recently highlighted by Lyew-Ayee (2004).

Geomorphologic model

The CHILD model (Channel–Hillslope Integrated Landscape Development) is a numerical model of landscape evolution (Tucker et al. 2001a, b). The evolution of topography over time is simulated through interaction and feedback between surface runoff, erosion and the transport of sediment. CHILD is used for modelling a large number of geomorphic processes at the scale of a drainage basin. In the CHILD model, the evolution of topography (z) over time (t) results from the combination of several factors:

$$\frac{\partial z}{\partial t} = \frac{\partial z}{\partial t}\Big|_{creep} + \frac{\partial z}{\partial t}\Big|_{erosion} \\ + \frac{\partial z}{\partial t}\Big|_{tectonic} + \frac{\partial z}{\partial t}\Big|_{dissolution}. \quad (1)$$

These various factors are numerically solved using a particular spatial framework (Fig. 1): Voronoï polygons (Braun & Sambridge 1997). The first three factors are present in all landscape evolution models mentioned above. A detailed description of the spatial framework can be found in Tucker et al. (1997, 1999, 2001a, b). The last factor concerns dissolution and this is precisely what is original about the present study.

Mass transfer onto slope

The transport of sediments is modelled through a diffusion equation (McKean et al. 1993; Kooi & Beaumond 1994; Braun & Sambridge 1997):

$$\frac{\partial z}{\partial t}\Big|_{creep} = k_d \Delta^2 z \quad (2)$$

where $k_d (m^2/s)$ is the diffusion coefficient of sediment transport. Beaumont et al. (1992) describe k_d in terms of physics by associating it with both lithology and climate. Thus, the values of k_d tend to decrease in the case of resistant rocks, and to increase under harsh climates. The values of k_d vary from 0.0001 to 0.01 m^2/y (e.g. Martin & Church 1997). Given the geological and climatic conditions of the cockpit karst typesite in Jamaica, we should expect low values for k_d, but since data on specific values is lacking, we have put several values to the test.

Mechanical erosion and deposition

Mechanical erosion and deposition are modelled using a continuity equation (Tucker et al. 2001a, b):

$$\frac{\partial z}{\partial t}\Big|_{erosion} = -\Delta q_s \quad (3)$$

where q_s (m^2/s) is the sediment flux per unit width. In the case of limited transport fluvial processes, the transported flux of sediments q_s is expressed as a power law (Willgoose et al. 1991; Tucker & Bras 1998):

$$q_s = \frac{1}{\rho_{CaCO_3}(1-\omega)} k_f (k_t Q_{sur}^{m_f} S^{n_f} - \tau_c)^p \quad (4)$$

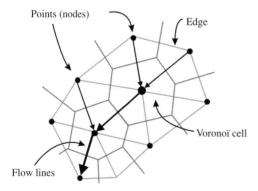

Fig. 1. Schematic illustration of the spatial framework in the CHILD model (after Tucker et al. 1999) and illustration of steepest-descent flow routing in TIN framework.

where ρ_{CaCO_3} (kg/m^3) is sediment mass density – here carbonate rock, ω is sediment porosity, S is topographic gradient and τ_c (N/m^2) is critical shear stress; k_f (kg m/N/y), k_t (N y/m^4), m_f, n_f and p are parameters.

The values of the equation (4) parameters can be estimated using the Einstein–Brown transport equation (Bogaart et al. 2003). Equation (4) may be written as follows:

$$q_s = 40\rho_{CaCO_3} V \sqrt{g(s-1)d_{50}^3} \times \left(\frac{1}{(s-1)d_{50}}\right)^3 Q_{sur}^{1.8} n^{1.8} S^{2.1} \quad (5)$$

where $s = \rho_{CaCO_3}/\rho_w$ is the specific density of sediment grains, d_{50} (m) is the median sediment grain size, n is Manning's resistance coefficient and V is the fall velocity of the sediment grain,

$$V = \sqrt{\frac{2}{3} + \frac{36\mu^2}{gd_{50}^3(s-1)}} - \sqrt{\frac{36\mu^2}{gd_{50}^3}}(s-1). \quad (6)$$

with $\rho_w = 1000$ kg/m^3, $\rho_{CaCO_3} = 2700$ kg/m^3, $d_{50} = 10^{-3}$ m, $n = 0.03$ as specific values for these parameters, we have:

$$q_s = 0.02 Q_{sur}^{1.8} S^{2.1}. \quad (7)$$

Correspondence to equation (4) gives $k_f = 1.0$ kg m/N/y, $k_t = 0.125$ N y/m^4, $m_f = 1.8$, $n_f = 2.1$, $\tau_c = 0.0$ N/m^2 and $p = 1.0$.

Tectonic movements

Movements resulting from tectonic processes in general can be simulated using a simple model:

$$\frac{\partial z}{\partial t}\bigg|_{tectonic} = U \quad (8)$$

where U (m/s) is the uplift velocity or the base level change rate.

In the present case, we will consider that tectonic movements are constant in each grid of the model. In view of parameterization of this equation, the history of our study site should be looked at.

After the deposit of limestone sediments during the Oligocene and Eocene, the area apparently underwent a tectonic lift, resulting in its plateaus rising to a height of 600 m above sea level (Lyew-Ayee 2004). Thus, karstification would have started to occur only at the beginning of the Pliocene, when climatic conditions were favourable (Versey 1972; Pfeffer 1997). Because of the lack of information, we can consider here that the impact of tectonics happened before karstification. As a result, equation (8) can be simplified, by supposing that $U = 0$ m/s.

Dissolution processes

Carbon dioxide (CO_2) contained in the atmosphere is dissolved by precipitation and can therefore be found in soil, as well as surface water and groundwater. The dissolution of carbon dioxide produces carbonic acid, H_2CO_3, which is the major factor for the dissolution of calcite ($CaCO_3$) contained in carbonate rocks.

Such dissolution may occur in two totally different contexts: open systems where carbonates are in direct contact with atmospheric phenomena; and closed systems, within tight fractures, for example. In open systems, precipitation comes into contact with carbonate rocks and the dissolution of the latter will then depend on the quantity of water in contact and on the hydrodynamic nature of the flows (Dreybrodt 1988).

The calcite dissolution reaction in an open system can be expressed as follows (Trudgill 1985):

$$CaCO_3 + CO_2 + H_2O \Leftrightarrow \\ CaCO_3 + H_2CO_3 + Ca^{2+} + 2HCO_3^-. \quad (9)$$

It is essential to determine the carbon dioxide flux F (mol/m^2/s) since it helps to determine the quantity of dissolved calcite and consequently the quantity of eroded carbonate rocks. This flux can be expressed as follows (Buhmann & Dreybrodt 1985; Kaufman & Braun 1999):

$$F = k_1 \left(1 - \frac{[Ca^{2+}]}{[Ca^{2+}]_{eq}}\right) \text{ if } [Ca^{2+}] \leq [Ca^{2+}]_s \quad (10)$$

$$F = k_2 \left(1 - \frac{[Ca^{2+}]}{[Ca^{2+}]_{eq}}\right)^4 \text{ if } [Ca^{2+}] > [Ca^{2+}]_s \quad (11)$$

where k_1 (mol/m^2/s) and k_2 (mol/m^2/s) are kinetic constants of the dissolution and $[Ca^{2+}]_{eq}$ (mol/m^3) is the equilibrium calcium concentration, that is when the solution is saturated with calcium. $[Ca^{2+}]_s$ represents the threshold of calcium concentration at which the flux F shifts from low-order kinetics to high-order kinetics. Dreybrodt and Eisenlohr (2000) show that, when the solution concentration is close to equilibrium, the carbonate dissolution rate decreases because of impurities (phosphates and silicates). Therefore, beyond a threshold value $[Ca^{2+}]_s$, the value of the flux F

significantly decreases in a non-linear way (Fig. 2). The value of this threshold may vary by $0.7 \leq [Ca^{2+}]_s \leq 0.9$ (Kaufman & Braun 1999).

In the case of low-order kinetics, hydration of CO_2 and ion transport through molecular diffusion are two rate-limiting factors. The value of the coefficient k_1 is (e.g. Buhmann & Dreybrodt 1985):

$$k_1 = k_0 \left(1 + \frac{k_0 \delta}{6 D_{Ca} [Ca^{2+}]_{eq}}\right)^{-1} \quad (12)$$

where k_0 (mol/m²/s) is a constant value, D_{Ca} (m²/s) is molecular diffusion coefficient of calcium and δ (m) is the diffusion boundary layer.

In the case of high-order kinetics, the rate-limiting factor is the dissolution of calcite at the interface between rock and the solvent. The value of the constant k_2 is (e.g. Dreybrodt 1996):

$$k_2 = k_1 \left(1 - [Ca^{2+}]_s\right)^{-3}. \quad (13)$$

The coefficient k_2 is several orders higher than k_1.

The case of time-space isotropic dissolution

Dissolution of carbonate rocks is also called denudation. Several denudation models have been proposed (e.g. Corbel 1959; Ford 1981). White's (1984) maximum denudation model is the most comprehensive (Dreybrodt 1988; White 1988):

$$\frac{\partial z}{\partial t}\Big|_{dissolution} = DR = -\frac{1000\, m_{Ca}(P-E)}{\rho_{CaCO_3}} \\ \times \left(\frac{K_1 K_C K_H P_{CO_2 open}}{4 K_2 \gamma_{Ca} \gamma^2 HCO_3}\right)^{\frac{1}{3}} \quad (14)$$

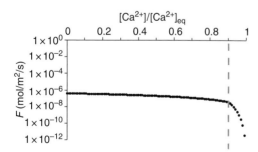

Fig. 2. Dependence of dissolution rates on saturation ratio $[Ca^{2+}]/[Ca^{2+}]_{eq}$. Owing to inhibition, the non-linear surface rates close to equilibrium are so slow that they become rate-limiting. The vertical line separates the region of linear kinetics from that of non linear kinetics $([Ca^{2+}]_s = 0.9[Ca^{2+}]_{eq})$.

where DR (m/s) is the denudation rate, spatially uniform over all the study area, E (m/s) is the evapotranspiration, $(P - E)$ (m/s) is the effective precipitation, m_{Ca} (kg/mol) is the atomic weight of the calcite and ρ_{CaCO_3} (kg/m³) is the volumic weight of the calcite, K_1 (mol/m³), K_2 (mol/m³), K_C (mol²/m⁶) and K_H (mol/m³/atm) are equilibrium constants which only depend on temperature, and $P_{CO_2}{}^{open}$ (atm) is the pressure of carbon dioxide in open system.

The values of carbon dioxide pressure are governed by local factors and may thus be directly related to climatic conditions (Brook et al. 1983):

$$\log P_{CO_2} = -3.47 + 2.09\left(1 - e^{-0.00172 ER}\right). \quad (15)$$

In Jamaica for example, $ER \approx 1000$ mm/a, which gives $P_{CO_2} \approx 0.0176$ atm. This value is consistent with the values Smith & Atkinson (1976) gave for tropical areas: $P_{CO_2} \approx 0.01$ atm.

Equation (14) requires no adjustment and combines three climatic variables: temperature, carbon dioxide pressure and precipitation. When White's (1984) model is applied, it is assumed that denudation is maximal and constant all over the studied area. Variations in topography over time, as given in equation (1), will also be constant. Therefore, this model is isotropic in space and time. As an example, the denudation value of cockpit karst is 0.13 m/k years approximately.

The case of time–space anisotropic dissolution

Karstification is a complex process, and it is difficult to establish clear and quantifiable relations between its various components: calcite dissolution, surface runoff and groundwater flow. Many authors assume that there is a close relation between surface denudation, hence a decrease in the topographic elevation, and subsurface dissolution processes.

Dissolution of carbonates in a network of fractures results in their disappearing into the depth of the karst. This complete dissolution of materials results in the topographic surface sinking and the resulting depressions filling with soil. A description of such processes can be found in the epikarst concept (e.g. Ford & Williams 1989). Consequently, the higher the subsurface dissolution of the fractured medium is, the quicker the overlying topography subsides. This epikarst concept is the starting point of the introduction to the anisotropic dissolution of carbonates.

Epikarst functions as follows (Williams 1985): take a carbonate rock with a gently sloping

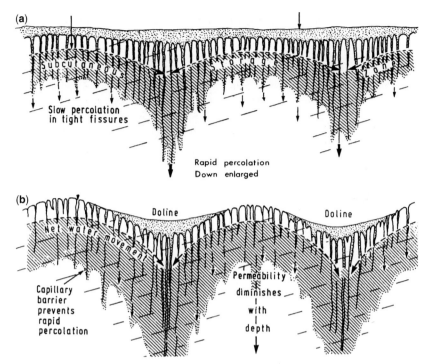

Fig. 3. Illustration of the subcutaneous (or epikarstic) zone in karstified rocks (after Williams 1985). Limestone terrains inherit a three-dimensional permeability from the structural features of the bedrock (**a**). The interconnected fissure network provides solute pathways. Fissures are considerably widened by solution. Preferred flow paths develop at the base of the epikarstic zone. These are enhanced by solution and become increasingly dominant (**b**).

topography and a network of fractures (Fig. 3a). Precipitation will both run across surface and infiltrate the network of fractures. Since fractures narrow rapidly as they get deeper, water located in the subcutaneous zone is perched and percolates downward, from high points towards lower points. The shape of this hydrodynamic flow globally corresponds to that of the overlying topography. Consequently, water in the subcutaneous zone usually flows down to lower points of the perched water table, that is to say to the lower points of topography. This large inflow of water results in a quicker dissolution in the fractures located at the lower points of the perched water table and consequently enhancement of the topography (Fig. 3b). Positive feedback can be noted between the water flow in the subcutaneous zone and dissolution in the fractures (Bauer *et al.* 2005). The present approach is a good example of the time–space anisotropy of carbonate dissolution: the rate of dissolution depends on the initial position of rocks with regard to slope and hence water flow.

Consequently, it can be noted that the implementation of the hydraulic functioning of the epikarst in our model requires coupling surface flow, subsurface flow and the underground dissolution of fractured carbonates.

Surface and sub-surface flows

The CHILD model allows several patterns of surface runoff: either the Hortonian runoff or runoff as a result of excessive saturation of the soil. Such procedures are used to simulate surface runoff but they do not take into account local infiltration in each cell of the model. In the present case, the point is to implement the hydrodynamics of an epikarst, that is to say being able to determine the quantity of runoff water and infiltrating water into each cell of the model. For this purpose, we chose a simple flow procedure which considers hydrodynamics of our system at cell scale (Figs 1 & 4):

$$Q_{in} = Q_{insur} + Q_{insub} + P \times A_i$$
$$Q_{sub} = \min(Q_{in}, Q_{subcapacity}) \quad (16)$$
$$Q_{sur} = \max(Q_{in} - Q_{sub}, 0)$$

where Q_{in} (m^3/s) is sum of the surface (Q_{insur}, m^3/s) and the subsurface (Q_{insub}, m^3/s) flows into the cell under consideration. P (m) is precipitation and A (m^2) surface of the cell. The volume

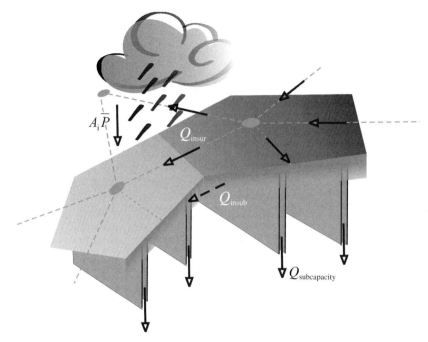

Fig. 4. Using a flow routing described in the text, surface runoff and subsurface flow are computed by means of a fracture network.

of infiltrating water ($Q_{subcapacity}$, m³/s) varies according to the hydraulic conductivity of the associated cell. The volume of water that does not infiltrate (Q_{sur}, m³/s) is the surface runoff. Such a procedure requires that hydraulic conductivity is assigned to each cell. It will then be possible to simulate the hydrodynamic functioning of the epikarst.

Hydrodynamics in one fracture

In order to determine the hydraulic conductivity of a fracture network, first the hydraulic conductivity of a single fracture should be expressed. We assume that each fracture in the network is represented by two parallel planes separated by a distance of a_0 (m), corresponding to the fracture aperture.

In the case of laminar flow, the flow rate within the fracture is expressed using Poiseuille's equation (Beek et al. 1999):

$$Q_{fracture} = \frac{g}{12\mu} a_0^3 b_0 M \frac{h}{L} \quad (17)$$

where g (m/s²) is the gravity, μ (m²/s) the dynamic viscosity, b_0 (m) the fracture width, h (m) the hydraulic head, L (m) the fracture depth and $M = -0.575(a_0/b_0) + 1$ a geometric parameter.

The dissolution of calcite on the inner surfaces of the fracture results in the fracture widening, and an increasing flow rate. Beyond a determined value, the flow rate becomes turbulent and the equation (17) cannot be applied any longer. The threshold between laminar and turbulent flow is computed using the Reynolds number (Marsily 1986):

$$Re = \frac{2 \times 10^6 Q_{fracture}}{a_0 + b_0}. \quad (18)$$

Flow rate is turbulent when $Re > 2000$. The linear relation between flow rate and the hydraulic load no longer exists and the flow rate is computed using the Darcy–Weissbach equation (Dreybrodt 1988):

$$Q_{fracture} = \sqrt{\frac{4g(a_0 b_0)^3 h}{f(a_0 + b_0)L} \frac{h}{|h|}} \quad (19)$$

where f is the friction factor,

$$\frac{1}{\sqrt{f}} = -2 \log_{10}$$
$$\times \left(\frac{r(a_0 + b_0)}{7.42 a_0 b_0} + 2.51 \nu \frac{(a_0 + b_0)^3 L}{16 g(a_0 b_0)^3 h} \right). \quad (20)$$

The roughness value of the fracture r (m) is approximately 2% that of its aperture (Romanov et al. 2003).

Consequently, the maximum flow rate which can infiltrate into a fracture with simple geometry is known. We can then deduce the maximum flow rate ($Q_{subcapacity}$) which can infiltrate into a network of parallel fractures of a cell (Bear et al. 1993; Singhal & Gupta 1999):

$$Q_{subcapacity} = \frac{Q_{fracture} N_{fracture}}{s} \quad (21)$$

where s (m) is the mean distance between fractures and $N_{fracture}$ the number of fractures of the cell.

Dissolution inside a fracture

As previously noted, the calcite present on the fracture walls dissolves, thus widening the fractures over time. Such a process is a typical profile of the evolution of fracture widening. This profile is the result of differential kinetics along the fracture (Dreybrodt 1996; Dreybrodt & Gabrovsek 2000):

$$F_{lower}(x) = k_1 \left(1 - \frac{[Ca^{2+}]_{inclose}}{[Ca^{2+}]_{eqclose}}\right) e^{-\frac{2k_1(a_0+b_0)x}{Q[Ca^{2+}]_{eqclose}}}$$

for $x \leq x_s$

(22)

$$F_{higher}(x) = k_2 \left[\frac{6k_1(a_0+b_0)(x-x_s)}{Q[Ca^{2+}]_{eqclose}}\right.$$

$$\left. + \left(1 + \frac{[Ca^{2+}]_{sclose}}{[Ca^{2+}]_{eqclose}}\right)^{-3}\right]^{-\frac{4}{3}}$$

for $x > x_s$

$$x_s = \frac{Q[Ca^{2+}]_{eqclose}}{2k_1(a_0+b_0)} \ln \left(\frac{1 - \frac{[Ca^{2+}]_{inclose}}{[Ca^{2+}]_{eqclose}}}{1 - \frac{[Ca^{2+}]_{sclose}}{[Ca^{2+}]_{eqclose}}}\right)$$

where $[Ca^{2+}]_{inclose}$ is the calcium concentration at the entrance of the fracture, $[Ca^{2+}]_{eqclose}$ is the calcium concentration at equilibrium and *close* refers to a closed system, a deep fracture.

Beyond a certain distance x_s (m), the flow related to low-order kinetics quickly decreases (Fig. 5). Consequently, if high-order kinetics were not taken into account beyond x_s, dissolution would be insignificant. The evolution of fracture widening can then be computed (Dreybrodt et al. 1999; Kaufmann 2003):

$$\frac{\partial a}{\partial t} = \frac{m_{Ca}}{\rho_{CaCO_3}} F_{higher}(L). \quad (23)$$

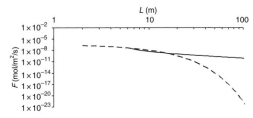

Fig. 5. Dissolution rates along the uniform fracture. Fractures widening take place and is controlled by a positive feedback between dissolution and flow into the fracture. The dashed line represents the dissolution rates for linear kinetics valid up to equilibrium. Note the steep decay in dissolution rates by ten orders of magnitude in the first 70 m of the fracture. The solid line shows the dissolution rates for non-linear kinetics. Here the rates drop much more slowly in a hyperbolic manner, and sufficient dissolution is still active at the exit to widen the fracture.

In the present study, equations expressing the dissolution along the inner walls of the fracture and those expressing fracture hydrodynamics have been taken into account. These coupled equations represent the hydraulic functioning of the epikarst and correspond to our anisotropic time–space dissolution model.

Results

The purpose of the present study is to understand the evolution of cockpit karst terrains. Firstly, we have described the integration processes of carbonate rock dissolution into the CHILD model. We have developed two types of possible scenarios relating to cockpit karst formation. The first scenario involves isotropic time–space dissolution. The second one is an anisotropic time–space dissolution model and it corresponds to an epikarst concept.

In order to know whether the processes involved in both approaches could recreate the geomorphometry of cockpit karst terrains, it is necessary to compare the simulated landscapes with the real ones. To do so, we will use Lyew-Ayees's (2004) in-depth study, carried out over 84 km² in the cockpit karst area of Jamaica. Data result from geostatistical calculations based on a high-resolution digital elevation model. The GIS digital tools were used to calculate a large number of morphometric parameters linked to these typical landscapes, e.g. roughness, summit and depression density. In his study, Lyew-Ayee (2004) put forward two morphometric criteria typical of cockpit karst landscapes: average slope and relative relief. Results from morphometric analysis reveal

Table 1. *Morphometric properties of the Cockpit Country study area (Lyew-Ayee 2004)*

Name	Slope (deg)	Relative relief (m)
Barbecue Bottom	32.63	74.2
Quickstep	26.04	54.11
Windsor	27.45	57.08
Elderslie	23.46	47.46
Nassau Valley	22.99	45.97
Queen of Spain's Valley	21.86	43.54

that the average slope of the cockpit karst country (Jamaica) varies from 21.86 to 32.63°. Similarly, the relative relief varies from 43.54 to 77.02 m (Table 1). Slopes are calculated from the DEM using a moving window – a 3 × 3 cell kernel – which moved throughout the grid. To calculate slopes, the method is a third-order finite difference estimator which uses the eight outer points of the moving window.

Determination of relative relief needs to identify sinks and summits of the landscape. Then relative relief may be determined using sink and summit points. A cluster analysis of both the average slope and the average relative relief was used (Lyew-Ayee 2004) as a crucial measure in distinguishing cockpit and non-cockpit karst landscapes.

Reference simulations

Reference simulations are implemented without taking carbonate dissolution into account; the last component of the equation (1) is then nil $\frac{\partial z}{\partial t}|_{\text{dissolution}} = 0$ m/s. Such reference simulations aim to provide comparison criteria with those simulations using isotropic and anisotropic dissolution approaches, as described above.

Measurement data obtained by Lyew-Ayee (2004) from a DEM gave the initial geometry of our system. Consequently, we have chosen to study a limestone plateau discretized into 4600 Voronoi cells. The dimensions of our system correspond to average measurements carried out by Lyew-Ayee (2004) in the cockpit area (Fig. 6):

- a star-shaped distribution of depressions towards which runoff is initially directed on a 1% hillslope;
- a 220 m distance between these depressions.

As mentioned above, we carried out the present study at single cockpit scale, in order to test the processes under consideration. Once the processes are validated, it will be possible to extend the scale of the study to a whole cockpit country.

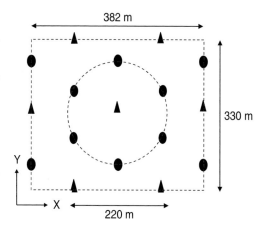

Fig. 6. Design of the simulated area: depressions (circles) are star-shaped around summits (triangles). Distance between summits and depressions is 110 m, which corresponds to an average value from Lyew-Ayee's (2004) morphometric data.

Five reference simulations were performed (Fig. 7) with varying annual precipitation P (m/a) and diffusion coefficient k_d (m²/a). Each simulation should spread over the whole period available for karstification, which is 10 million years (Pliocene era and part of the Pleistocene era). During each simulation, the variations of the morphometric parameters were recorded, in order to compare them with those of the real landscape.

With a constant value of the diffusion coefficient k_d (Fig. 8), an increase in the average precipitation improves simulations. However, with such an increase, a satisfactory adjustment of the real data can never be obtained. With a constant precipitation value, we varied the diffusion coefficient k_d to determine its order of magnitude. It seems that $k_d \approx 10^{-3}$ m²/s is the closest value to the real data.

These reference simulations lead to a major conclusion: whatever the values of the model parameters, the simulated cockpits never have the same average morphometric features as those observed in real landscapes. Consequently, mechanical erosion and diffusion of sediments should not be the only processes taken into account.

Simulations using time–space isotropic dissolution

This application of the CHILD model takes into account carbonate rock dissolution, and considers that it is isotropic over time and space (see above). Therefore, the last component of equation (1) is not nil but equal to $\frac{\partial z}{\partial t}|_{\text{dissolution}} = DR$ m/s,

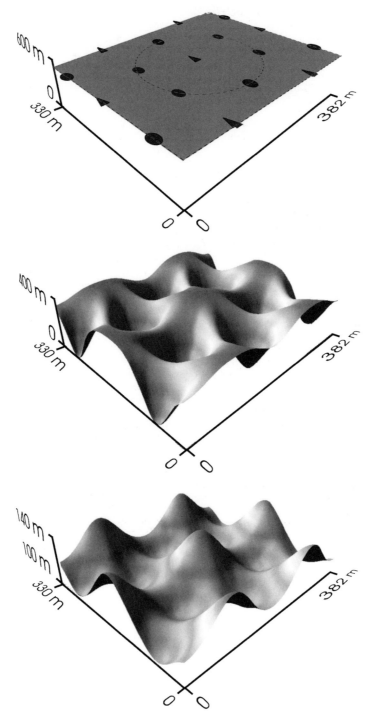

Fig. 7. According to the initial geometry and boundary conditions of the system, CHILD is running to simulate a cockpit karst shape (upper picture). Surface runoff is initially directed towards depressions (middle picture) and erosion is computed until the end of the simulation (lower picture).

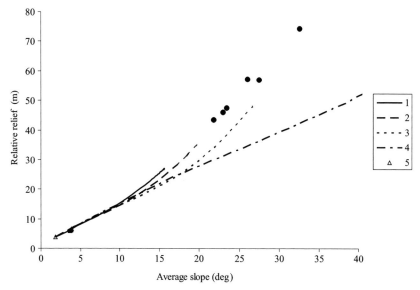

Fig. 8. Reference simulations of the evolution of average slope v. relative relief for various parameters. Here, dissolution is not taken into acount. Real data from Lyew-Ayee's (2004) study are black circles and represent average values of both the relative relief and the slope in the Cockpit Country, Jamaica. (1) $P = 1$ m/a and $k_d = 10^{-3}$ m^2/a; (21) $P = 2$ m/a and $k_d = 10^{-3}$ m^2/a; (3) $P = 3$ m/a and $k_d = 10^{-3}$ m^2/a; (4) $P = 2$ m/a and $k_d = 10^{-4}$ m^2/a; 5, $P = 2$ m/a and $k_d = 10^{-2}$ m^2/a.

where DR is a constant value provided by White's (1984) maximum denudation model.

Initial and boundary conditions are similar to those of the reference simulations. In the case of the present simulations, the DR value had to be worked out. As previously noted, this value depends on a significant number of variables, including precipitation and carbon dioxide pressure. The order of magnitude of

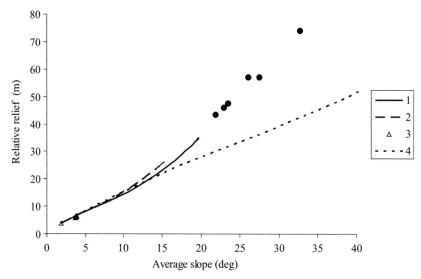

Fig. 9. Simulations using the isotropic dissolution model: evolution of average slope v. relative relief for various parameters. (1) $P = 2$ m/a, $k_d = 10^{-3}$ m^2/a and $P_{CO_{2open}} = 10^{-2}$ atm; (2) $P = 1$ m/a, $k_d = 10^{-3}$ m^2/a and $P_{CO_{2open}} = 10^{-2}$ atm; (3) $P = 2$ m/a, $k_d = 10^{-2}$ m^2/a and $P_{CO_{2open}} = 10^{-2}$ atm; (4) $P = 2$ m/a, $k_d = 10^{-4}$ m^2/a and $P_{CO_{2open}} = 10^{-2}$ atm.

Table 2. *Chemical parameters of the model*

Parameter	Description	Unit	Value
$[Ca^{2+}]_{eqopen}$	Concentration of calcium at equilibrium (opened system)	mol m^{-3}	4.1
$[Ca^{2+}]_{eqclose}$	Concentration of calcium at equilibrium (closed system)	mol m^{-3}	1.9
$[Ca^{2+}]_{sopen}$	Threshold concentration (opened system)	mol m^{-3}	3.69
$[Ca^{2+}]_{sclose}$	Threshold concentration (closed system)	mol m^{-3}	1.71
$[Ca^{2+}]_{inclose}$	Concentration of calcium (entrance of the fracture)	mol m^{-3}	0.0–0.9
k_0	Low-order rate coefficient	mol m^{-2} s^{-1}	4×10^{-7}

Table 3. *Hydrodynamical parameters of the model*

Parameter	Description	Unit	Value
g	Gravitational acceleration	m s^{-2}	9.81
h/L	Gradient in fractures		1.0
μ	Kinematic viscosity	m^2 s^{-1}	1.14×10^{-6}
r	Fracture wall roughness	m	$2 \times a_0/100$
$D_{Ca}^{laminar}$	Calcium molecular diffusion (laminar flow)	m^2 s^{-1}	10^{-9}
$D_{Ca}^{turbulent}$	Calcium molecular diffusion (turbulent flow)	m^2 s^{-1}	10^{-5}

carbon dioxide pressure in the karst cockpit region is $P_{CO_2} \approx 0.01$ atm (Smith & Atkinson 1976; Miotke 1975, in White 1984, p. 195).

In these simulations, we only varied precipitation in order to change DR values (Fig. 9). The other parameters depending on the assigned value of DR are given in Tables 2–4.

The findings of these simulations, which assume that dissolution is isotropic over time and space, are almost similar to those of the reference simulations. This approach only adds a constant component to the evolution of topography over time. Consequently, the decrease in the topographic elevation z is simply quicker than that of the reference simulations. The topographic elevation reaches the water table level within only 4 million years, whereas such a level is not reached in the reference simulation in 10 million years.

Thus, the conclusions drawn from the time–space isotropic dissolution model are quite similar to those of the reference simulations. Assuming a constant denudation in space and time does not allow us to simulate a cockpit karst morphometry identical to that observed in the field.

Simulations using time–space anisotropic dissolution

This application of the CHILD model takes into account carbonate dissolution, but considers that it is anisotropic over time and space (see above). Therefore, the last component of equation (1) is not a constant anymore but equal to $\frac{\partial z_i}{\partial t}|_{dissolution} = \alpha_i(z_i,t)\, DR$ m/s, where DR is a constant value provided by White's (1984) maximum denudation model, and $\alpha_i(z_i,t)$ is a function depending on space and time.

The function $\alpha_i(z_i,t)$ should help us recreate the links between hydrodynamics and dissolution, which should be as close to reality as possible. This is developed through the use of the epikarst

Table 4. *Environmental parameters of the model*

Parameter	Description	Unit	Value
P	Precipitation	m/a	1–3
$P_{CO_{2open}}$	Carbon dioxyde pressure (opened system)	atm	10^{-2}
$P_{CO_{2open}}$	Carbon dixoyde pressure (closed system)	atm	10^{-3}
ρ_{CaCO_3}	Calcite density	kg/m^3	2400
ρ_ω	Water density	kg/m^3	1000
m_{Ca}	Calcium atomic mass	kg/mol	0.04
m_{CaCO_3}	Calcite atomic mass	kg/mol	0.1

concept described above. We have also described the relation between surface runoff, subsurface flow and the dissolution of fractured carbonate rocks.

The dissolution of fractured carbonate rocks is the result of water infiltration into the fractures and thus depends on the position of fractures along the cockpit hillslope. In fact, the higher the dissolution, the wider the fracture opens, and the greater water infiltration there is.

In the case of cockpit karst, surface denudation seems directly linked to the dissolution of underlying fractured rocks (Williams 1985; Lyew-Ayee 2004), and combining both processes seem to be relevant:

$$\frac{\partial z_i}{\partial t} = \alpha_i(z_i, t) DR = \frac{\frac{\partial a}{\partial t}|_i}{\frac{\partial a}{\partial t}|_{max}} DR \quad (24)$$

where $\frac{\partial a(t)}{\partial t}|_i$ represents the fracture aperture variation in the i cell according to the time step dt, and $\frac{\partial a(t)}{\partial t}|_{max}$ is the maximum aperture variation obtained, during a similar time step, over the whole simulated area.

It is thus obvious that function $\alpha_i(z_i, t)$ varies over time and space since it depends on both the position of the cell along the cockpit slope and the aperture of the fracture which widens over time. Moreover, this function lies between 0 and 1, and denudation will never exceed White's (1984) maximum denudation. Equation (24) also shows that denudation reaches its maximum where fracture dissolution is the highest, that is to say on the lower slopes of the cockpits. Conversely, denudation is limited on the cockpit summits where the amount of infiltrated water is less and the rate of dissolution slower.

The present time–space anisotropic dissolution approach requires that hydraulic conductivity is assigned to each cell of the studied area. This has been implemented through the development of a network of parallel fractures over the whole area of study. Fractures are separated by space $s = 1$ m and their initial aperture is $a_0 = 10^{-4}$ m. Initial hydraulic conductivity for each cell is (Gabrovsek & Dreybrodt 2001):

$$k = \frac{g}{12\mu} \frac{a_0^3}{s}. \quad (25)$$

Initial hydraulic conductivity is then 10^{-7} m/s approximately, this value being typical of initial karstification process (Singhal & Gupta 1999). All the other variables involved in simulations are listed in Tables 2–4.

Here are the various calculation stages used in simulations of time–space anisotropic dissolution:

1. Starting from the initial geometry of a fracture, its hydraulic conductivity and that of the corresponding Voronoï cell are worked out.
2. Surface runoff and subsurface flow is calculated using equation (17).
3. If the flow into fractures is laminar, the flow rate is computed using equation (17), the molecular diffusion coefficient $D_{Ca}^{laminar}$, and the diffusion boundary layer $\delta = a_0/2$.
4. If the flow is turbulent, the flow rate is computed using equation (19), the molecular diffusion coefficient $D_{Ca}^{turbulent}$ and the diffusion boundary layer $\delta = a_0/S_h$. Indeed, when the flow is turbulent, the fluid is separated from the mineral surface by a diffusion boundary layer which is a_0/S_h thick. The transfer between fluid and mineral depends on its thickness (Incropera & Dewitt 1996). S_h is the Sherwood number:

$$S_h = \frac{\frac{f}{8}(Re - 1000)Sc}{1 + 1.27\sqrt{\frac{f}{8}(Sc^{\frac{2}{3}} - 1)}} \quad (26)$$

where Sc is the Schmitt number and is about 1000 when the fluid is water.

5. The various variables linked to dissolution are computed: k_1 (equation (12)), k_2 (equation (13)) and the dissolution rate at the lower extremity of the fracture $F_{higher}(L)$ (equation (23)).
6. Fracture aperture is worked out, assuming that the dissolution inside the fracture is uniform along its whole profile (equation (23)).
7. The topographic denudation is calculated, assuming that it is related to dissolution in the fractures (equation (24)).
8. Since fractures have widened, the new values of hydraulic conductivity are computed as well as those of the cells.
9. Back to stage 2.

We have very little scope for manipulating the values of the simulation variables, given that we have previously examined the influence of precipitation P and diffusion coefficient k_d. Concerning closed systems such as fractures, the value of carbon dioxide pressure is much lower than in open systems, and is approximately 10^{-3} atm (Dreydbrot 1988; Dreybrodt et al. 1999; Kaufman & Braun 2001).

The only adjustable parameter is the concentration of incoming calcium into the fractures ($[Ca^{2+}]_{in}$). This parameter only depends on the properties of the overlying soil. When the soil is very thin or almost nonexistent, $[Ca^{2+}]_{in}$ tends to zero, but when the soil is thicker, $[Ca^{2+}]_{in}$ approaches $[Ca^{2+}]_s$ (Smith et al. 1972).

Simulations taking into account the space–time anisotropic dissolution are presented in Figures 10

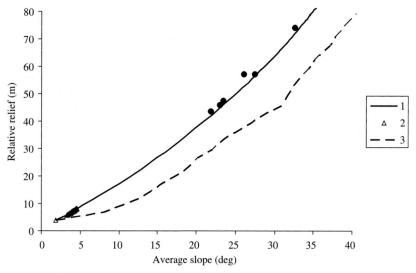

Fig. 10. Simulations using the anisotropic dissolution model: evolution of average slope v. relative relief for various parameters. (1) $P = 2$ m/a, $k_d = 10^{-3}$ m^2/a, $P_{CO_{2open}} = 10^{-2}$ atm and $P_{CO_{2close}} = 10^{-3}$ atm; (2) $P = 2$ m/a, $k_d = 10^{-2}$ m^2/a, $P_{CO_{2open}} = 10^{-2}$ atm and $P_{CO_{2close}} = 10^{-3}$ atm; (3) $P = 2$ m/a, $k_d = 10^{-4}$ m^2/a, $P_{CO_{2open}} = 10^{-2}$ atm and $P_{CO_{2close}} = 10^{-3}$ atm.

& 11. In general, this model simulates cockpit karst landscapes with a morphometry quite similar to that observed in the field. More particularly, simulation 1 in Fig. 10 shows variables whose values are very consistent with literature on that subject, where real data are $P = 2$ m/a, $k_d = 10^{-3}$ m^2/a $P_{CO_{2open}} = 10^{-2}$ atm and $P_{CO_{2close}} = 10^{-3}$ atm.

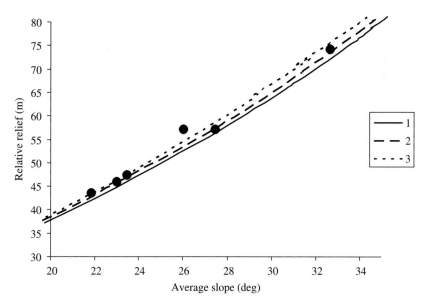

Fig. 11. Simulations using the anisotropic dissolution model: evolution of average slope v. relative relief for various parameters. 1, $P = 2$ m/a, $k_d = 10^{-3}$ m^2/a, $P_{CO_{2open}} = 10^{-2}$ atm, $P_{CO_{2close}} = 10^{-3}$ atm and $[Ca^{2+}]_{inclose}/[Ca^{2+}]_{eqclose} = 0$ (this simulation is actually the same that simulation in Fig. 10); (2) $P = 2$ m/a, $k_d = 10^{-3}$ m^2/a, $P_{CO_{2open}} = 10^{-2}$ atm, $P_{CO_{2close}} = 10^{-3}$ atm and $[Ca^{2+}]_{inclose}/[Ca^{2+}]_{eqclose} = 0.5$; (3) $P = 2$ m/a, $k_d = 10^{-3}$ m^2/a, $P_{CO_{2open}} = 10^{-2}$ atm, $P_{CO_{2close}} = 10^{-3}$ atm and $[Ca^{2+}]_{inclose}/[Ca^{2+}]_{eqclose} = 0.9$.

If we go into the detail, it can be noted that the quality of simulations may be improved by adjusting the $[Ca^{2+}]_{in}$ parameter, corresponding to calcium concentration at the entrance of the fracture (Fig. 11). Today, the cockpit region of Jamaica has thin soil (Lyew-Ayee 2004) and it is reasonable to assume that this type of soil already existed in this region, at least during certain periods, when karstification occurred.

The cockpit karst landscapes which are simulated in this space–time anisotropic dissolution model have very similar morphometric features to those observed in the field. Such simulations are based on the epikarst concept, which describes the coupling of surface runoff, subsurface flow and the dissolution of carbonate rocks. In fact, our simulations provide numerical validation of the epikarst concept.

Conclusion

The aim of the present study was to simulate the morphological formation processes of cockpit karst landscapes in Jamaica. The CHILD model was used to simulate landscape evolution. Two applications of the model were implemented, one based on the hypothesis of time–space isotropic dissolution and the other based on a space–time anisotropic dissolution.

Karst landscapes, simulated on the scale of a cockpit, were compared with reference simulation where only mechanical erosion was taken into account, and henceforth, where rock dissolution was sidestepped.

Morphological features and their variations in time were deduced from these simulated cockpit karst landscapes. Two morphological features based on Lyew-Ayee's (2004) work were chosen to compare simulated karst landscapes and real karst landscapes: average hillslope and relative relief.

The adequacy between the simulated karst morphometry and that of real karst landscapes constitutes an essential stage in the validation of the hypotheses put forward on their formation. With regard to the findings of the present study, the following main points should be emphasized:

- Mechanical erosion, occurring through surface runoff, cannot be the only process accounting for the shape of cockpit karst landscapes. This is consistent with karstification which highlights the central part of dissolution processes in mechanical erosion (White 1984).
- The hypothesis of space–time isotropic dissolution of carbonate rocks, combined with mechanical erosion, does not improve the quality of simulations. In fact, the evolution of the topographic level is similar to that of mechanical erosion, within one constant. The topographic level decreases faster, but with the same spatial variability.
- Space–time anisotropic dissolution was introduced by modelling the processes described in the epikarst concepts. This application highlights the fact that surface runoff, subsurface flow and carbonate rock dissolution have a combined effect. Therefore, a fracture system corresponding to the cells within the model was developed. The lowest fractures on the hillslope received the largest quantity of infiltration water from the perched water table (sub-cutaneous or epikarstic zone). Consequently, the lowest fractures along the hillslope dissolved more rapidly and thus infiltrated more water, and so on. There is positive feedback between runoff into fractures and their enlargement through dissolution. The more active the dissolution of fractured rocks in the subcutaneous zone is, the more extensive the denudation of the topographic surface appears. In fact, denudation involves various complex processes which are not discussed here (e.g. driving of particles towards the bottom of fractures, and collapse of the fractured zones into underlying caves).

The present study highlights how essential it is to take space–time anisotropic dissolution processes into account when studying karstification. Spatial anisotropy was modelled using the heterogeneity of hydraulic conductivity, observed in the karstic systems. Taking heterogeneity into account allows us to recreate the hydrodynamic functioning of a karstic system. It is then possible to combine runoff and dissolution processes more accurately. Temporal anisotropy is modelled using the positive feedback resulting from the combining of runoff and dissolution. Such positive feedback proves to be essential to the understanding of the overall functioning of cockpit karst terrains in Jamaica.

We thank Parris Lyew-Ayee for his constructive discussions on the cockpit karst country in Jamaica, Anne Bouillon for English review and two referees have helped to improve and to clarify this manuscript.

References

AHNERT, F. & WILLIAMS, P. W. 1997. Karst landform development in a three-dimensional theoretical model. *Zeitschrift für Geomorphologie, Suppl.*, **108**, 63–80.

BAUER, S., LIEDL, R. & SAUTER, M. 2005. Modeling the influence of epikarst evolution on karst aquifer genesis: a time-variant recharge boundary condition for joint karst–epikarst development. *Water Resources Research*, **41**(9), W09416.

BEAR, J., TSANG, C. F. & DE MARSILY, G. (eds) 1993. *Flow and Contaminant Transport in Fractured Rock*. Academic Press, San Diego, CA.

BEAUMONT, C., FULLSACK, P. & HAMILTON, J. 1992. Erosional control of active compressional orogens. *In*: MCCLAY, K. R. (ed.) *Thrust Tectonics*. Chapman and Hall, New York, 1–18.

BEEK, W. J., MUTTZALL, K. M. K. & VAN HEUVAN, J. W. 1999. *Transport Phenomena*, 2nd edn. Wiley, Chichester.

BOGAART, P. W., TUCKER, G. E. & DE VRIES, J. J. 2003. Channel network morphology and sediment dynamics under alternating periglacial and temperate regimes: a numerical simulation study. *Geomorphology*, **54**, 257–277.

BRAUN, J. & SAMBRIDGE, M. 1997. Modelling landscape evolution on geological time scales: a new method based on irregular spatial discretization. *Basin Research*, **9**, 27–52.

BROOK, G. A., FOLKOFF, M. E. & BOX, E. O. 1983. A global model of soil carbon dioxide. *Earth Surface Processes and Landforms*, **8**, 79–88.

BUHMANN, D. & DREYBRODT, W. 1985. The kinetics of calcite dissolution and precipitation in geologically relevant situations of karst areas. 1. Open system. *Chemical Geology*, **48**, 189–211.

CLEMENS, T., HÜCKINGHAUS, D., SAUTER, M., LIEDL, R. & TEUTSCH, G. 1997. Modelling the genesis of karst aquifer systems using a coupled reactive network model. *In*: Hard Rock Hydrosciences, Proceedings of Rabat Symposium S2, Vol. **241**, IAHS Publications.

CORBEL, J. 1959. Vitesse de L'erosion. *Zeitschrift für Geomorphologie*, 31–28.

COULTHARD, T. J. 2001. Landscape evolution models: a software review. *Hydrological Processes*, **15**, 165–173.

COULTHARD, T. J., KIRKBY, M. J. & MACKLIN, M. G. 1998. Modelling the 1686 flood of Cam Gill Beck, Starbotton, upper Warfedale. *In*: HOWARD, A. & MACKLIN, M. G. (eds) *The Quaternary of the Eastern Yorkshire Dales: Field Guide*. Quaternary Research Association, London, 11–18.

DAY, M. J. 1979a. The hydrology of polygonal karst depressions in northern Jamaica. *Zeitschrift für Geomorphologie*, **32**, 25–34.

DAY, M. J. 1979b. Surface roughness as a discriminator of tropical karst styles. *Zeitschrift für Geomorphologie*, **32**, 1–8.

DREYBRODT, W. 1988. *Processes in Karst Systems*. Springer, Berlin.

DREYBRODT, W. 1996. Principles of early development of karst conduits under natural and man-made conditions revealed by mathematical analysis of numerical models. *Water Resources Research*, **32**, 2923–2935.

DREYBRODT, W. & EISENLOHR, L. 2000. Limestone dissolution rates in karst environments. *In*: KLIMCHOUK, A., FORD, D. C., PALMER, A. N. & DREYBRODT, W. (eds) *Speleogenesis: Evolution of Karst Aquifers*. Natural Speleological Society, USA, 136–148.

DREYBRODT, W. & GABROVSEK, F. 2000. Dynamics of the evolution of a single karst conduit. *In*: KLIMCHOUK, A., FORD, D. C., PALMER, A. N. & DREYBRODT, W. (eds) *Speleogenesis: Evolution of Karst Aquifers*. Natural Speleological Society, USA, 184–193.

DREYBRODT, W., GABROVSEK, F. & SIEMERS, J. 1999. Dynamics of the early evolution of karst. *In*: PALMER, A. & PALMER, M. (eds) *Karst Modelling*. Karst Waters Institute: Charlestown, WV, 184–193.

FORD, D. C. 1981. Geologic structure and a new exploration of limestone cavern genesis. *Transactions of the Cave Research Groups of Great Britain*, **13**, 81–94.

FORD, D. C. & WILLIAMS, P. W. 1989. *Karst Geomorphology and Hydrology*. Unwin Hyman, London.

GABROVSEK, F. & DREYBRODT, W. 2001. A comprehensive model of the early evolution of karst aquifers in limestone in the dimensions of length and depth. *Journal of Hydrology*, **240**, 206–224.

GROVES, C. G. & HOWARD, A. D. 1994. Early development of karst systems 1. Preferential flow path enlargement under laminar flow. *Water Resources Research*, **30**, 2837–2846.

GUNN, J. 1981. Hydrological processes in karst depressions. *Zeitschrift für Geomorphologie*, **25**, 313–331.

INCROPERA, F. P. & DEWITT, D. P. 1996. *Fundamentals of Heat and Mass Transfer 889*. Wiley, New York.

KAUFMANN, G. 2003. A model comparison of karst aquifer evolution for different matrix-flow formulations. *Journal of Hydrology*, **283**, 281–289.

KAUFMANN, G. & BRAUN, J. 1999. Karst aquifer evolution in fractured rocks. *Water Resources Research*, **35**, 3223–3238.

KAUFMANN, G. & BRAUN, J. 2000. Karst aquifer evolution in fractured, porous rocks. *Water Resources Research*, **36**, 1381–1392.

KAUFMANN, G. & BRAUN, J. 2001. Modelling karst denudation on a synthetic landscape. *Terra Nova*, **13**, 313–320.

KOOI, H. & BEAUMONT, C. 1994. Escarpment evolution on high-elevation rifted margins: insights derived from a surface processes model that combines diffusion, advection, and reaction. *Journal of Geophysics Research*, **99**, 12191–12209.

LYEW-AYEE, P. 2004. Digital topographic analysis of cockpit karst: a morpho-geological study of the Cockpit Country region, Jamaica, PhD thesis, University of Oxford.

MARTIN, Y. & CHURCH, M. 1997. Diffusion in landscape development models: on the nature of basic transport relations. *Earth Surface Processes and Landforms*, **22**, 273–279.

MARSILY DE, G. 1986. *Quantitative Hydrology*, Elsevier, Oxford.

MCKEAN, J. A., DIETRICH, W. E., FINKEL, R. C., SOUTHON, J. R. & CAFFEE, M. W. 1993. Qualification of soil production and downslope creep rates from cosmogenic 10Be accumulations on a hillslope profile. *Geology*, **21**, 343–346.

PFEFFER, K. H. 1997. Paleoclimate & tropical karst in the West Indies. *Zeitschrift tür Geomorphologie, Suppl.*, **108**, 5–13.

ROMANOV, D., GABROVSEK, F. & DREYBRODT, W. 2003. The impact of hydrochemical boundary conditions on the evolution of limestone karst aquifers. *Journal of Hydrologie*, **276**, 240–253.

SIEMERS, J. & DREYBRODT, W. 1998. Early development of karst aquifers on percolation networks of fractures in limestone. *Water Resources Research*, **34**, 409–419.

SINGHAL, B. & GUPTA, R. 1999. *Applied Hydrogeology of Fractured Rocks*. Kluwer Academic Press, Dordrecht.

SMITH, D. I. & ATKINSON, T. C. 1976. Process, landforms and climate in limestone regions. *In*: DERBYSHIRE, E. (ed.) *Geomorphology and Climate*. Wiley, London, 367–409.

SMITH, D. I., DREW, D. P. & ATKINSON, T. C. 1972. Hypotheses of karst landform development in Jamaica. *Transactions of the Cave Research Group of Great Britain*, **14**, 159–173.

SWEETING, M. M. 1972. *Karst Landforms*. Macmillan, London.

TUCKER, G. E. & BRAS, R. L. 1998. Hillslope processes, drainage density, and landscape morphology. *Water Resources Research*, **34**, 2751–2764.

TUCKER, G. E. & SLINGERLAND, R. L. 1994. Erosional dynamics, flexural isostasy, and long-lived escarpments: a numerical modeling study. *Journal of Geophysical Research*, **99**, 12229–12243.

TUCKER, G. E., GASPARINI, N. M., LANCASTER, S. T. & BRAS, R. L. 1997. An integrated hillslope and channel evolution model as an investigation and prediction tool. Technical Report prepared for the United States Army Corps of Engineers Construction Engineering Research Laboratories.

TUCKER, G. E., GASPARINI, N. M., BRAS, R. L. & LANCASTER, S. L. 1999. A 3D computer simulation model of drainage basin and floodplain evolution: theory and applications. Technical report prepared for the United States Army Corps of Engineers Construction Engineering Research Laboratory.

TUCKER, G. E., LANCASTER, S. T., GASPARINI, N. M. & BRAS, R. E. 2001a. The channel–hillslope integrated landscape development model (CHILD). *In*: HARMON, R. S. & DOE, W. W. (eds) *Landscape Erosion and Evolution Modeling*. Kluwer Academic Press, Dordrecht, 349–388.

TUCKER, G. E., LANCASTER, S. T., GASPARINI, N. M., BRAS, R. E. & RYBARCZYK, S. M. 2001b. An object-oriented framework for hydrologic and geomorphic modeling using triangulated irregular networks. *Computers and Geosciences*, **27**, 959–973.

TRUDGILL, S. T. 1985. *Limestone Geomorphology*. Longman, London.

VERSEY, H. R. 1972. Karst in Jamaica. *In*: HERAK, M. & SPRINGFIELD, V. T. (eds) *Karst: Important Karst Regions of the Northern Hemisphere*, Elsevier, Amsterdam, 445–466.

WHITE, W. B. 1984. Rate processes: chemical kinetics and karst landform development. *In*: LA FLEUR, R. G. (ed.) *Groundwater as a Geomorphic Agent*. Allen & Unwin, London, 227–248.

WHITE, W. B. 1988. *Geomorphology and Hydrology of Karst Terrains*. Oxford University Press, New York.

WILLGOOSE, G. R., BRAS, R. L. & RODRIGUEZ-ITURBE, I. 1991. A physically based coupled network growth and hillslope evolution model, 1, theory. *Water Resources Research*, **27**, 1671–1684.

WILLIAMS, P. W. 1985. Subcutaneous hydrology and the development of doline and cockpit karst. *Zeitschrift für Geomorphologie*, **29**, 463–482.

Debris flows as a factor of hillslope evolution controlled by a continuous or a pulse process?

ERIC BARDOU & MICHEL JABOYEDOFF

Institute of Geomatics and Risks Analysis, Faculty of Geosciences and Environment, University of Lausanne, CH-1015 Lausanne, Switzerland (e-mail: eric.bardou@idealp.ch)

Abstract: Flood effectiveness observations imply that two families of processes describe the formation of debris flow volume. One is related to the rainfall–erosion relationship, and can be seen as a gradual process, and one is related to additional geological/geotechnical events, those named hereafter extraordinary events. In order to discuss the hypothesis of coexistence of two modes of volume formation, some methodologies are applied. Firstly, classical approaches consisting in relating volume to catchments characteristics are considered. These approaches raise questions about the quality of the data rather than providing answers concerning the controlling processes. Secondly, we consider statistical approaches (cumulative number of events distribution and cluster analysis) and these suggest the possibility of having two distinct families of processes. However the quantitative evaluation of the threshold differs from the one that could be obtained from the first approach, but they all agree in the sense of the coexistence of two families of events. Thirdly, a conceptual model is built exploring how and why debris flow volume in alpine catchments changes with time. Depending on the initial condition (sediment production), the model shows that large debris flows (i.e. with important volume) are observed in the beginning period, before a steady-state is reached. During this second period debris flow volume such as is observed in the beginning period is not observed again. Integrating the results of the three approaches, two case studies are presented showing: (1) the possibility to observe in a catchment large volumes that will never happen again due to a drastic decrease in the sediment availability, supporting its difference from gradual erosion processes; (2) that following a rejuvenation of the sediment storage (by a rock avalanche) the magnitude–frequency relationship of a torrent can be differentiated into two phases, the beginning one with large and frequent debris flow and a later one with debris flow less intense and frequent, supporting the results of the conceptual model. Although the results obtained cannot identify a clear threshold between the two families of processes, they show that some debris flows can be seen as pulse of sediment differing from that expected from gradual erosion.

Observations of debris flow in small alpine catchments show at first sight the existence of two 'types' of events: ones that are frequently produced by sediment sources that are near and/or within the stream and that do not significantly change the morphology of the stream bed; and ones that incise the stream bed deeply, changing the catchment in the long term, which could be related to rare, but major, events. The flood effectiveness (in the sense of Wolman & Gerson 1978) resulting from these two 'types' of event differs strongly. Observations related to alpine debris flows made by other authors show the existence of extraordinary processes of erosion (at least at a local scale) very different from the gradual erosion behaviour on a short time scale (e.g. Haeberli *et al.* 1991; Tonanzi & Troisi 1996). Is this apparently unsteady behaviour of the system reflecting a continuous process, the existence of two kinds of processes separated by threshold crossing, or the existence of a gradual and an extraordinary behaviour of catchments?

The geomorphological systems we consider here range from dismantling cliffs to the alluvial fans, referred to hereafter as the alpine catchment system. The component on which this paper focuses is the volume of sediments exported out of the system or stored on the fans, mainly by debris flows. The relative magnitude of the events (e.g., 'normal' or 'catastrophic' events) will be defined according to whether the catchment exhibits steady or unsteady behaviour, a distinction which is made according to the effectiveness of the event (Newson 1980). A system in a steady-state is defined as one in which the evolution of the slope proceeds by erosion (in the broad sense) that is proportional to the intensity of the tectonic and climatic agents and consequently that can be correlated with them (Schumm 1977; Burbank & Anderson 2001). A system which is not in a steady-state is defined as one where disruptive erosive events occur, which strongly deviate from the erosion–climatic agent relationship. This could be induced by widespread erosion over the basin (Nouvelot 1990; Meunier

1991; Cannon et al. 2001), delivery of sediment to the stream system by landslides (Ellen & Fleming 1987; Jakob et al. 1997), internal erosion of the stream bed (Koulinski 1993; Zimmermann et al. 1997) or a combination of these factors. When generalized to a major part of the catchment, these various erosive events can be viewed as differing from the 'normal operating' conditions of the catchment. This unsteady-state behaviour of erosion was recognized some time ago as a possible evolutive process of the geomorphological system and has recently been modelled in geomorphological studies (e.g. Tucker & Slingerland 1996; Pinter & Brandon 1997).

To study the sediment dynamics of the alpine catchment system, several approaches can be followed. In the first step, data analysis (at a regional scale and at a local scale) is performed to see if some classes could be defined, reflecting the existence of two 'types' of events. Then a simple conceptual model was built in order to look at the relationship between the components of the alpine catchment system. Finally, case studies analysis was used to verify and discuss the results obtained.

Data sources

In the Alps only a few catchments have been systematically surveyed and instrumented. Thus, available time series of sediment fluxes are generally incomplete and inhomogeneous for a given catchment. For this reason, analysis is often done at a regional scale, in order to increase the amount of data. The data used in this study comes from the Swiss Alps and was collected mainly during the last 25 years. Repeated rain storm events and intensification of land use during this period have highlighted the problems caused by sediment transfer. The data comes from Swiss official syntheses (OFEE 1991; Haeberli et al. 1992; Rickenmann & Zimmermann 1993; Zimmermann et al. 1997; OFEG 2000; BWG 2002; Petraschek & Hegg 2002), as well as unpublished operational reports and a survey carried out by the authors.

Exploratory data analysis

On the basis of the available data, a descriptive approach relating the event volume to the sole catchment area was performed in order to characterize the catchments and the associated events. This general approach gives a scattered result which is difficult to analyse. In a second step, it was interesting to relate the event volume to the specific difference of level, a property of the catchment that is more characteristic for sediment production.

Methods

Volume of one event v. catchment area. One of the basic relations used to study a debris flow is to relate its volume to the catchment area. This kind of representation has been used in past studies attempting to predict debris flow volume (Kronfellner-Kraus 1984; D'Agostino 1996; Rickenmann 1999). As the data sources were different, and because the reported volume possibly was only a fraction of the total volume (especially for small tributaries that reach the main river when it is in flood), we tried to estimate the potential error on the total volume of sediment.

Volume of one event v. specific difference of level. In order to take into account the morphometry of the catchment, which has a great influence on the sediment dynamics (Evans 1997), the specific difference of level, D_s (Melton 1965; Marchi & Brochot 2000), was used,

$$D_s = \frac{alt_{max} - alt_{min}}{\sqrt{A}}$$

where alt_{max} and alt_{min} are the minimum and the maximum altitudes of the catchment, expressed in masl, and A is the catchment area, expressed in square metres.

Based on the reports, we tried to define the causes of the magnitude of the events. If the events seemed to be influenced by other causes than the rainfall–erosion relationship, such as a lake outburst or deep erosion, we classed them under the appellation 'extraordinary geological causes evident' (cf. Fig. 2).

Results

Figure 1 plots the volumes of debris flows as a function of the catchment area in which they occurred. This graph shows a very large dispersal of the data, and it should be noted that these data only represent values from events sufficiently extraordinary to be listed by the Swiss cantonal and federal services. This scattering of volume data is observed all over the European Alps (Kronfellner-Kraus 1984; Franzi 2001), but the overall trend is to have bigger debris flows in wider catchments. Comparison of the data with formulas that infer event volume from the catchment area showed that they give results one or more orders of magnitude higher or lower than the reported data. As a consequence, the area of the catchment alone cannot explain the variability of the observed volumes: the sediment supply is a function of the area, but is also related to other parameters.

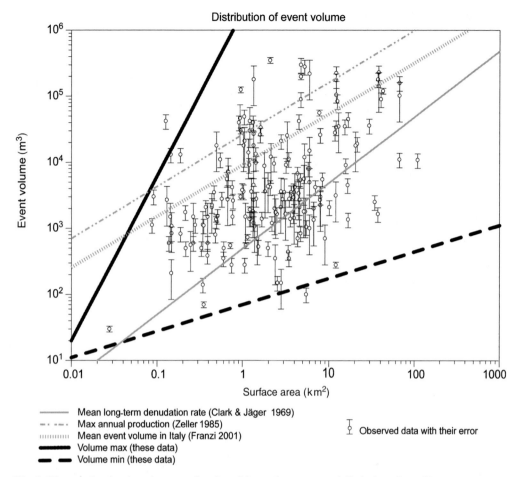

Fig. 1. Diagram showing the volume as a function of the catchment area and displaying other volume–area relations found in the literature.

However, it is interesting to consider the formula of Zeller (1985) depicting the maximal annual volume (in the Swiss Alps) and that of Franzi (2001) representing the mean volume of an event (in the Italian Alps). The two formulas have the same increasing trends that differ from the trend obtained with the long-term denudation rate (e.g. Clark & Jäger 1969).

In spite of this difference between the theoretical long term denudation rate and real event volume, the data seems to present – at least on a regional basis – a continuum. However, a closer look at the data base showed that, for the same catchment, the volume ranged over two orders of magnitude. Some of the reported events showed that catchments already known for the debris flow production (i.e. with data concerning 'normal' events) could be hit by intense mass wasting or extraordinary erosion processes resulting in debris flow volume significantly higher than that observed before. Table 1 shows that the operating of the catchment could differ from the classical rainfall–event relationship that gradually erodes the storage of sediments. These causes may then be viewed in terms of disequilibrium. The main causes of disequilibrium of the system reported in the Alps were glacial lake outbursts, blockage by rock fall and extreme erosion (cf. Table 1). Erosion could be seen as extreme when, for example, a moraine bastion was entrenched (a value of 250 m^3/m on a short reach was reported by Tonanzi & Troisi 1996) or when the torrent bed was deeply eroded, that is, by several metres – 5–7 m are reported from operational surveys. This represents two to three times the height of the former cross-section of the torrent (Petraschek, pers. comm., Chambon et al. 2005, and authors' observation). Such erosion led to an average sediment production of 50–70 m^3/m

Table 1. *Causes of some events with high magnitude that occurred during the nineteenth century in the Canton of Valais*

Name	Area (km^2)	Date	Volume (m^3)	Causes
Baltschiederbach	42.64	14 October 2000	120,000	Very heavy rainfall over a long period + no previous event for a long period
Saxé	0.95	10 November 1939	125,000	Hydrogeological enhancement in conjunction with very mobile material deposited upstream
St Barthélémy	12.05	20 July 1930	125,000	Rock avalanche
Täschbach	37.23	15 June 2001	150,000	Lake outburst
St Barthélémy	12.05	11 August 1927	160,000	Rock avalanche
Illgraben	4.73	3 October 1995	180,000	Unknown (but possible effect of a rockfall dam)
Saasbach	5.22	24 July 1987	200,000	Very heavy rainfall over a long period leading to important mass wasting all over the catchment[a]
Saltina	66.01	24 September 1993	250,000	Very heavy rainfall over a long period leading to important mass wasting all over the catchment[a]
Illgraben	4.73	6 June 1961	300,000	Lake outburst (lake dammed by a Rock avalanche)
Bossay	2.08	15 October 2000	350,000	Man-made debris flow due to ruin of a pipe

[a]During these events, many other catchments were hit by mass wasting and no detailed survey was done at the time of the event.

on the whole range of the torrent, which is a very high value. These very high values were comparable with the maximum values of bank erosion (15–30 m^3/m), before considering sediment input as point source (landslide), as proposed by Hungr *et al.* (1984).

In a sense, the catchment area integrated too many variables, some of which were not linked to the sediment production, hindering refinement of the analysis. The specific difference of level depicted the potential energy of the catchment. Figure 2 shows that the data could be separated in two major groups: one with high density of small events (i.e. a relative high frequency of occurrence) and one with events of high magnitude apparently more influenced by geological/geotechnical causes (when they are known). If there was overlapping of the events influenced by geological causes other than the rainfall–erosion relationship, and if there were uncertain causes for some events, it seems that at least two zones could be differentiated at the 7.5×10^4 m^3 level. Moreover, in Figure 2 the population of small and frequent events without additional geological cause represents a non-exhaustive population, whereas the data for events of high magnitude, mainly influenced by geological causes, came from an exhaustive dataset. It follows that, below 2×10^4 m^3, the real density of small events is greater and thus the separation between the two zones becomes more evident.

From the regional analysis of the volume v. the catchment area, as well as from its refinement with the specific difference of level, it was difficult to see if the events of higher magnitude were catastrophic events or if they were the tail of the distribution of gradual erosion of watersheds. Site-specific analysis supported the separation of the events into two families on the basis of the processes of erosion.

Analysis with statistical tools

Even if the data were incomplete, it was interesting to see if they could really be separated into two groups and where the threshold lay between the two groups. Two statistical methods were used to look at this hypothesis.

Methods

Cumulative number of events. The goal of this analysis was to see if the event volume came from a single population or more. This analysis was based on the distribution of the cumulative number of events exceeding a given volume. It had already been successfully used to analyse rockfalls (Wieczorek *et al.* 1998; Hungr *et al.* 1999) as

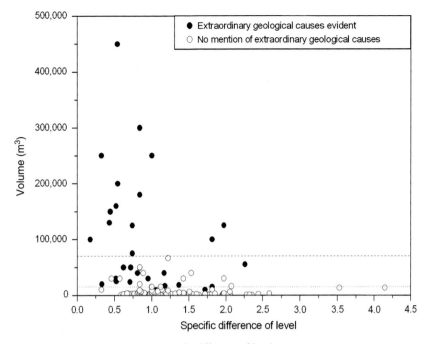

Fig. 2. Volume distribution according to the specific difference of level.

well as other kinds of mass wasting processes (Guzzetti et al. 2002; Brardinoni et al. 2003). As these authors have shown, part of the distribution can be described by a power law:

$$N(V) \approx V^{-b}$$

where V is the volume of the events, $N(V)$ the number of events that have a volume exceeding V and the exponent b is a constant parameter.

Wards' cluster analysis. The goal of the cluster analysis was to extract groups of data (Davis 1986). Here the idea was to look for the level of volume at which two families (one with a small and frequent debris flow and the other with a large and rare debris flow) could be distinguished. The difference between the sample elements was measured by an iterative procedure. The elements were aggregated and put into clusters. The higher the level of aggregation, the less similar were the members in the respective cluster. Among the various possibilities to compute aggregation rules (i.e. if the neighbouring cluster could be aggregated together in the next step), the following procedures were retained:

- The measuring of the 'distance' between data (here the volume) was done using the most common type of procedure: the Euclidean distance. It refers to the geometric distance in the multidimensional space and is computed as:

$$\text{distance}(x,y) = [\Sigma_i (x_i - y_i)^2]^{1/2}$$

where x and y are the coordinates of the object.
- As the aggregation rule, Ward's method (Ward 1963) was applied. To evaluate the distances between the clusters, the variance was analysed. This method attempted to minimize the sum of squares (SS) of any two (hypothetical) clusters that could be formed at each step.

Results

From the cumulative number of events analysis (cf. Fig. 3), it appeared that the data distribution was separated into two parts. For the high volume (on the right of the graph) the data could be fitted to a power law. The exponent b was 0.423 ± 0.007, which is surprisingly near the value of c. 0.4 found by Dussauge et al. (2003) for rockfalls. For this type of approach, the incompleteness of the data, especially for small volumes, can be put forward as an explanation for the presence of two families. Nevertheless, as already proposed by other authors, the observed knick point on the curve was much higher than if it was caused by the incompleteness of the data (Hovius et al. 2000; Guzzetti et al. 2002; Guthrie

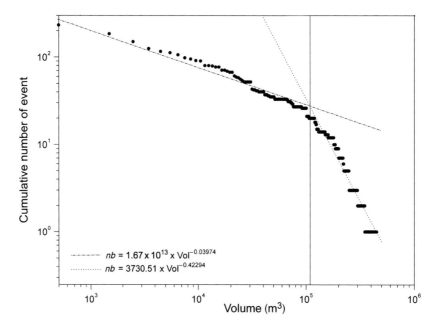

Fig. 3. Cumulative distribution of volume values.

& Evans 2004). This approach makes it possible to set a limit between two classes of volumes at approximately 1×10^5 m^3.

Another way of looking at the limit between the small and frequent events resulting from gradual erosion and catastrophic events of high magnitude was the cluster analysis. It was performed until two clusters were built. Once the two groups were constituted, the limiting branch of the hierarchical tree was reported. The data could be drawn as two box-and-whisker plots representing the distribution of the event within the two families. Again the limit appeared towards 10^5 m^3, as it can be seen from Figure 4.

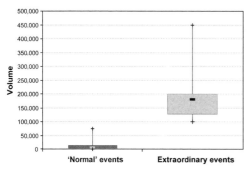

Fig. 4. Box-and-whiskers plot of the distribution of the population of the two families formed by the cluster analysis.

Conceptual modelling

Given the environment of alpine catchment and the data available, it was interesting to investigate the sediments dynamic resulting from different mechanisms of volume formation. As differentiation between gradual and catastrophic events can appear to be possible only for a short time scale (the one used for natural hazard assessment), a longer time scale was envisaged. If one takes a geological scale of time into account, it is possible for a sufficient number of events to exist on a site to build a steady statistical population. At a shorter time-scale, the evolution of the landscape fluctuates, partly due to the variability of the sediment fluxes (Pratt-Sitaula et al. 2004). To avoid analyses biased by the smoothing of these diverse and complex phenomena, a simple conceptual stochastic model was built.

Methods

For this study a model was developed to estimate the behaviour of catchments suffering debris flow, balancing the volume available. This made it possible to describe the sedimentary budget dynamics of an alpine catchment system on time scales under 10,000 years. If landscape evolution models already exist, the integration of a physical difference in the sediment response according to the magnitude and to the temporal evolution of the sediment availability is new (Tucker & Slingerland 1996).

The model is made up of two modules that stochastically generate: (1) the volume of sediment available at a given time according to erosion and landslide sediment production; and (2) the triggering conditions – if they are met, the whole available sediments are supposed to form a debris flow.

The components of the model are described in Figure 5. The model assumes a catchment area (S) from which only a part (S_A) could deliver sediments to the torrent by erosion (E) during one event (as observed in many cases e.g. Haeberli et al. 1992). The catchment contains an initial stock of sediment having a volume (V_i) such as moraine and others sediments. On average, V_i is assumed on the entire surface S. Only part of the store of sediment is easily available (R), and presents a random variable (D) with time. This random variability reflects the complexity of the processes that deliver sediments to the gully.

The debris flows are assumed to be triggered by precipitation above a given intensity threshold Th_p (expressed in mm/h), following a normal distribution (m_p, σ_p). The time frequency of precipitations follows a Poisson distribution with a mean frequency (t_p). Landslides, including rockfalls, are assumed to be an important potential sediment supply (Iverson et al. 1997; Sandersen et al. 2001). An average volume V_l and a return period T_l, both following Poisson distributions as indicated on Figure 5, control the alimentation of the gullies by landslides. This variability in the initiating conditions of sediment supply induces a threshold in erosion, which is an important issue in landscape erosion modelling (Tucker & Whipple 2003).

Some other factors that influence the sediment yield, like vegetation (Cerdan et al. 2002; Istanbulluoglu & Luce 2004), tectonic (Tucker & Slingerland 1996; Hovius et al. 1997) and thermal

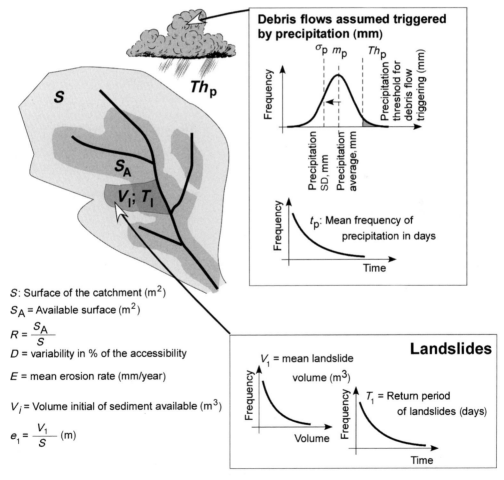

Fig. 5. Description of the conceptual model and its components.

Table 2. *Parameters of the long-term simulations*

Variable	Units	Simulation 1	Simulation 2	Simulation 3
S	m^2	10,000,000	10,000,000	10,000,000
V_i	m^3	30,000,000	30,000,000	—
e_i	m	3	3	—
E	mm/a	0.5	—	0.5
V_1	m^3	10,000	—	10,000
T_1	years	3	—	—
m_p	mm	5	5	5
σ_p	mm	3	3	3
t_p	days	10	10	10
Th_p	mm	13	13	13
R	%	15	15	15
D	%	20	20	20

effects (i.e. passage through the freezing point, Bogaart *et al.* 2003; Bardou & Delaloye 2004), are not modelled here. As this model is a first attempt to conceptualize the variation of the sediment dynamics at the event scale, simplicity was preferred. Nevertheless, if quantification of the sediment volume is researched, a more complex model should be built.

The initial parameters of the model were as follows. An erosion rate of 0.5 mm/a and a three-year return period for a landslide with an average volume of 10,000 m^3 were used, which are relevant for an alpine region subject to debris flows. This represents the sedimentary input (cf. Table 2). On the other hand, a stock of sediment of 30 Mm^3 was considered (roughly evaluated in analogy with typical alpine catchments). Precipitations were supposed to fall on average every 10 days. Around 1% of these precipitations was supposed to lead to debris flows. The time step adopted thus was 1000 days and the total duration of the simulation was approximately 10,000 years.

Results

Initially, three catchment behaviours were modelled (cf. Fig. 6), each one having a different origin for the sediment: (1) a basin with a mixed origin (storage + weathering, column A); (2) another with only a primary storage of sediments (column B); and (3) a third with supply only by weathering (column C). For each of these simulations, the climatic factors were considered to be identical (cf. Fig. 6). Only the availability of the sediments varied.

Starting the simulation for a catchment with storage (e.g. moraines) and sediment supply (erosion and landslides), the debris flow volumes and frequency were higher at the beginning of the period, before reaching steady-state behaviour after 5000 years, that is to say half of the reference period (cf. Fig. 6, column A). For a catchment that possessed only a limited input of fresh sediments (i.e. the supply by weathering is equal to zero, column B), large debris flows occurred during the first 2000 years, that is to say one-fifth of the reference period. After that period, a progressive decrease of the debris-flows volumes to zero after 4000 years could be seen. On the other hand, a catchment without pre-existing sediment storage (i.e. the storage was zero, column C), but possessing a constant input of fresh sediments, reached a steady-state behaviour after 2000 years, that is to say one-fifth of the reference period.

Following the type of formation of the debris flow volume, there could be a great difference between sediment volumes available at the beginning and the end of the period. After 20–50% of the time of the reference period, the volume never reached the high value simulated in the first part of the period again. This reflects the tendency of the catchment to reach its equilibrium.

However, this simulation could not simply be extrapolated to real data, due to the limitations of the various assumptions made, i.e. the climatic scheme was considered to be identical for the whole period. In order to reach a more 'realistic' modelling, the model was run over a 150 year period. This time horizon corresponded to the last important generation of sediment storage in the European Alps (the Little Ice Age). Over this 150 year period, the assumption of stationarity of the parameters was less strong. At this time-scale, the existence of different classes of events volumes still appeared. The same trend (i.e. proportion of high magnitude events at the beginning of the series) can be seen (Fig. 7). The events were due to the presence of easily mobilized sediments, which could, for example, correspond to the deglaciation phases (Evans 1997) or the alteration of moraine bastion (Tonanzi & Troisi 1996; Delaloye pers. comm.).

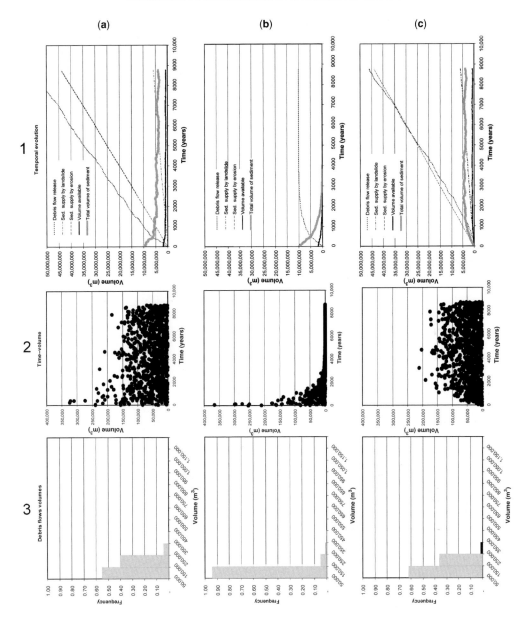

Fig. 6. Results of the simulation for three different origins of sediments.

Discussion

Distinction between gradual erosion and catastrophic events

The results obtained from the data analysis led to the differentiation between small and frequent events resulting from gradual erosion and larger but rarer events resulting from catastrophic geological causes in the catchment. Although the separating range between the events judged as catastrophic and normal events was computed using several methods, it should be noticed that the values fall in a wide range. This is not only due to the quality of the data and to the different shortcomings inherent to each method, but probably also to the variability of nature (e.g. there is surely an overlap between the higher volume supposed to come from gradual behaviour and the lower volume produced by noticeable geological processes).

Table 3. *Summary of the limits found with the various approaches*

Method	Volume threshold
Geomorphologic (according to specific difference of level)	75,000 m^3
Cumulative number of event	100,000 m^3
Cluster analysis	100,000 m^3

Table 3 summarizes the different thresholds of volumes separating the two possible behaviours of a catchment. For the Central European Alps, depending in part on the predisposition of the catchment, this threshold is around 5×10^4 m^3. This distinction into two families of events is based on differences in physical processes related to volume formation of debris flow (linked with peaks of sediments supply, e.g. the one given in Table 1).

The defined thresholds have to be taken as indicative, and not as well-determined values. They are merely order of magnitude estimates. They could be moderated by the predisposition of the catchment and/or by human changes in the watershed (e.g. building of safety works that reduce the sediment supply, at least for a certain time).

Availability of sediments as a cause of the two distinct behaviours

A simple stochastic model suggests that the distribution of debris flow volumes depends strongly on both the supply of fresh sediment and the initial storage of sediment. For catchments only supplied by weathering, the events show a quite regular distribution. The presence of sediment storage (e.g. Quaternary deposits) can induce large debris flows at the beginning of the period, with volume significantly higher than can be restored by the rate of fresh sediment supply. As a consequence, the storage will be progressively emptied. Thus, debris flows decrease in volume and in frequency. The present distribution of large debris flows is therefore incomplete, because it has to be correlated with the early period of the catchment history. At present, most of the catchments contain Quaternary deposits. Large debris flows are still available, but they will not represent the standard volume. Owing to the gradual erosion processes acting on the catchment, smaller events will be more frequent. In other words, the highest volumes belong to same statistical distribution. The law of large numbers is not applicable, and thus extraordinary events can appear discontinuous with the background activity of the catchment. This transient behaviour at the beginning of the activity in the catchment that appears for each mode of sediment supply is consistent with other models (Tucker & Slingerland 1996) and field measurement in comparable environments (Evans 1997).

At least two process types of debris-flow formation coexist within a given stream: one implying a gradual erosion (steady-state process) and the other a pulse of erosion (process in disequilibrium with its environment). The pulse events, producing large volumes, are due to peaks in sediment supply and/or change in triggering factors. They can be followed by periods of slight activity in the gully. Therefore, the sediment fluxes can be described by a variable pattern. This variability has been the rule during Holocene (Martin 2000; Hinderer 2001).

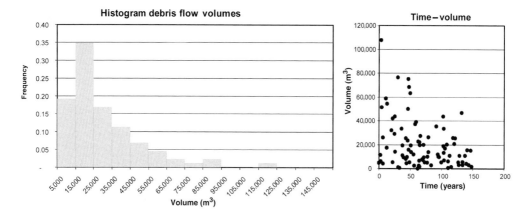

Fig. 7. Simulation for a catchment with stock and recharge by weathering over a period of 150 years (i.e. approximately from the end of the Little Ice Age to the present).

Thus the generation of a sediment volume forming the debris flow is a complex function of various phenomena having neither the same time scale nor the same magnitude. The output of this function (i.e. the description of the behaviour of the catchment) is not continuous, but shows thresholds. The level of these thresholds may vary with time (Baker 2002). In the case of the alpine catchment system, the magnitude of an event results from at least two different behaviours of the catchment: the gradual one is related to the generalized erosion (Meunier 1991), and the catastrophic one is influenced by the extent of the geological and the hydrogeological phenomena occurring in the upper part of the drainage system.

Case studies

Two case studies are presented in order to illustrate how and why the sediment supply can change with time. This enables us to consider how the results presented above (although descriptive and linked to strong hypotheses) are relevant to field observations.

The Illgraben catchment

The active area of the Illgraben catchment reaches 4.4 km^2 and elevations range from 900 to 2720 masl. The study conducted in the Illgraben torrent since 1997 have shown that it is prone to sudden floods. Several texts dating from the end of the Middle Ages mention these floods. In the analysis of the severe floods which struck Switzerland during the end of the nineteenth century, the Illgraben appears to be a torrent prone to causing problems to human infrastructures (Culmann 1864). In the scope of the present study, the distribution of debris flow linked with a rock avalanche event will be looked at in detail.

In 1961, a 5 Mm^3 large rock avalanche occurred in the upper part of the catchment (Lichtenhahn 1971; Eisbacher & Clague 1984). This event produced important deposits along the axis of the valley, which became a new storage of easily available sediments. This deposit dammed a lateral channel of the Illgraben's torrent, resulting in the creation of a lake. The failure of this dam triggered the most extreme debris flow recorded in a century. It was triggered by low-magnitude rainfall. Figure 8 shows that, after this geological event, an increase in debris flow frequency was observed on the main fan (approximately 3–4 a year up to 1963), followed by a long period of relatively low debris flow activity (Zimmermann 2000; Bardou et al. 2003).

This case study agrees with the results of the conceptual modelling. The increase of debris flows after 1961 can be linked to the high availability of sediment provided by the rock avalanche deposit (i.e. a new sediment storage). The frequency of the debris flows increased over some years, until the rock avalanche deposit talus reached equilibrium. Afterwards the frequency decreased over some years, probably due to the protection work built in reaction of the 1961 event. Then the torrent recovered a dynamic that corresponded to its new equilibrium, until a new geological event happens and changed the sedimentary predisposition of the catchment. The first event, the one that occurred when the dam was breached, could be considered as a catastrophic (it is the largest event listed in the 'history'). The subsequent events were readjustment of the bed cross-section to the new storage of sediment. It could be considered that these events differed from the gradual dynamic of erosion of the torrent.

The Saxé and Métin catchments

The areas of the Saxé and Métin catchments are respectively 0.93 and 1.01 km^2. The two torrents are an average of 275 m apart (cf. Fig. 9). Both catchments are made from Granodiorites, which are surmounted by Limestones belonging to the Nappe of Morcles (cf. Fig. 10). In the middle part of the Saxé catchment, an important mass of deposited sediments (from not clearly identified palaeo processes) provides a mass of loose sediments. This is not the case in the Métin catchment, where availability of sediments is low.

In November 1939, the settlement of Saxé was inudated by a 125,000 m^3 large debris flow, flowing through the village over 5 h, when the Métin's torrent produced only several hundred cubic metres (Montandon 1940). An important part of the deposited volume was eroded from the sediments located in the middle part of the hillslope, as no other scars are visible. From the mapped inundated area and from the estimation done in 1939, we attempted to assess the potential volume of in-place sediment that was removed (accounting for water content, 10%, and bulking, 30%). This estimation gave 86,000 m^3. The present estimation of the missing volume in the current gully gave a value of 88,000 m^3. That is a variation of only 2000 m^3 on the best estimate of volume in place potentially mobilized during the event of 1939. The magnitudes of the values of the missing volume (surprisingly similar, considering that some events implying a few thousand cubic meters have been recorded since 1939) and the current morphology (allowing a reconstitution of that of 1939) show that the gully was only notched slightly to have

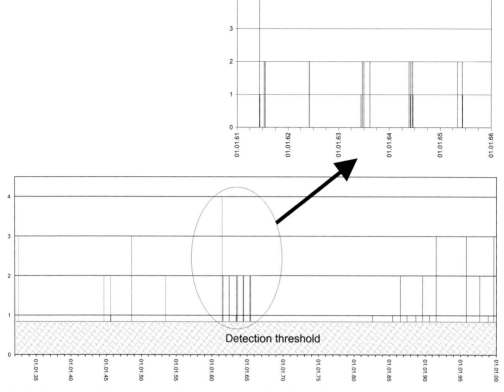

Fig. 8. Distribution of debris flow exceeding a survey threshold and classified according to their magnitude (assessed from the damages and flooded area) in the Illgraben torrent.

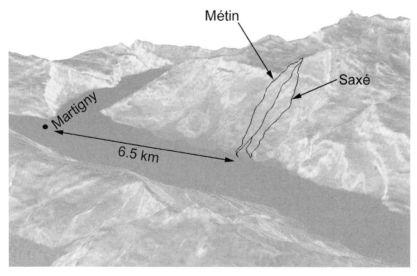

Fig. 9. Perspective of the catchments of Saxé and Métin.

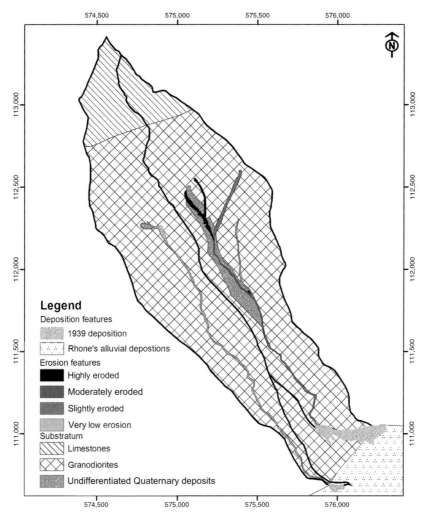

Fig. 10. Geological disposition and geomorphology of the present gullies in the Saxé and Métin's catchment. The reference is the Swiss coordinate given in metres.

sufficient sediments at disposal (cf. Fig. 11). According to the traces currently mapped, the storage of the necessary sediments had to be in the gully (the proximity of the old torrent-bed and the easily movable sediments, which corresponds to the parameter S_A in the model). As a result, the quantity of available sediments decreased. The estimation of probable maximum volumes today made according to the method of Hungr et al. (1984) shows that, with the observed predisposition, an event like that of 1939 is unlikely to occur. Figure 11 indicates that the catchment tends to pass from a transport-limited condition to a supply-limited one. This kind of change would influence the magnitude–frequency relationship. This has already been observed in other areas with significant relief (Newson 1980). The conclusions of Montandon, although careful, show that the event already appeared extraordinary at the time when the traces were still fresh: 'As far as we know, localities situated on these fans have not suffered ... serious damage due to debris flow ... Otherwise, it seems that we would have been informed by local historians ... It is certain that in Saxé, going back for four generations – to 100 or 120 years – one has no memory of a similar misfortune to that of November 1939 – which does not say ... that nothing serious did happen there 200, 500 or 1000 years ago' (translated from French, Montandon 1940).

Analysis of the rainfall compared with the average water necessary to the mobilization of

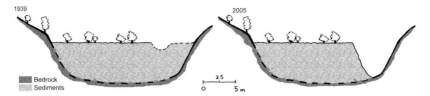

Fig. 11. Cross-section of the torrents at an altitude of 1200 masl, where reconstitution of the probable morphology of the event of 1939 is the most precise (the linear production of 17 m^3/m at this place). The estimate of the current potential delivery of sediments is of 6.5 m^3/m. If the ratio (approximately 1 to 2) is supposed to be similar throughout the channel and with an estimate of current probable maximum volume around 70,000 m^3, i.e. a little more than half of the event of 1939 (estimated by other sources, Marquis pers. comm.) one can admit that the reconstitution is coherent.

such a large debris flow (evaluated from rheologic measurements, Bardou, 2002) shows that the one-day rainfall with a 100-year return period applied for 3 consecutive days would have been necessary. The rainfall recorded at the raingauge at Martigny, situated 6.5 km southeastwards, and the witness reports did not indicate this kind of exceptional precipitation (cf. Fig. 9). Furthermore, old pictures show that, at the end of the event, snow replaced rainfall (Kunz 1939). This inconsistency between the rainfall, the runoff transport capacity and the effective volume led us to consider an additional contribution of water. This contribution has to be sought among hydrogeological effects (e.g. the temporary outbreak of a local spring, a fact observed in this area by the authors during field work in 2001), as is often the case in such an environment (Hungr et al. 1984).

This case study shows that the conditions leading to a trigger could change and in some case perhaps could not be put together. From a physical point of view, the events that occurred in Saxé's torrent come from two distinct geomorphological families. In this case, the emptying of the sediment storage after the 1939 event was so drastic that, without other changes, the present torrent seems to be unable to produce a volume similar to that of 1939. Furthermore, the unknown origin of the water that produces the debris flow changes the possibility to assist, or not, such an event.

What the case studies illustrate

Detailed surveys of historical events make it possible to link the statistical and conceptual models to reality. The two case studies presented support the hypothesis of coexistence of continuous and catastrophic erosion processes on a catchment already observed by others authors (e.g. Jakob et al. 1997). The behaviour of the Saxé catchment was related to the change in triggering factors, drastic emptying of the sediment storage and/or the temporary apparition of a local spring, that no longer enables events with a magnitude similar to that one of 1939. In the case of the Illgraben torrent, complex geological processes formed the sediment supply. These can evolve rapidly or in a discrete way. When one of these processes unbalances the catchment operating (here the rock avalanche), other processes can take place in turn (e.g. natural dam failure), causing a series of large and frequent debris flows differing from the gradual erosion expected in such a catchment. After that crisis, the catchment tends to reach its equilibrium-production again.

Conclusions

Different approaches, made at different spatial and temporal scales, indicate that at least two different families of events can be distinguished. Depending on the processes involved in the catchment, an event in either steady-state (gradual erosion) or in non-steady-state (extraordinary behaviour) can take place. However, it is difficult, and in a certain way of little importance, to set a limit between the families. Nevertheless, the different methods used made it possible to set a threshold at around $7.5-10 \times 10^4$ m^3 for the western Swiss Alps. This threshold is only valid for the environmental settings and there is probably an overlap between the two families. The study of the frequency distribution of debris flow volumes, including an event that took place in volcanic settings, Jakob (2005), show that there is a knick point in the distribution at 1×10^6 m^3 (linked with a change in the process of available sediment production).

This coexistence of two different behaviours has to be integrated into hazard assessment procedures. This could be done by a comprehensive analysis of the relevant catchment. The new conceptual model, developed for analysing the variability of the sediment fluxes in a given catchment, made it possible to clarify the time scale necessary for statistically representative distribution. This model also gave new insights into the causes of catastrophic events. The predisposition of the catchment can be enhanced by geological events that change the magnitude of triggering factors. As a consequence, the identification of the possible supply of sediments, their dynamic of renewal and possible triggering factors, is of great importance to asses the general sedimentary behaviour of the catchment. Each catchment should be conceptualized as a system yielding complex interactions. In particular, distinctions should be made between the different geomorphological processes that participate either in the gradual erosion of the catchment or in catastrophic events.

The authors wish to thank the road and river service of the 'Canton du Valais' (SRCE) for giving us access to their archives and to the engineers and geologists with whom we shared some interesting discussions. We also wish to thank two anonymous reviewers whom made substantial improvements of the early paper possible. We thank also our two colleagues F. Baillifard and T. Oppikofer for their comments.

References

BAKER, V. R. 2002. High-energy megafloods: planetary settings and sedimentary dynamics. *In*: MARTINI, I. P., BAKER, V. R. & GARZÓN, G. (eds) *Flood and Megaflood Processes and Deposits: Recent and Ancient Examples*. Special Publication **32** of the International Association of Sedimentologists, 3–15.

BARDOU, E. 2002. Méthodologie de diagnostic des laves torrentielles sur un bassin versant alpin. Thesis, no. 2479. EPFL, Lausanne.

BARDOU, E. & DELALOYE, R. 2004. Effects of ground freezing and snow avalanche deposits on debris flows in alpine environments. *Natural Hazards and Earth System Sciences*, **4**, 519–530.

BARDOU, E., FOURNIER, F. & SARTORI, M. 2003. Paleofloods reconstruction on Illgraben torrent (Switzerland): a need for today frequency estimation, *International Workshop on Paleofloods, Historical Data & Climatic Variability: Applications in Flood Risk Assessment*, Barcelona.

BOGAART, P. W., TUCKER, G. & DE VRIES, J. J. 2003. Channel network morphology and sediment dynamics under alternating periglacial and temperate regimes: a numerical simulation study. *Geomorphology*, **54**, 257–277.

BRARDINONI, F., SLAYMAKER, O. & HASSAN, M. A. 2003. Landslide inventory in a rugged forested watershed: a comparison between air-photo and field survey data. *Geomorphology*, **54**, 179–196.

BURBANK, D. W. & ANDERSON, R. S. 2001. *Holocene Deformation and Landscape Responses, Tectonic Geomorphology*. Blackwell Science, Malden, 159–173.

BWG, 2002. *Hochwasser 2000 – Les crues 2000*. OFEG-BWG, Bern.

CANNON, S. H., KIRKHAM, R. M. & PARISE, M. 2001. Wildfire-related debris-flow initiation processes, Storm King Mountain, Colorado. *Geomorphology*, **39**, 171–188.

CERDAN, O., LE BISSONNAIS, Y., SOUCHÈRE, V., MARTIN, P. & LECOMTE, V. 2002. Sediment concentration in interrill flow: interactions between soil surface conditions, vegetation and rainfall. *Earth Surface Process and Landforms*, **27**, 193–205.

CHAMBON, G., ESCANDE, S., LAIGLE, D. & TACNET, J. M. 2005. Commune des Contamines-Montjoie; Département de la Haute-Savoie; Torrent du Nant d'Armancette. Evènement du 22 août 2005. *Compte rendu de la visite de terrain du 31 août 2005*, Cemagref, Grenoble.

CLARK, S. P. & JÄGER, E. 1969. Denudation rate in the Alps from geochronologic and heat flow data. *American Journal of Science*, **267**, 1143–1160.

CULMANN, K. 1864. Bericht an den hohen schweizerischen Bundesrath über die Untersuchung der schweiz. Wildbäche, vorgenommen in den Jahren 1858, 1859, 1860 und 1863, Zürich.

D'AGOSTINO, V. 1996. Analisi quantitativa e qualitativa del transporto solida torrentizio nei bacini montani del Trentino Orientale, I problemi dei grandi comprensori irrigui. Associazione Italiana di Ingegneria Agraria, Novara, 111–123.

DAVIS, J. C. 1986. *Statistics and Data Analysis in Geology*. Wiley, Chichester.

DUSSAUGE, C., GRASSO, J.-R. & HELMSTETTER, A. 2003. Statistical analysi of rockfall volume ditributions: implications for rockfall dynamics. *Journal of Geophysical Research*, **108**, 2286.

EISBACHER, G. H. & CLAGUE, J. J. 1984. Destructive mass movement in high mountains: hazard and management, 84-16. Geological Survey of Canada.

ELLEN, S. D. & FLEMING, R. W. 1987. Mobilization of debris flows from soil slips, San Franscico Bay region, California. Debris flow and avalanche: process recognition and mitigation. *Reviews in Engineering Geology*, **VII**, 31–40.

EVANS, M. 1997. Temporal and spatial representativeness of alpine sediment yields: Cascade Mountains, British Columbia. *Earth Surface Processes and Landforms*, **22**, 287–295.

FRANZI, L. 2001. A statistical method to predict debris-flow volumes deposited on a debris fan. *Physics and Chemistry of the Earth, Part B: Hydrology, Oceans, Atmosphere*, **26**, 683–688.

GUTHRIE, R. H. & EVANS, S. G. 2004. Analysis of landslide frequencies and characteristics in a natural system, coastal, British Columbia. *Earth Surface Processes and Landforms*, **29**, 1321–1339.

GUZZETTI, F., MALAMUD, B. D., TURCOTTE, D. L. & REICHENBACH, P. 2002. Power-law correlations of landslide areas in central Italy. *Earth and Planetary Sciences Letters*, **195**, 169–183.

HAEBERLI, W., RICKENMANN, D. & ZIMMERMANN, M., 1991. Murgänge, Ursacheanalyse der Hochwasser 1987. Ergebnisse der Untersuchung. Mitt. no. **14**. Landeshydrologie und -geologie, Bern.

HAEBERLI, W., RICKENMANN, D., RÖSSLI, U. & ZIMMERMANN, M. 1992. *Murgänge 1987, Dokumentation und Analyse*. 97.6. VAW, Zürich.

HINDERER, M. 2001. Late Quaternary denudation of the Alps valley and lake fillings and modern river loads. *Geodinamica Acta*, **14**, 231–263.

HOVIUS, N., STARK, C. P. & ALLEN, P. A. 1997. Sediment flux from a mountain belt derived by landslide mapping. *Geology*, **25**, 231–234.

Hovius, N., Stark, C. P., Hao-Tsu, C. & Jiun-Chan, L. 2000. Supply and removal of sediment in a landslide-dominated mountain belt: Central Range, Taiwan. *Journal of Geology*, **108**, 73–89.

Hungr, O., Morgan, G. C. & Kellerhals, R. 1984. Quantitative analysis of debris torrent hazards for design of remedial measures. *Canadian Geotechnical Journal*, **21**, 663–677.

Hungr, O., Evans, S. G. & Hazzard, J. 1999. Magnitude and frequency of rock falls and rock slides along the main transportation corridors of southwestern British Columbia. *Canadian Geotechnical Journal*, **36**, 224–238.

Istanbulluoglu, E. & Luce, C. H. 2004. Modeling of interactions between forest vegetation, disturbances, and sediment yields. *Journal of Geophysical Research*, **109**, F01009.

Iverson, R. M., Reid, M. E. & LaHusen, R. G. 1997. Debris-flow mobilization from landslides. *Annual Review of Earth and Planetary Science*, **25**, 85–138.

Jakob, M. 2005. A size classification for debris flows. *Engineering Geology*, **79**, 151–161.

Jakob, M., Hungr, O. & Thomson, B. 1997. Two debris flow with anomalously high magnitude. *In*: Chen, C.-I. (ed.) *First International Conference on Debris-Flow Hazards Mitigation, Mechanics, Prediction and Assessment*. ASCE, San Fransico, 382–394.

Koulinski, V. 1993. *Etude de la formation d'un lit torrentiel par confrontation d'essais sur modèle réduit et d'observation de terrain*; série Etude **15**. CEMAGREF, Grenoble.

Kronfellner-Kraus, G. 1984. Extreme Festofffrachten und Grabenbildungen von Wildbächen, Interpraevent. VHB, Villach, 109–118.

Kunz, T. 1939. *Coulée de Saxé*. Collection Kern, Mediatèque Valais, Images et Sons.

Lichtenhahn, C. 1971. Zwei Betonmauern: die Geschieberückhaltsperre am Illgraben (Wallis). *In*: Hochwasserbekämpfung, F. F. V. (ed.) *International Symposium on Interpraevent*, 451–456.

Marchi, L. & Brochot, S. 2000. Les cônes de déjection torrentiels dans les Alpes françaises; morphométrie et procesus de tranbsport solide torrentiel. *Revue de géographie alpine*, **88**, 23–38.

Martin, Y. 2000. Modelling hillslope evolution: linear and nonlinear transport relations. *Geomorphology*, **37**, 1–21.

Melton, M. A. 1965. The geomorphic and paleoclimatic signifiance of alluvial deposits in southern Arizona. *Journal of Geology*, **73**, 1–38.

Meunier, M. 1991. *Eléments d'Hydraulique torrentielle*. CEMAGREF, Grenoble.

Montandon, F. 1940. *La coulée d'éboulis de Saxé*. Globe, 79.

Newson, M. D. 1980. The geomorphological effectiveness of floods – a contribution stimulated by two recent events in Mid-Wales. *Earth Surface Processes*, **5**, 1–16.

Nouvelot, J. F. 1990. *Erosion mécanique, transport solide, sédimentation dans le cycle de l'eau*. Orstom, Grenoble.

OFEE, 1991. Analyse des causes des crues de l'année 1987, rapport final, no. **5**, OFEE.

OFEG, 2000. *Hochwasserereignisse in den Kantonen Wallis und Tessin im Oktober 2000*. OFEG, Bern.

Petraschek, A. & Hegg, C. 2002. Ereignisanalyse Hochwasser 2000. *Wasser, Energie, Luft – eau, énergie, air*, **94**, 105–106.

Pinter, N. & Brandon, M. T. 1997. How erosion builds mountains. *Scientific American*, **276**, 74–80.

Pratt-Sitaula, B., Burbank, D. W., Heimsath, A. & Ojha, T. 2004. Landscape disequilibrium on 1000–10,000 year scales Marsyandi River, Nepal, central Himalaya. *Geomorphology*, **58**, 223–241.

Rickenmann, D. 1999. Empirical relationships for debris flows. *Natural Hazards*, **19**, 47–77.

Rickenmann, D. & Zimmermann, M. 1993. The 1987 debris flows in Switzerland: documentation and analysis. *Geomorphology*, **8**, 175–189.

Sandersen, F., Bakkehøi, S., Hestnes, E. & Lied, K. 2001. The influence of meteorological factors on the initiation of debris flows, rockfalls and rock slides in Norway. *In*: Kühne, M., Einstein, H. H., Krauter, E., Klapperich, H. & Pöttler, R. (eds) *International Conference on Landslides, Causes, Impacts and Countermeasures*. Verlag Glückauf Essen, Davos, 199–208.

Schumm, S. A. 1977. *The Fluvial System*. Wiley, New York.

Tonanzi, P. & Troisi, C. 1996. Gli eventi alluvionali del settembre–ottobre 1993, Piemonte.

Tucker, G. & Slingerland, R. 1996. Predicting sediment flux from fold and thrust belts. *Basin Research*, **8**, 329–349.

Tucker, G. & Whipple, K. X. 2003. Topographic outcomes predicted by stream erosion models: sensitivity analysis and intermodel comparison. *Journal of Geophysical Research*, **107**, ETG 1, 1–16.

Ward, J. H. 1963. Hierarchical grouping to optimize an objective function. *Journal of the American Statistical Association*, **58**, 236.

Wieczorek, G. F., Morrissey, M. M., Iovine, G. & Godt, J. W. 1998. Rock-fall hazard in the Yosemite Valley. United States Geological Survey Open-File Report **98-467**.

Wolman, M. G. & Gerson, R. 1978. Relative scales of time and effectiveness in watershed geomorphology. *Earth Surface Proceses*, **3**, 189–208.

Zeller, J. 1985. Festoffmessung in kleinen Gebirgseinzugsgebieten. *Wasser, Energie, Luft – eau, énergie, air*, **77**, 246–251.

Zimmermann, M. 2000. *Geomorphologische Analyse des Illgraben*. GEO 7, Bern.

Zimmermann, M., Mani, P. & Gamma, P. 1997. *Murganggefahr und Klimaänderung – ein GIS-basiert Ansatz*. vdf, Hochschulverlag AG, Zürich.

Limits to resolving catastrophic events in the Quaternary fluvial record: a case study from the Nene valley, Northamptonshire, UK

REBECCA M. BRIANT[1], PHILIP L. GIBBARD[1], STEVE BOREHAM[1], G. RUSSELL COOPE[2] & RICHARD C. PREECE[3]

[1]*Quaternary Palaeoenvironments Group, Department of Geography, University of Cambridge, Downing Place, Cambridge CB2 3EN, UK (e-mail: b.briant@qmul.ac.uk)*

[2]*Centre for Quaternary Research, Department of Geography, Royal Holloway, University of London, Egham, Surrey TW20 0EX, UK*

[3]*Department of Zoology, Downing Street, Cambridge CB2 3EJ, UK*

Abstract: Flood events within rivers are responsible for much erosion and deposition. Thus, deposits laid down during floods could potentially comprise the bulk of the Quaternary fluvial record. However, it is difficult to detect individual flood events, as effectively illustrated by the Middle Devensian (Weichselian) to Holocene fluvial sequence from the Nene Valley, Northamptonshire, described in this paper. This is due to limits in the resolution of sedimentological, palaeontological and geochronological techniques. Geochronological techniques have the highest resolution, but error bars of c. 50 years (radiocarbon) and up to 2 ka (optically stimulated luminescence) in the Late-glacial do not allow detection of floods lasting only a few weeks or less. Geochronology is, however, essential for linking periods of fluvial deposition to climatic phases at the marine isotope substage scale. Thus, multiple age determinations show remnant Middle Devensian deposits within a facies association mainly of Younger Dryas age, showing similar fluvial response to climate during both time periods. Palaeontological assemblages suggest that climate was also similar, although with some subtle differences. Determining 'average' fluvial activity in response to broad climate phases improves understanding of how rivers behave over long time periods, even though determination of the role of flood events in the Quaternary fluvial record remains elusive.

Large-scale or 'catastrophic' events are important to the evolution of the landscape because they are often the most geomorphologically effective (e.g. Thorn 1988; Knox 1993). They are particularly important over longer time periods such as the Quaternary, since the geological record is highly discontinuous, consisting mostly of sediments deposited during short, high-energy events (e.g. Gibbard & Lewin 2002). Their extent will vary depending on the type of river and the dominant depositional processes operating, but is likely to be pronounced within the braided systems typical of much Quaternary cold-stage deposition (e.g. van Huissteden *et al.* 2001). These had a hydrological regime characterized by catastrophic annual snowmelt floods, as confirmed by quantitative estimates of temperature from palaeontological evidence (Coope 2000).

In addition to individual flood events, climatic phases are important in the interpretation of fluvial system behaviour over long time-scales. These 'phases' are characterized by a dominant hydrological regime with associated magnitude and frequency relationships. On a small scale, stacking of larger flood events may show trends over tens of years. Phases may also last for hundreds of years, for example changes in flooding frequency and deposition have been linked to cooler climatic phases during the Holocene (Macklin & Lewin 2003). On a larger scale flood frequency and hydrological regime change significantly both between interglacial and glacial phases and within glacials, and it is often possible to link fluvial activity to successive climate phases (e.g. during the Late-glacial, Kasse *et al.* 1995; the Middle Devensian (Weichselian), Briant *et al.* 2004b or the Early Devensian, Briant *et al.* 2005).

Recognizing individual flood events on time-scales longer than a few hundred years is harder, and indeed even in the Holocene may be restricted to particularly favourable locations (e.g. Baker 1984). Thus the formation of the Strait of Dover, probably in the Anglian (Smith 1985, 1989), may not have been a single catastrophic event. Gibbard (1995) argues that the sequence at Cap Gris Nez is a single record of multiple events that cannot convincingly be separated. A similar explanation is also plausible for the 'event' breaching the Wight–Purbeck ridge in the former Solent system. This seems to have occurred due to both fluvial downcutting in the Devensian and marine transgression at the start of the Holocene, but it cannot be dated with any certainty (e.g. Velegrakis et al. 1999). Furthermore, the potential impact of flood events may be enhanced by previous periglacial breakdown of rock material that is often 'invisible' in the rock record (e.g. West 1991). Despite this, if it were possible to detect individual flood events, this would provide valuable extra detail on river activity during climatic phases.

Quaternary fluvial sediments date largely from cold stages and were therefore deposited in flood-dominated braided, periglacial rivers with either an arctic or subarctic nival regime dominated by snowmelt flooding (e.g. Woo 1990). Thus, deposition potentially only occurred during single annual flood events, the sedimentary record representing only the largest, which may or may not be separately detectable. This paper uses a case study from a Middle Devensian to Holocene sequence in the middle reaches of the Nene Valley in Northamptonshire to investigate fluvial response to both climatic phases and individual flood events. Although faunal evidence suggests large-scale flood events probably occurred annually, resolving individual flood events is beyond the power of current sedimentological, palaeontological and geochronological methodology.

The Nene Valley

The middle and upper reaches of the Nene valley are incised into Jurassic Oolitic Limestone and underlain by Lias clay (Horton 1989). Quaternary deposits comprise those mapped as the First Terrace (Horton 1989) and designated as the Ecton Member of the Nene Valley Formation (Maddy 1999). These underlie the Holocene alluvium and have traditionally been assigned to the Middle to Late Devensian Substages on the basis of various radiocarbon dates (e.g. Shotton et al. 1969, 1970; Brown et al. 1994). Downstream, gravel deposits have been mapped at various altitudes above the floodplain and, as the river emerges from the limestone uplands, it fans out to form a gravel fan shape.

Previous research in this section of the river has concentrated on organic material within the sequence, and the development of the river in the Late-glacial and early Holocene (e.g. Bell 1968; Morgan 1969; Holyoak & Seddon 1984; Brown et al. 1994). The most detailed sedimentological description of River Nene First Terrace gravels is from Thrapston, which is similar to the sequence described below. In addition to horizontally bedded gravel with organic beds, Bell (1968) noted organic 'erratics' preserved within the gravel and a poorly exposed 'older, coarser gravel' (Bell 1968, p. 40) underlying the main succession. In addition, gravels are recorded 'overlying beds with post-glacial fossils' (Bell 1968, p. 41) and fine-grained sediments overlie the main succession throughout.

The depositional succession

The exposures at Stanwick, Northamptonshire, span grid references SP943701 to SP954703 (Figs 1 & 2). The full succession comprises between 1.5 and 3 m of gravels, overlain by up to 2 m of clays and silts. Four facies associations (ST-1 to ST-4) are recognized at the site on the basis of distinctive textures, facies associations, and laterally continuous erosional surfaces. These are described and interpreted below (Table 1 and Fig. 3), using the facies code scheme of Miall (1996).

Samples of the finer-grained material were collected for radiocarbon and optically stimulated luminescence (OSL) dating, from both 'tie-point' (i.e. samples for each technique came from the same channel fill) and 'framework' locations (where OSL and radiocarbon samples were not directly associated). The locations of these samples are shown in Figure 2 and listed in Table 6. Methods applied to geochronological and associated palaeontological samples are described in detail below.

Facies association ST-1

Facies association ST-1 consists of up to 2 m of laterally discontinuous matrix-rich clast-supported fine to very coarse ironstone-rich gravel with a dark brown sand and silt matrix, reaching maximum thickness in the northeast of the quarry (Fig. 3). ST-1 is dominated by horizontal beds of gravel approximately 50 cm thick (Gh), with occasional sand and silt drapes and rare lenses of planar cross-bedded gravels (Gp, Sp). A number of wedge forms with secondary fills and adjacent downturned strata occur within ST-1 (Fig. 2), truncated by the upper contact of the facies association. The lower contact of facies association ST-1 is not seen and the upper contact is usually erosional.

Following Miall's (1996) bounding surface classification, the lower contact of facies association ST-1 is interpreted as representing a period of channel migration, downcutting prior to deposition. The sediments of facies association ST-1 are dominated by architectural element GB – gravel bars and bedforms (cf. Miall 1996). This is typical of a shallow gravel-bed braided system with multiple shallow ephemeral channels and

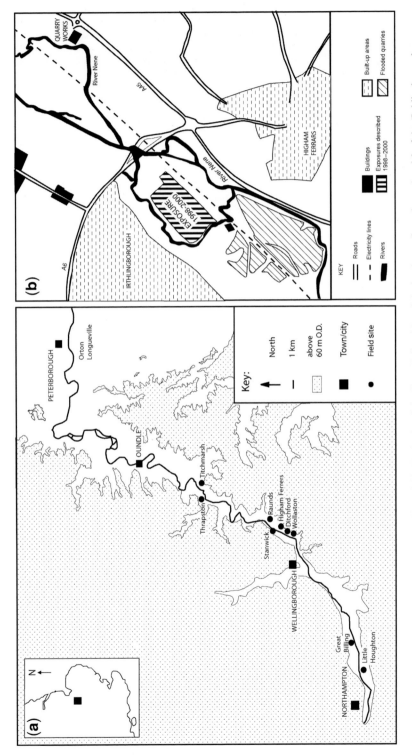

Fig. 1. Location of exposures at Stanwick, Northamptonshire, within England and the Nene Valley (a) and the local area (b – phases 1 and 2 of the 'Irthlingborough extension' to the ARC/Hanson Stanwick quarry).

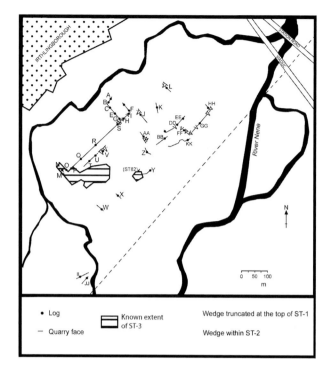
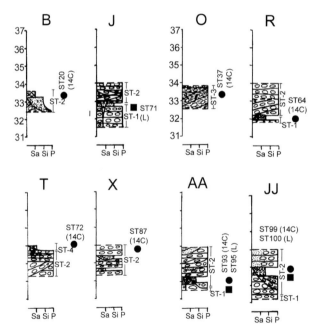

Fig. 2. Exact stratigraphic and spatial position of samples taken from exposures in phases 1 and 2 of the 'Irthlingborough extension' to the ARC/Hanson Stanwick quarry between December 1998 and July 2000. Only those logs from which samples were taken are shown. Vertical scale in metres O.D., with values quoted where log heights were obtained by surveying, and blank where they were extrapolated.

Table 1. *Sedimentology of Quaternary fluvial sequence at Stanwick. Sedimentological characteristics and interpreted environment of deposition of facies associations 1–4 at Stanwick, Northamptonshire*

Facies association	Facies assemblage	Lower bounding surface (and order)	Geometry	Architectural element(s)	Interpretation	Depositional environment
ST-4	Fsm	Gradational, horizontal	Massive silty clay with abundant molluscs and woody debris in lowest 0.5–1 m	FF	Overbank fines	System dominated by vertical accretion (e.g. Brown *et al.* 1994)
ST-3	Gp, Gh, Gs	Erosional, concave-upward (fourth-order)	Clay-rich gravels with shells and wood fragments, largely cross-bedded	HO	Scour hollows infilled by clay matrix material during falling stage flows	Scour-dominated braided river
ST-2	Gh, Gp, Sp, Gs, Ss, Fsm, Fl	Erosional, both sub-horizontal and concave-upward (fifth-order)	Horizontally bedded gravel with organic 'rip-up' clasts. Scour forms infilled with gravel and organic material	GB, HO	Both gravel bars and bedforms and scour hollows (infilled during high and falling stage flows)	Transitional form between shallow gravel-bed and scour-dominated braided rivers
ST-1	Gh, rare Gp, Sp	Not observed, erosionally overlies bedrock (fifth-order)	Matrix-rich horizontally bedded cobble gravel. Rare sand and silt drapes and cross-bedded gravel lenses	GB	Gravel bars and bedforms. Rare bar-top drapes and scour-fills	Shallow gravel-bed braided river

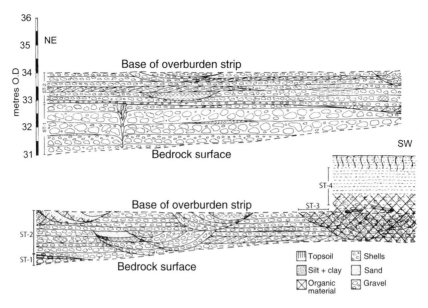

Fig. 3. Schematic cross-section through the Pleistocene deposits at Stanwick, Northamptonshire based on exposures seen between December 1998 and July 2000, showing the nature of and relationships between the four facies associations identified at the site. Exact sample locations are given in Fig. 2 and Tables 6 and 7.

only barform sediments preserved (Miall 1996). The discharge regime was probably an arctic nival regime dominated by a single snowmelt flood (Church 1974; Woo 1990). The features of the wedge forms described above suggest that they are ice-wedge casts (Black 1976). Given their single stratigraphic position, they probably developed during a prolonged period of subaerial exposure and permafrost development post-dating deposition of the facies association. It is not clear whether permafrost also formed during deposition of the facies association.

Facies association ST-2

Facies association ST-2 consists of 1-2 m of laterally continuous fine to coarse gravel with a coarse sand matrix, thickest in the southwest of the quarry where facies association ST-1 thins (Fig. 3). Sedimentary structures are highly variable and include horizontal bedding (Gh), planar cross-bedded lenses (Gp, Sp) and scour-fill structures (Gs, Ss), the latter more common and thicker in the southwest of the quarry. 'Rip-up' clasts of peat-like material, similar to those seen at Thrapston (Bell 1968), or organic-rich silt and clay (Fsm) occur occasionally, as does laminated sand and silt with plant beds (Fl), both as a 'bar-top' structure (log R, Fig. 2) and within scour-fill structures (logs B, X, AA and JJ, Fig. 2). Wedge forms are observed rarely within the facies association, terminating at various levels, and concentrated in the northeast of the quarry (Fig. 2). The lower contact of the facies association is usually erosional and the upper contact, where seen beneath facies association ST-3, is erosional and concave-upward.

The lower contact of facies association ST-2 truncates ice-wedge casts and sedimentary features, and is laterally continuous, representing a fifth-order bounding surface (Miall 1996) and therefore a further period of channel migration. The main architectural elements in the sediments of facies association ST-2 are GB – gravel bars and bedforms (facies Gh, Gp and Gs) – and HO – scour hollows (facies Gs, Ss, Fl and Fsm). It therefore has some similarities to the shallow gravel-bed braided style of ST-1 (after Miall 1996). However the scour-fill structures are not typical of such systems. For this reason it probably represents a transitional form between a shallow gravel-bed and a scour-dominated braided style. Periods of low flow are important, possibly in a subarctic nival discharge regime (Church 1974; Woo 1990) in which finer-grained facies were deposited and preserved. The wedge forms in facies association ST-2 are interpreted as ice-wedge casts for the same reasons as those in facies association ST-1. The rarity of ice-wedge casts within the sediments probably reflects fluvial activity. Significant flood activity and flood-plain reworking would decrease ice-wedge casting both because opportunities for development were fewer, and because preservation potential was lower. This is more likely than decreasing severity of climate given the valley-confined nature of the reach and the coleopteran faunal evidence for climatic severity (described below) during deposition of this facies association.

Facies association ST-3

ST-3 is laterally discontinuous (Fig. 2) and stratigraphically overlies ST-2 in the southwest of the quarry (Fig. 3). It has a maximum thickness of 1 m, and consists of fine to coarse clast-supported ironstone-rich pebbles with a matrix dominated by dark grey clay. Abundant shell and wood fragments are also present, as are occasional peat-like 'rip-up' clasts. Common sedimentary structures include planar cross-bedding (Gp), horizontal bedding (Gh) and scour-fill structures (Gs). The lower contact of ST-3 is erosional and concave-upward at the base of several scour-fill structures. The upper contact is gradational and sub-horizontal beneath facies association ST-4.

The discontinuity of the lower contact of facies association ST-3 means that this is interpreted as a fourth-order bounding surface (Miall 1996), and thus a small-scale shift of form in the system, representing the basal scour of minor channels. The sediments of facies association ST-3 are similar to Miall's (1996) architectural element HO – scour hollows – as this contains facies Gh and Gs, which are the main features of facies association ST-3. Thus the facies association is thought to represent a scour-dominated braided style, with some stabilization of shallow gravel channels but limited overbank deposition. The presence of temperate fossils (see below) indicates temperate climatic conditions. The bimodal particle-size distribution (gravel and clay) suggests a peaked hydrological regime dominated by high and low stage flows with limited moderate flows. The good preservation of many shells indicates that individual clasts were not transported far. This probably means that low-stage flows were dominant with occasional peak floods entraining gravel.

Facies association ST-4

Facies association ST-4 consists of up to 2 m of laterally continuous massive silty clay. In the basal 0.5–1 m, abundant molluscs and woody debris occur, with rarer vertebrate material. The upper 1–1.5 m of the clay is non-fossiliferous. The lower and upper contacts are gradational and sub-horizontal.

The gradational lower contact of ST-4 indicates overtopping of previous deposits, without erosion. The association is interpreted as representing Miall's (1996) FF – overbank fines – with a blanket-like geometry and facies Fsm. This suggests a system dominated by vertical accretion processes, such as in the stable bed aggrading banks (SBAB) model of Brown et al. (1994). In this model, based on observations of Holocene sediments in the Nene valley, channel courses remain stable and these and the surrounding floodplain gradually accrete fine sediment. The abundant fossils in the base of the facies association suggest that it was deposited under temperate climatic conditions. Flows must have been low-energy, as the particle size of the facies association is fine, and the fossils are frequently intact. The discharge regime may have been similar to the present day, which is non-flashy with a mean gauged flow of c. 9 m^3/s (Brown et al. 1994).

Palaeontology

Plant macrofossils

Plant macrofossil material was prepared for radiocarbon dating by wet sieving with distilled water to 250 μm, followed by separation and identification under a low-power stereo microscope. Samples were dried overnight at 105°C after identification to avoid fungal growth following the recommendations of Wohlfarth et al. (1998). Many of the macrofossils were broken, but sufficiently well-preserved to allow identification.

Preservation was poor in samples from facies association ST-2, with broken leaves and frequently degraded seeds (Table 2). All samples show a similar range of taxa, although with variation, possibly due to local taphonomic processes. For example, ST93 is dominated by aquatic elements and its assemblage is slightly sparser, as would be expected from an organic silt accumulating at some distance from active channels (West et al. 1993; West 2000). Assemblages are also similar to previous floras reported from Great Billing and Little Houghton and radiocarbon dated to the Middle Devensian (Morgan 1969; Holyoak & Seddon 1984). Dwarf birch (Betula nana) and various Salix (willow) taxa are present, including Salix cf. repens. Alnus sp. (alder) wood and fruits in ST20 is the only taxon which is not typical of Devensian floras (Godwin 1975; West 2000). Thus it might be a reworked element. Alternatively, it could have been a part of the contemporary flora, since Bennett and Birks (1990, p. 127) state that, although not abundant until the mid-Holocene, Alnus was 'probably present as early as 10,000 BP and conceivably present even earlier'. The number of herbs present is restricted and dominated by members of the Asteraceae. Carex (sedge) seeds are limited, and the aquatic flora includes Scirpus (sedge) and Potamogeton sp. (broad-leaved

Table 2. Plant macrofossils from Stanwick

Sediment type	Facies association ST-2				Facies association ST-3	Facies association ST-4
	ST20	ST64	ST93	ST99	ST37	ST72
	Plant bed in sand	Plant bed in sand	Organic silt	Plant bed in sand	Gravel with clay	Plant bed in sand
Taxa						
Trees and shrubs						
Corylus avellana L. nut	Many				1	
Alnus sp. wf	5					
Alnus sp. fr						
Betula nana L. lf	Few	2		Several		
Betula nana L. fr	25	6		37		
Betula nana L. ccs	7			5		
Salix herbacea L. lf				Several		
Salix polaris Wahl. lf			2	Several		
Salix cf. *repens* lf	Few	Many				
Salix sp. lf		Many		Many		2
Salix sp. cap						1
Salicaceae cf. *Salix* wf		Many				Many
Salix/Betula lf	Few					
Sambucas nigra L. s					3	
Herbs						
Achillea millefolium L. a			1			
Armeria maritima (Mill.) Willd. c				18		
Caryophyllaceae s				7		
Cerastium sp. s			4			
Compositae s	4	42		31		7

Taxon	ST20	ST64	ST99	ST72	ST93	ST37
Draba sp. cap	1					
Ericaceae wf						Several
Linum perenne agg. s						1
Polygonum aviculare agg. fr		1				
Potentilla sp. a	2					
Silene sp. s			1			
Thalictrum sp. a		1		3		
Umbelliferae fr	2					2
Grasses/sedges						
Carex aquatilis-type n	2	5	3	59		6
Carex trigonous-type n	2	2		43		5
Juncus sp. s			3			
Aquatics						
Chara o			239			
Eleocharis cf. *palustris* n			2			
Nuphar lutea (L.) Sm. s					2	
Potamogeton sp. fst	47		20		25	2
Ranunculus (subg. *Batrachium*) a					4	7
Scirpus sp. n		1		8		
Zannichellia palustris L. a	1					16
Unidentified wf						Many

Plant macrofossil assemblages from radiocarbon dating samples from Stanwick, Northamptonshire between 1998 and 2000 (Fig. 2). Taxa used for dating listed in Table 6. Nut, nut; wf, wood fragments; fr, fruit; lf, leaf fragment; csc, conescale; cap, capsule; s, seed; a, achene; c, calyx; n, nutlet; o, oospore; fst, fruitstone. Values given are actual counts. Sample weights and material processed: ST20, ST64, ST99, ST72 sieved >1 kg, picked >1 mm fraction; ST93 sieved >1 kg, picked >250 μm fractions; ST37 sieved 1.5 kg, picked >1 mm fraction.

pondweed). The floodplain environment indicated during deposition of facies association ST-2 supported a mixture of herbs and grassland, with a local growth of aquatic taxa. Greater detail of interpretation is limited, since it was not possible to differentiate taxa within the Asteraceae, which is a large, cosmopolitan family. Climatically, the occasional presence of *Salix* cf. *repens* may suggest less severe conditions than those indicated by *Salix herbacea* or *Salix polaris*, because it has a lower latitudinal distribution at the present day.

The assemblage from facies association ST-3 is small in comparison to those from facies association ST-2, and contains many broken elements. However, a number of significant taxa are present which are not in the flora from facies association ST-2 or other previously reported assemblages, i.e. *Corylus avellana*, *Sambucus nigra* and *Nuphar lutea*. *Corylus avellana* (hazel) is significant because it shows the presence of shrubs close to the floodplain. That this is the only arboreal macrofossil found is perhaps consistent with the dominance of *Corylus* pollen in the sample (Table 3). The other two are significant because they are temperate in character. *Nuphar lutea* (yellow water-lily) is found in lakes and pools as far south as North Africa (though rarely) at the present day (Clapham *et al.* 1981). *Sambucus nigra* (elder) is a shrub common in Europe, southwards from Scandinavia and northern Scotland (Clapham *et al.* 1981). The local environment is hard to reconstruct from such a sparse flora, but may have included patches of disturbed ground with stands of hazel and elder scrub.

The flora from ST72 in facies association ST-4 is also difficult to interpret because it is sparse and poorly preserved. However, *Salix* taxa may have been shrub rather than dwarf forms because of the presence of wood fragments in addition to leaves. The restricted herb flora is dominated by Asteraceae (daisy family). The aquatic flora includes *Carex* and *Scirpus*, *Potamogeton* sp. and *Nuphar lutea*. The floodplain environment indicated during deposition of this facies association was a mixture of shrubs and grassland, with local

Table 3. *Pollen analyses from Stanwick*

Sediment type	Facies association ST-2	Facies association ST-3
	ST99	ST37
	Plant bed in sand	Gravel with clay
Taxa		
Trees and shrubs		
Betula	2.6	1.4
Pinus	27.6	4.2
Picea	1.3	0
Ulmus	0	0.3
Quercus	0	3.9
Tilia	0	2.3
Alnus	0	22.5
Fraxinus	0	0.6
Acer	0	0.3
Picea	0	0.6
Corylus	0	22.8
Salix	0	0.3
Juniperus	1.3	0
Rhamnus	0	0.3
Herbs		
Poaceae	50.0	23.9
Cyperaceae	5.3	2.3
Other herbs	7.9	9.3
Main sum (excluding aquatics and pre-Pleistocene types)	76	355
Aquatics (as % main + aquatics)	6.2	1.1
Pre-Pleistocene types (as % main + pre-Pleistocene)	3.8	0.6

Summary palynological assemblages of selected samples from Stanwick, Northamptonshire (Fig. 3). Data are presented as percentages of total land pollen and spores excluding aquatic taxa and pre-Pleistocene types. Taxonomic arrangement follows Stace (1991).

growth of aquatic taxa. Climatically, the presence of the fully temperate species *Nuphar lutea* suggests that this facies association was deposited during a temperate event.

Palynology

Pollen preparation was carried out using the standard procedures employed by the Quaternary Palaeoenvironments Group, Department of Geography, University of Cambridge (adapted from Bennett *et al.* 1990). The pollen concentration was low in ST99 and the pollen count was too small to give a statistically significant result for this sample.

The assemblage from ST99 (facies association ST-2) contained low pollen concentrations and was dominated by grasses (Poaceae) and herbs (Table 3). Arboreal pollen included *Betula* (birch), *Juniperus* (juniper), *Pinus* (pine) and *Picea* (spruce), with *Pinus* reaching 27%. This assemblage is similar to those from Thrapston (Bell 1968), Titchmarsh (lower channel), Orton Longueville (lower channel) (Holyoak & Seddon 1984), and both the basal and upper channels at Ditchford (Brown *et al.* 1994), although percentages of tree pollen are often lower. *Juniperus* is a shade-intolerant early colonizer only previously reported at Orton Longueville (Holyoak & Seddon 1984).

The broad picture of the local environment is of a grass- and herb-dominated landscape with occasional *juniper* and *birch* scrub, and scattered stands of boreal woodland. The assemblage is similar to those seen from the Late-glacial Interstadial sequence at Windermere, although lacking in *Salix* and *Empetrum* (Coope & Pennington 1977). There are few similarities with the Upton Warren Interstadial, which was treeless, and shows only an increase in the diversity of the tall herb component (Kerney *et al.* 1982; Coope *et al.* 1997).

Pollen was more abundant and better-preserved in ST37 (from ST-3) and the sample shows a very different assemblage to ST99 (Table 3). The dominant taxa present are trees and shrubs. *Alnus* (alder) formed a significant component of the tree pollen, together with *Corylus*, *Quercus* (oak), *Ulmus* (elm) and *Tilia* (lime), making it different from other assemblages previously reported from the Nene. However, *in situ Alnus* roots found growing in the top of the upper channel at Raunds have been dated to 5195 ± 65 BP (SRR-3606, Brown *et al.* 1994), approximately contemporaneous with deposition of ST-3. Important shrubs are *Salix* and *Rhamnus* (alder buckthorn). Grass and herb percentages are lower than in ST99, but still up to *c.* 34% of the total. The overall picture of the environment is of a mature mixed-oak woodland, with local stands of alder carr in damp areas, and patches of grassland on the floodplain. The presence of *Tilia* suggests a mid-late Holocene age, since this taxon did not reach Britain until approximately 8000 cal. BP (e.g. at Hockham Mere; Bennett 1983). The dominance of *Corylus* suggests correlation with zone VII of Godwin's (1975) Holocene pollen zonation; and zones IIa-c from West (1980), with an age of *c.* 5700 BP.

Mollusca

The molluscan assemblages from the radiocarbon dated samples within facies association ST-2 are sparse and poorly preserved, particularly in ST20 and 93, which yielded only a few specimens of *Lymnaea peregra* and *Pisidium* sp. (Table 4). In contrast, ST99 has a slightly richer fauna. The aquatic fauna is dominated by *Pisidium* sp. and species indicative of slow-flowing water bodies or small ponds and ditches (*Gyraulus laevis*, *G. crista*, *Valvata piscinalis*). Other species (e.g. *Lymnaea truncatula*) inhabit damp ground or marshland, consistent with living on a river floodplain. The presence of *Carychium tridentatum*, *Punctum pygmaeum* and *Cochlicopa* sp., none of which are common in Devensian stadial faunas, may suggest slight warming during this time period. *Carychium tridentatum* is unknown in the British Devensian stadials (Holyoak 1982). These are different from assemblages reported from Little Houghton (Holyoak & Seddon 1984), which are less diverse and contain *Pupilla muscorum*, but similar to those from the Titchmarsh lower channel (Holyoak & Seddon 1984) and Thrapston (Bell 1968, with unusual species such as *Punctum pygmaeum*, *Gyraulus crista* and *Pisidium milium*, although *Carychium minumum* occurs at Titchmarsh and Thrapston, in contrast with *Carychium tridentatum* at Stanwick.

In contrast, the fauna from facies association ST-3 is more diverse (Table 4) and dominated by aquatic taxa. The assemblage is dominated by *Bithynia tentaculata*, *Valvata piscinalis*, *Gyraulus albus* and *Pisidium* sp. that are not indicative of any one fluvial regime, as they show preferences for both slow-flowing water (e.g. *Bithynia tentaculata*, *Valvata piscinalis*, *Lymnaea stagnalis* and *Lymnaea auricularia*) and shallow ponds (e.g. *Lymnaea truncatula*, Planorbidae). Additionally, *Theodoxus fluviatilis*, *Ancylus fluviatilis* and some species of *Unio* favour hard substrates. Thus the local environment suggested is one of slow-flowing river channels with a hard substrate, surrounded by a floodplain in which shallow pools have developed. This is consistent with the discharge regime determined above on the basis of sedimentology. With the exception of *Vallonia excentrica*, the terrestrial species prefer damp ground of various

Table 4. Molluscs from Stanwick

Species	Facies association ST-2			Facies association ST-3	Facies association ST-4
	ST20	ST93	ST99	ST37	ST72
Aquatic					
Theodoxus fluviatilis				24	
Valvata cristata				7	
Valvata piscinalis			2	36	
Bithynia tentaculata				99	1
Bithynia leachii				2	
Lymnaea truncatula			1	6	
Lymnaea stagnalis				1	
Lymnaea auricularia				8	
Lymnaea peregra	1	4	2	12	
Bathylomphalus contortus				1	
cf. Gyraulus laevis			2		
Gyraulus albus				48	
Gyraulus crista			1	6	
Hippeutis complanatus				2	
Ancylus fluviatilis				21	
Unio sp.				15	
Sphaerium corneum				4	
Pisidium amnicum				3	
Pisidium cf. casertanum					
Pisidium henslowanum				1	
Pisidium sp.	14	1	20	84	
Aegopinella nitidula				2	
Terrestrial					
Carychium tridentatum			1		
Oxyloma pfeifferi			4		
Succinea/Oxyloma				7	
Vallonia pulchella				2	
Vallonia excentrica				1	
Vallonia pulchella/excentrica				2	
Punctum pygmaeum			1		
Cochlicopa sp.			1	1	
Cepaea nemoralis				2	

Molluscan assemblages from radiocarbon dating samples from Stanwick, Northamptonshire between 1998 and 2000 (Fig. 2). Values given are actual counts. Sample weights and material processed: ST20, ST99, ST72 sieved 1 kg, picked >1 mm fraction; ST93 sieved >1 kg, picked >250 μm fractions; ST37 sieved 1.5 kg, picked >1 mm fraction. *Pisidium* species includes: ST20 – *Pisidium nitidum, Pisidium subtruncatum, Pisidium henslowanum*; ST99 – *Pisidium subtruncatum, Pisidium milium*; ST37 – *Pisidium nitidum, Pisidium subtruncatum, Pisidium henslowanum, Pisidium moitesserianum*.

sorts, suggesting that they are derived from close to the river channel. *Cepaea nemoralis* prefers well-vegetated habitats, which can be damp, as indicated by plant macrofossils and pollen from this sample (Tables 2 and 3).

Climatically, the indication is that the assemblage from ST37 represents a temperate event, since in addition to the diversity of the assemblage there are a number of species only recorded from the Holocene such as *Theodoxus fluviatilis* and *Cepaea nemoralis* (e.g. Holyoak 1983). The assemblage from ST37 is very similar to that found in Holocene deposits at Staines and dated to between c. 8700 and c. 6700 BP (Preece & Robinson 1982). Thus facies association ST-3 may date from the early Holocene onwards.

Coleoptera

The fossils were extracted by wet sieving over a 0.3 mm sieve and concentrated by the now standard paraffin separation technique (Coope 1986). Coleoptera were recovered from deposits of facies associations ST-2 and ST-3. Those from ST-2 were generally well preserved and yielded similar assemblages (Table 5).

Coleoptera from facies association ST-2 The local environment interpreted from the beetle fauna from ST-2 is dominated by pools of still water, as indicated by the various *Helophorus* species (especially in ST93 which is also dominated by aquatic plants) and all the species of Dytiscidae. *Cercyon marinus*

Table 5. *Coleoptera from Stanwick*

Coleoptera in taxonomic order	Facies association ST-2					Facies association ST-3
	ST20	ST64	ST87	ST93	ST99	ST37
Carabidae						
Carabus glabratus Payk.						
Elaphrus lapponicus Gyll.					1	
Elaphrus riparius (L.)					1	
Notiophilus aquaticus (L.)				2		1
Notiophilus sp.					1	
*Bembidion hasti Sahlb.				1	1	
*Bembidion dauricum (Motsch.)					1	
Bembidion aeneum Germ.				2		
Bembidion guttula (F.)	1					
Patrobus assimilis Chaud.			1			1
Pterostichus nigrita (Payk.)		1				
Amara quenseli (Schönh.)					1	
Dytiscidae						
Hydroporus sp.			1		2	
Agabus bipustulatus (L.)			1			
Agabus arcticus (Payk.)					1	
Agabus congener (Thunb.) group			1			
Ilybius fuliginosus (F.)			1			
Ilybius sp.	1					
Colymbetes sp.					1	1
Hydraenidae						
Ochthebius lenensis Popp.			9			
Ochthebius minimus (F.)						1
Ochthebius sp.	1				1	
*Helophorus obscurellus Popp.	1	1	1	1	2	
Helophorus 'aquaticus' (L.) = *aequalis* Thoms.	1		3	2	1	
*Helophorus sibiricus (Motsch.)	3	2	3			
*Helophorus glacialis Villa			1		2	
Helophorus granularis (L.)				15	2	
Helophorus sp.	1					1
Hydrophilidae						
Cercyon marinus Thoms.			1			
Cercyon sternalis Shp.					1	
Laccobius sp.			1			1

(*Continued*)

Table 5. Continued

Coleoptera in taxonomic order	Facies association ST-2					Facies association ST-3
	ST20	ST64	ST87	ST93	ST99	ST37
Sphaeriidae						
Sphaerius acaroides Waltl.					1	
Staphylinidae						
*Pycnoglypta lurida (Gyll.)		1	16		1	
Olophrum fuscum (Grav.)	2	1	9		4	
*Olophrum boreale					1	
Eucnecosum brachypterum (Grav.) or *norvegicum Munst.	3	4	8	1	14	
*Acidota quadrata Zett.					1	
Geodromicus cf nigrita (Müll.)	1		2	1		
*Boreaphilus henningianus Sahlb.	1	1	1	2	1	
Trogophloeus sp.			2			
Oxytelus rugosus (F.)			1			
Oxytelus nitidulus Grav.				1		
Platystethus nodifrons Mannh.						1
Bledius sp.						1
Stenus juno (Payk.)	1	1	2	1	2	1
Stenus sp.						
Tachinus sp.				1		
Philonthus sp.						
Aleocharinae Gen. et sp. indet.	10	7	9	5	21	
Elateridae						
Athous vittatus (F.)						1
Cantharidae						
Cantharis paludosa Fall.					1	
Helodidae						
Gen. et sp.indet.			2			1
Dryopidae						
Oulimnius tuberculatus Müll.					1	
Oulimnius troglodytus (Gyll.)						1
Normandia nitens (Müll.)						1
Riolus sp.		1				

Taxon	ST20	ST64	ST87	ST93	ST99	ST37
Byrrhidae						
Simplocaria semistriata (F.)					1	
Simplocaria metallica (Sturm)	1					1
Byrrhus sp.	1					
Coccinellidae						
Hippodamia arctica (Schneid.)	1					
Ceratomegila ulkii Crotch	1					
Anthicidae						
Anthicus sp.			1			
Scarabaeidae						
Aphodius affinis Panz.			1			
Aphodius sp.				2		2
Phyllopertha horticola (L.)				1		1
Chrysomelidae						
Donacia marginata Hoppe					1	1
Donacia simplex F.						
Plateumaris sericea (L.)						
Phytodecta cf *viminalis* (L.)					1	
Phyllodecta vitellinae (L.) or *polaris* (Schneid.)					1	
Galeruca tanaceti (L.)					1	
Chaetocnema transversa (Marsh.)					1	1
Crepidodera ferruginea (Scop.)					1	
Curculionidae						
Apion sp.	1					
Otiorhynchus rugifrons (Gyll.)					1	1
Otiorhynchus ovatus (L.)					1	
Sitona sp.						
Notaris aethiops (F.)	1					2
Hypera sp.					7	1
Limnobaris pilistriata (Steph.)						1
Phytobius sp.	1					
Phyllobius sp or *Polydrusus* sp.						
Rhynchaenus foliorum (Müll.) group						1

Coleopteran assemblages from samples taken for radiocarbon dating from Stanwick, Northamptonshire between 1998 and 2000 (Fig. 2). Values given are minimum numbers of individuals. Species marked with an asterisk are not native in Britain at the present day. Sample weights and material processed: ST20, ST64, ST87, ST93, ST99 sieved 1 kg, picked >250 μm fractions; ST37 sieved 1.5 kg, picked 250 μm fractions.

and *Cercyon sternalis* live in wet moss and mud at the margins of ponds. Accumulations of damp decomposing plant debris are the habitat of the predatory Staphylinidae. Reedy vegetation surrounded the pools, as indicated by *Notaris aethiops* and *Limnobaris pilistriata*. Vegetation on the floodplain also included mosses (*Simplocaria metallica* and other Byrrhidae are obligate moss-feeders). *Phytodecta viminalis* feeds on shrubby willows and the larvae of the minute weevil *Rynchaenus* of the *foliorum* group can feed on dwarf *Salix*. The ladybird *Hippodamia arctica* feeds on Aphida on *Salix* bushes and also on *Betula nana* (Strand 1946). These records are interesting given the concentration of *Salix* leaves in several of the samples (Table 2). Other parts of the floodplain must have been considerably drier, however, because the *Otiorynchus* species present are largely xerophilic, as are some of the scavenging species, such as *Notiophilus aquaticus*. *Amara quenseli* lives in dry unshaded situations where the soil is sandy and often unstable and the vegetation patchy and sparse.

The thermal climate indicated by all the beetle assemblages from facies association ST-2 is cold and continental, with 12 species not now native to the British Isles. The ranges of all these are exclusively boreal or boreomontane at the present day. The most unusual of these is *Ceratomegilla ulkei*, which is an Asiatic and North American species. *Helophorus obscurellus*, *Helophorus sibiricus* and *Hippodamia arctica* live today exclusively in arctic Europe and in Siberia. *Helophorus glacialis* is a northern and montane species found throughout Europe at high altitudes but at low levels in the extreme north, e.g. on the Kola Peninsula (Angus 1992). Quantitative estimates of the thermal climate have been made using the mutual climatic range (MCR) method (Atkinson *et al.* 1987) from the carnivorous and scavenging beetle species from each sample. The following figures were obtained where T_{max} is the mean temperature of the warmest month (July) and T_{min} the mean temperature of the coldest months (January and February). Actual mean monthly temperatures probably lay between these limits.

ST 20: T_{max} between 6°C and 11°C; T_{min} between −25°C and −13°C; 6 species used

ST 64: T_{max} between 7°C and 11°C; T_{min} between −23°C and −13°C; 7 species used

ST 87: T_{max} between 7°C and 12°C; T_{min} between −22°C and −5°C; 5 species used

ST 99: T_{max} between 7°C and 11°C; T_{min} between −23°C and −13°C; 13 species used

ST 93: T_{max} between 4°C and 12°C; T_{min} between −32°C and −14°C; 4 species used

Biostratigraphically, the coleopteran assemblages from samples ST20, ST64, ST87 and ST99 are very similar to one another. They include several species common in Late-glacial (Younger Dryas) faunas but rare in Middle Devensian ones. In particular the relative abundance of *Helophorus glacialis* should be noted as it is a common and ubiquitous member of the Younger Dryas fauna (e.g. Coope & Pennington 1977; Walker *et al.* 1993, 2003) and is extremely rare in earlier Devensian contexts. On the basis of their coleopteran assemblages therefore, samples ST20, ST64, ST87 and ST99 all seem to have been deposited during Younger Dryas times. It is perhaps significant that in sample ST93, which has a Middle Devensian date (see below) and slightly lower reconstructed temperatures (see above), *Helophorus glacialis* was absent. However, species that are common members of the Middle Devensian faunas in Britain are absent from all facies association ST-2 samples. These species include *Diacheila polita*, *Pterostichus kokeili*, *Holoboreaphilùs nordenskioeldi*, *Aphodius holdereri* and many others. This is in contrast to beetle faunas from Standlake Common (dated to 29,200 ± 300 BP – Birm-334; Briggs *et al.* 1985); Brandon (dated to 29,000 ± 500 BP – NPL-87; Coope 1968); Great Billing (dated to 28,225 ± 330 BP – Birm.75; Morgan 1969; Shotton *et al.* 1969); and Thrapston (dated to 25,780 ± 870 BP – Birm-113; Shotton *et al.* 1970; Coope, unpublished coleopteran data). All these sites have radiocarbon dates close to that obtained from ST93 (see below).

Coleoptera from facies association ST-3
The coleopteran assemblage from sample ST37 (i.e. from facies association ST-3) is very different from those in facies association ST-2 (Table 5). It has fewer beetle species (only 20 taxa in all) and the fossils are very fragmentary.

A different fluviatile environment is indicated by the coleoptera from ST-3. The running water species *Oulimnius troglodytus* and *Normandia nitens* live in well-oxygenated, usually shallow water in streams that flow continuously throughout the year. They suggest that at this time there was a single channel locally present. *Colymbetes* and *Helophorus* indicate the presence of still or slowly flowing water either as floodplain pools or backwaters. Reedy vegetation grew beside the river, as indicated by *Donacia marginata* (a monophage feeding on *Sparganium ramosum*) and *Donacia simplex* (an oligophage feeding on such plants as *Glyceria*, *Carex*, *Sparganium* and *Typha* – Koch 1992). Away from the river margin open, drier grassland is indicated by *Phyllopertha horticola* and *Athous vittatus*, the larvae of which live underground on roots. The only, rather tenuous, evidence for the local presence of trees is the weevil *Phyllobius/Polydrusus*, many of which feed on the leaves of deciduous trees.

Climatically, the assemblage from ST37 suggests temperate conditions similar to, or warmer than, those of the present day. For example, *Phyllopertha horticolis* has never been found in Devensian deposits. In this assemblage only *Aphodius affinis* is not native to Britain at the present day. Its modern range is central and southern Europe, suggesting thermal conditions perhaps warmer than those of the present day. Because of the small number of individuals present, it is not possible to obtain an MCR estimate of the palaeotemperature.

Vertebrate fossils

Two samples of vertebrate material were taken from the base of facies association ST-4 and are now archived in the Peterborough Museum. Sample ST57 (log M, Fig. 2) was identified by Dr D. Schreve and consisted of parts of two antlers from a fully mature Red Deer (*Cervus elaphus*). Sample ST82 (Fig. 2) was identified by Dr J. Stewart and consisted of a fish head from a gadid (Cod family, possibly burbot). Neither of these species is diagnostic of particular climatic conditions or time period.

Geochronology

Radiocarbon

Samples were submitted for accelerator mass spectrometric (AMS) radiocarbon dating from seven locations within the sequence, with three replicate sub-samples from the 'tie-point' sample ST99 within ST-2. Submitted samples consisted of plant macrofossils chosen carefully to avoid potential problems of hard-water error or reworking. They comprised terrestrial taxa which could have formed part of the contemporaneous flora and varied between samples depending on abundance (Tables 2 and 6).

Most of the radiocarbon dates from Stanwick are consistent with both stratigraphy and palaeontology. However, ST93 is considerably older than the other dates from facies association ST-2 (Fig. 6). Hard-water error is a possible problem with this sample because, in contrast with the other samples, the species dated was aquatic rather than terrestrial and the surrounding gravels contain a high percentage of limestone (Briant 2002). However, the magnitude of the difference (*c.* 17,000 years) is too great for hard-water error, which usually gives an offset of only 1500–2000 years (e.g. Day 1996). Even if old carbon from hard water contributed 50% of the total carbon in the sample, seeds with a real age of 11,500 BP would give an apparent age of only 17,000 BP (Olsen 1986). It is therefore likely that ST93 really is different, especially because it agrees with the OSL date ST95 (Table 6). ST93 was initially rejected for radiocarbon dating because of its lack of terrestrial macrofossils, but later dated because this associated luminescence date appeared anomalously old. Palaeontological assemblages (described above) may suggest some subtle differences between the environment during deposition of this sample and that of the rest of ST-2. However, the smallness of the sample means that these differences are not robust.

However, the younger radiocarbon dates place the bulk of facies association ST-2 between *c.* 9800 and *c.* 10,500 years BP (*c.* 11,200–12,800 cal. BP), i.e. within the Younger Dryas cold phase. This is broadly equivalent to the event-stratigraphically defined Greenland Stadial 1 (GS-1) (Walker *et al.* 1999). They are very similar to those from the Ditchford basal and upper channels (11,220 ± 45 years BP – SRR-4644, 10,280 ± 45 years BP – SRR-4642; Brown *et al.* 1994) elsewhere in the Nene. In addition, at Raunds a channel overlying the gravel succession was dated to between 11,395 ± 55 and 9375 ± 40 years BP (SRR-3604, SRR-3605) and one within the succession to between 12420 ± 60 and 10,870 ± 55 years BP (SRR-3610, SRR-3607a; Brown *et al.* 1994). The dominance of Younger Dryas deposits in this sequence may reflect the valley-confined nature of the river system in this reach, causing selective preservation.

Facies association ST-3 was deposited around 5100 years BP (*c.* 5900 cal. BP) and ST-4 from approximately 4000 years BP (*c.* 4400 cal. BP) onwards. As described above, these contain temperate palaeontological assemblages consistent with deposition during the Holocene. Comparison of the pollen spectra from ST37 with independently dated records suggests a correlation with pollen zones IIa–c (West 1980) *c.* 5700 BP, extremely close to the date from ST37. The radiocarbon dates from these associations (Table 6) are also consistent with the numerous Holocene radiocarbon dates obtained in the Nene by Brown *et al.* (1994).

Optically stimulated luminescence

Sand samples for OSL dating were taken in opaque plastic tubing and stored in light-tight bags until processed. Sample locations were chosen to maximize the likelihood of zeroing before deposition and were usually clean, well-sorted sand beds. Preparation to quartz followed the protocol outlined in Bateman & Catt (1996). Equivalent dose (D_e) was determined in the Godwin Laboratory,

Table 6. Radiocarbon and OSL dates from Stanwick, Northamptonshire

Facies association	Log	Radiocarbon sample	OSL sample	Species radiocarbondated	Radiocarbon laboratory code	Radiocarbon date	OSL date
ST-4	T	ST72		Salicaceae cf. *Salix* wood fragments	CAMS-75516	3980 ± 30 years BP (4420 ± 90 cal. BP)	
ST-3	O	ST37		Unidentified wood fragments	AA-40477	5145 ± 44 years BP (5910 ± 20 cal. BP)	
ST-2	B	ST20		*Alnus* sp. wood fragments	CAMS-75514	10,260 ± 40 years BP (11,770–12,316 cal. BP)	
ST-2	R	ST64		Salicaceae cf. *Salix* wood fragments	CAMS-75515	10,040 ± 40 years BP (11,305–11,685 cal. BP)	
ST-2	X	ST87		Salicaceae cf. *Salix* wood fragments	AA-41742	9825 ± 75 years BP (11,200 ± 50 cal. BP)	
ST-2	AA	ST93	ST95	*Potomageton* sp. fruitstones	AA-48182	27,290 ± 330 years BP (~32,000 cal. BP)	33.6 ± 1.8 ka
ST-2	JJ	ST99	ST100	*Betula nana* leaves	AA-41739	10,335 ± 92 years BP (11,782–12,597 cal. BP)	11.4 ± 0.6 ka
ST-2				*Salix herbacea* leaves	AA-41740	10,263 ± 94 years BP (11,754–12,334 cal. BP)	
ST-2				*Salix polaris* leaves	AA-41741	10,488 ± 80 years BP (12,189–12,812 cal. BP)	
ST-1	J		ST71				None calculated

Samples collected between 1998 and 2000 (see Fig. 2 for sample locations). Calibration of all dates after Stuiver *et al.* (1998), except ST93, for which Beck *et al.* (2001) was used to provide approximate calibration, although the lack of an internationally agreed scheme is noted (van der Plicht 2000). Uncertainties quoted as one standard deviation (SD) for radiocarbon and one standard error (SD/$\sqrt{n}$) for OSL.

Cambridge, by R. M. Briant, using the single aliquot regenerative (SAR) protocol of Murray & Wintle (2000), with modifications to preheat values (Briant 2002). Dose rates were determined by neutron activation analysis of representative surrounding sediments. Where sediments within a 30 cm radius of the sample were heterogeneous, dose rates were integrated, as explained in Appendix H of Aitken (1985). Cosmic dose rates were calculated using the equation of Prescott & Hutton (1994) and it was assumed that sediments had been buried to depth immediately after deposition. The water content used to attenuate dose rates was field moisture content with a 5% error. However, each sample was checked for its sensitivity to water content with reference to fully dry conditions and to saturation water content, as calculated from the density of packing within the tube. At this site, changes in water content values made little difference to the estimated ages (Briant 2002) and therefore field moisture content was used for all samples. Age was calculated by dividing the mean De by dose rate.

The results of the OSL measurements, water content values and dosimetry data are shown in Table 7. All OSL samples have been checked for preheat dependence after Bailey et al. (2001) and the equivalent dose calculated only from those aliquots with a recycling ratio within 10% of unity (Armitage 2001), located on the 'preheat plateau' and falling within the distribution defined using Chauvenet's criterion (Taylor 1997). Further details of luminescence behaviour may be found in Briant (2002). As a result of the rigorous acceptance criteria applied, it was not possible to calculate an age for ST71. None of the 10 measured aliquots passed the recycling ratio test, suggesting that sensitivity changes were too great for the SAR protocol to correct.

In contrast, the SAR protocol performed equally well on ST95 and ST100, with an 86.7% acceptance rate and mean recycling ratios very close to 1. In addition, both samples showed a preheat plateau at all temperatures, good behaviour and well-clustered normal equivalent dose (D_e) frequency distributions (Fig. 7). Both ST95 and ST100 agree well with their associated calibrated radiocarbon dates (Table 6), suggesting that the discrepancy between the dates represents the true spread of ages of sediments within facies association ST-2. Therefore the OSL dates confirm the radiocarbon dating and date the bulk of ST-2 to c. 11.4–11.6 ka with some older sediments preserved dating from c. 32–36 ka. Facies association ST-1 pre-dates this and ST-3 and ST-4 were deposited afterwards.

Fluvial response to climatic phases

The robust geochronological control on ST-2 dates most of this association to the Younger Dryas stadial. The climatic characteristics of this period are well known from other sites (e.g. Atkinson et al. 1987; Walker et al. 1993). It also allows differentiation of remnant Middle Devensian deposits within ST-2, which are not conclusively detectable using other evidence. For example, the type of organic material preserved within ST-2 is so variable (Figs 2 & 3) that the sedimentological differences between the two tie-point locations (Fig. 4a, b) are not significant. In addition, the differences in coleopteran faunas are subtle and may be merely a reflection of the relatively small samples used or taphonomic effects. The geochronological control on ST-3 and ST-4 is also important for linking river activity to specific climate phases, giving a more precise indication of exact timing within the Holocene. In contrast, the unsuccessful OSL dating of ST71 and lack of other suitable material from ST-1 means that this association cannot be linked to an externally defined climate phase.

The fact that remnant Middle Devensian deposits cannot be differentiated from those of Younger Dryas age on sedimentological grounds suggests similar fluvial behaviour during both phases. This is not surprising, given that analysis of coleopteran faunas suggests very similar climate at both times with the MCR technique estimating mean July temperatures between 7 and 12°C and mean January temperatures between −22 and −13°C for the Younger Dryas samples and slightly lower values from the older sample ST93. The braided style of this association probably represents

Table 7. *OSL dosimetry, equivalent dose and age estimates for samples from Stanwick, Northamptonshire*

Sample and laboratory code	Facies association and log	Field moisture (%)	Total dose rate (Gy/ka)	Mean D_e (Gy)	Age estimate (ka)
ST95	ST-2, log AA	17.09	1.32 ± 0.07	44.33 ± 0.94	33.62 ± 1.83
ST100	ST-2, log JJ	24.08	1.72 ± 0.08	19.60 ± 0.52	11.39 ± 0.60

Gy, grays; ka, thousands of years.

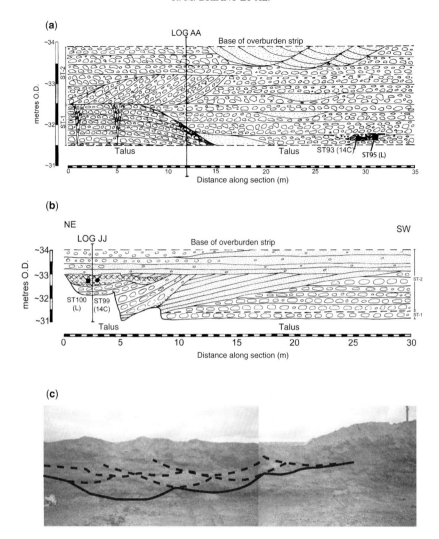

Fig. 4. Detailed sedimentology of key sections: section drawings from logs (**a**) AA and (**b**) JJ, where paired OSL and radiocarbon samples (ST93/95, ST99/100) were taken from single channel fills; (**c**) photograph from log U, showing sedimentology of channel fills within facies association ST-2.

a reaction to high discharge and sediment supply, typical of both climate phases. Coleopteran temperature reconstructions suggest that hydrological regimes during both phases were probably snowmelt-flood dominated, leading to high short-term discharges in the spring. In addition, precipitation is likely to have been higher in the Middle Devensian and Late-glacial than in the intervening parts of the Late Devensian. The Last Glacial Maximum appears to have been characterized by arid conditions over much of continental Europe (e.g. Bateman 1995; Kasse 1999; Briant *et al.* 2004*b*), probably due to ice-sheet-controlled anticyclonic systems causing a southwards shift of the Polar Front. Sediment supply would have been high during the Middle Devensian and Late-glacial because of sparse vegetation cover, as shown in the plant macrofossil and pollen assemblages from this site. Interestingly, no sediments have been found from the Late-glacial Interstadial. This is probably due to the low preservation potential of such material, especially given the valley-confined nature of the Nene at this point and the destructive fluvial style in ST-2. Incision and coarse-grained deposition is characteristic of fluvial activity in the Younger Dryas Stadial in Britain (Rose et al. 1980; Collins et al. 1996; Lewis et al. 2001) and northern France (e.g. Antoine 1994; et al. 2000). Similar patterns are also seen in other continental systems (e.g. Vandenberghe et al. 1994; van Huissteden & Kasse 2001; Schirmer 1995; Kozarski 1983; van Huissteden et al. 2001), although where these are lower-lying they are often sand braided rather than gravel braided.

During deposition of ST-3 and ST-4, the river was responding to a temperate climate with low sediment supply because continuous vegetation colonized the floodplain (grassland and alder) and valley sides (mixed-oak woodland), restricting runoff and sediment entrainment. Discharges were relatively low for the same reason, and also because there were no longer perennial snow banks concentrating discharge into a single season. However, the sediments in ST-3 suggest that, for a short time, the discharge regime included occasional peak flows sufficient to transport gravel a short distance, as many present-day temperate rivers will do. It is tempting to attribute this to increased precipitation associated with increased temperatures at the 'climatic optimum' (Iversen 1944) or to a period of higher precipitation recorded c. 5900 cal. BP from peat records (Hughes et al. 2000). Macklin (1999) suggests that a general increase in sedimentation occurred in the mid-Holocene, although recent synthesis of multiple fluvial records from Britain (Macklin & Lewin 2003) does not suggest widespread alluviation at this time. Without a more continuous record of Holocene fluvial development in the Nene, a specific climate link cannot be conclusively demonstrated, especially since gravel beds occur in the Welland dated to c. 4000 BP (French et al. 1992), and within predominantly fine-grained Holocene deposits in the Gipping valley (Rose et al. 1980). Thus periodic gravel deposition may be a common feature of Holocene fluvial systems in lowland southern Britain where suitable sediment is available.

Recognizing flood events in the Quaternary fluvial record – limits to resolution

Robust geochronological control therefore allows determination of fluvial behaviour in response to climate phases of hundreds to thousands of years' duration. However, it is significantly harder to detect individual flood events lasting only days or weeks due to limits in the resolution of sedimentology, palaeontology and geochronology.

Thus, whilst sedimentologically distinct packages of sediment (facies associations) bracketed by high-order bounding surfaces (Table 1) can be interpreted as representing distinct phases within the system, it is impossible to estimate how much time intervened between each event nor how events correlate between sections. When horizontal bedding is dominant, as in ST-1 (Figs 3 & 4a, b), this was probably laid down within bars, the cores of which often remain stable during floods, limiting identification of single events. In contrast, the scour-fill structures common in parts of ST-2 (Figs 3 & 4a, b) are more likely to relate to a single event, being scoured on the rising limb and filled on the falling limb. Even so, a group of intercutting scour-fill structures as in Figure 4c could be interpreted in two ways. The entire group could have been deposited during a single flood event with changes in flow level and energy within the event causing small-scale intra-group erosion and deposition. Alternatively, individual scour-fills might represent single flood events and the group comprise remnants from multiple events of varying magnitudes, possibly from significantly different time periods. This ambiguity shows that sedimentological description is often not sufficient to detect individual events because of the discontinuity of many bounding surfaces. Indeed, small-scale scour is typical of fluvial activity at various scales and not necessarily only of flood events. Furthermore, even in settings where events can be sedimentologically distinguished at individual locations, variable preservation means that the only reliable basis for correlating these is geochronology (Kochel & Baker 1988).

Palaeontological assemblages are likewise limited in their usefulness for detecting individual flood events, largely because they record average conditions over a year, with material 'swept' off adjacent surfaces during annual flooding (Holyoak 1984; West et al. 1993; West 2000). Whilst coleopteran faunas can provide quantitative climatic parameters within which to place sedimentological processes and do respond more quickly to rapid climatic changes than many other biological systems (Coope et al. 1997) this taphonomic process means that even they do not record seasonal differences. Furthermore, a wide range of flood magnitudes can occur within any climate regime and the subtle differences in temperature and precipitation between these are unlikely to be recorded by palaeontological data. Indeed, at this site it seems that the climate was sufficiently similar in the Middle Devensian and Younger Dryas that palaeontological assemblages within ST-2 are very similar to each other (Tables 2, 4 & 5), despite the fact that

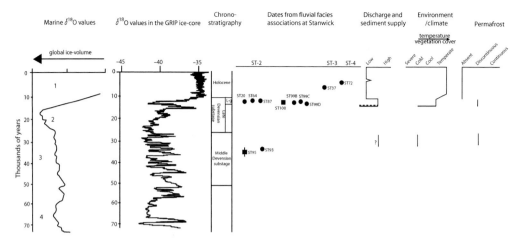

Fig. 5. Summary of fluvial activity in response to climate as recorded in the sequence from Stanwick, Northamptonshire during the Middle and Late Devensian, compared with the independent marine isotope record (Martinson et al. 1987) and GRIP ice-core record (Johnsen et al. 1997). Age estimates used to determine sedimentation periods include error bars, although these are too small to be depicted.

elsewhere coleopteran faunas from these time periods are much more distinct.

Geochronology has been shown to be the most effective way of dating fluvial sediments to specific climate phases, especially where it can be linked to good palaeontological data. Thus it might be thought to be the most promising for detecting individual events. However the dating techniques available within the Pleistocene times-scale are not yet sufficiently precise to differentiate individual flood events. For example, AMS radiocarbon dating of terrestrial macrofossils, as carried out in this study, is widely recognized as the most precise type of radiocarbon dating (Törnqvist et al. 1992; Lowe & Walker 2000; Turney et al. 2000). For this reason, and because they are relatively young, the Younger Dryas-age dates have

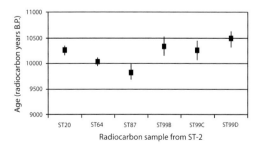

Fig. 6. Comparison of uncalibrated radiocarbon dates of Younger Dryas age from facies association ST-2, Stanwick, Northamptonshire. Error bars are shown at two standard deviations (see Table 6 for exact values).

relatively small error bars of c. 50 years (Fig. 6), although even these mean that an individual scour-fill could fall within a c. 100 year period, or longer after calibration due to factors such as the radiocarbon plateau during the Late-glacial (Table 6). In addition, the dates overlap (Fig. 6), meaning that individual events cannot even be separated from each other. Even the three age estimates from different species within the same channel fill (ST99) are not consistent. This phenomenon appears widespread in other samples as well, suggesting that radiocarbon dating may be sensitive to the type of macrofossil material dated (e.g. Walker et al. 1999; Turney et al. 2000).

Lack of precision is even more problematic with OSL dating, which is the main dating technique suitable for fluvial sediments older than c. 35,000 years BP (Roberts et al. 1994; Briant et al. 2004a). At this site, the 5% error bars that are typical of the technique lead to relatively 'small' error ranges of c. 0.5–2 ka that are again too large to detect deposition within a single flood event. In the Early Devensian, error bars are even larger (c. 5–10 ka, e.g. Briant et al. 2004a, 2005; Törnqvist et al. 2000), making it hard to link fluvial deposition even to single climate phases. Indeed, error bars are often larger than the duration of marine isotope substage-scale climate phases that might drive fluvial activity. Similar problems are encountered when comparing older OSL-dated deposits to the longer marine isotope stages (Bates et al. 2004; Briant et al. 2006), because error bars are even larger beyond the Devensian (Weichselian) Stage.

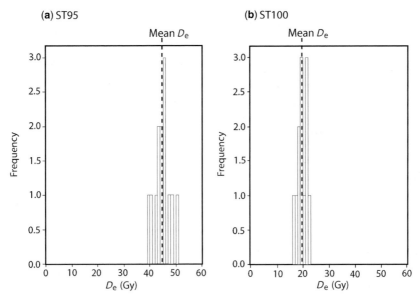

Fig. 7. Frequency distributions of equivalent dose distributions (D_e) in Grays (Gy) for the two OSL samples in facies association ST-2, Stanwick, Northamptonshire. Dotted line shows the mean D_e calculated after testing for preheat (PH) plateaux.

Conclusion

Many of the limitations outlined above with regard to detecting flood events within the Quaternary fluvial record will probably remain intractable. Increases in precision are most likely within geochronological techniques, but even here it seems unlikely that it will be possible to detect deposition from within a single year, let alone a flood event within that year. Nonetheless, geochronological control is sufficient, at least during the later parts of the Devensian (Weichselian) Stage, to link periods of fluvial deposition to climatic phases at the marine isotope substage scale. The improved chronological control on such sequences is novel, being part of a study applying the first large-scale OSL dating to British Devensian river deposits (e.g. Briant 2002; Briant et al. 2004a, b, 2007). It is valuable because it allows the determination of 'average' fluvial activity in response to broad climate phases, for example, detecting similarities between the Middle Devensian and Younger Dryas phases at this site. This will improve our understanding of how rivers behave over long time periods and how sensitive they are to climate changes on various scales. From the evidence available from this site and others (e.g. Briant et al. 2004b) it seems likely that river systems record responses only to large-scale changes within the climate system such as those described above. The dominance of younger deposits within this valley-confined part of the Nene Valley suggests that the record of fluvial activity at this site is probably highly discontinuous. Whilst this is probably an extreme case because there are less confined sequences where Early Devensian deposits are preserved (e.g. Briant et al. 2004a, 2005), the Quaternary fluvial record elsewhere is also likely to be 'mostly gaps' (Gibbard & Lewin 2002), even though the specific role of flood events within that record remains elusive.

This work was undertaken whilst R.M.B. was in receipt of a Mary Anne Ewart Research Studentship from Newnham College, Cambridge. The cooperation of Hanson Ltd is appreciated. Training in plant identification was given by C. Turner and luminescence procedures by M. Bateman, L. Zhou and A. Wintle. Radiocarbon determinations were carried out by the NERC Radiocarbon Laboratory in East Kilbride, Scotland under grant number 854/0500, with thanks to C. Bryant. Neutron Activation Analyses were undertaken by Becquerel Laboratories, Australia. The paper forms a contribution to the Fluvial Archives Group (FLAG) Focus 3.

References

AITKEN, M. J. 1985. *Thermoluminescence Dating.* Studies in Archaeological Science, London.
ANGUS, R. 1992. *Insecta, Coleoptera, Hydrophilidae, Helophorus.* Süsswasserfauna von Mitteleuropas. Gustav Fischer, Stuttgart, 20/19-2, 1–144.

ANTOINE, P. 1994. The Somme valley terrace system (northern France); a model of river response to Quaternary climatic variations since 800,000 BP. *Terra Nova*, **6**, 453–464.

ANTOINE, P., LAUTRIDOU, J. P. & LAURENT, M. 2000. Long-term fluvial archives in NW France: response of the Seine and Somme rivers to tectonic movements, climatic variations and sea-level changes. *Geomorphology*, **33**, 183–207.

ARMITAGE, S. J. 2001. SAR De(t) plots for Aeolian quartz from Inhaca Island, Mozambique. Abstract, Luminescence & ESR Dating Meeting, University of Glasgow.

ATKINSON, T. C., BRIFFA, K. R. & COOPE, G. R. 1987. Seasonal temperatures in Britain during the last 22,000 years, reconstructed using beetle remains. *Nature*, **325**, 587–592.

BAKER, V. R. 1984. Flood sedimentation in bedrock fluvial systems. *Memoirs of the Canadian Society of Petroleum Geologists*, **10**, 87–98.

BAILEY, S. D., WINTLE, A. G., DULLER, G. A. T. & BRISTOW, C. S. 2001. Sand deposition during the last millenium at Aberffraw, Anglesey, North Wales as determined by OSL dating of quartz. *Quaternary Science Reviews*, **20**, 701–704.

BATEMAN, M. D. 1995. Thermoluminescence dating of the British coversand deposits. *Quaternary Science Reviews*, **14**, 791–798.

BATEMAN, M. D. & CATT, J. A. 1996. An absolute chronology for the raised beach deposits at Sewerby, East Yorkshire, U.K. *Journal of Quaternary Science*, **11**, 389–395.

BATES, M. R., WENBAN-SMITH, F. F., BRIANT, R. & MARSHALL, G. 2004. Palaeolithic Archaeology of the Sussex/Hampshire Coastal Corridor. Final report: English Heritage ALSF project 3279.

BECK, J. W., RICHARDS, D. A. ET AL. 2001. Extremely Large Variations of Atmospheric 14C Concentration During the Last Glacial Period. *Science*, **292**, 2453–2458.

BELL, F. G. 1968. Weichselian glacial floras in Britain. Unpublished PhD thesis, University of Cambridge.

BENNETT, K. D. 1983. Devensian late-glacial and Flandrian vegetational history at Hockham Mere, Norfolk, England. 1. Pollen percentages and concentrations. *New Phytologist*, **95**, 457–487.

BENNETT, K. D. & BIRKS, H. J. B. 1990. Postglacial history of alder (*Alnus glutinosa* (L.) Gaertn.) in the British Isles. *Journal of Quaternary Science*, **5**, 123–133.

BENNETT, K. D., FOSSITT, J. A., SHARP, M. J. & SWITZUR, V. R. 1990. Holocene vegetational and environmental history at Loch Lang, South Uist, Western Isles, Scotland. *New Phytologist*, **114**, 281–298.

BLACK, R. F. 1976. Periglacial features indicative of permafrost: ice and soil wedges. *Quaternary Research*, **6**, 3–26.

BRIANT, R. M. 2002. Fluvial responses to rapid climate change in eastern England during the last glacial period. PhD thesis, University of Cambridge.

BRIANT, R. M., COOPE, G. R., PREECE, R. C. & GIBBARD, P. L. 2004a. The Upper Pleistocene deposits at Deeping St James, Lincolnshire: evidence for Early Devensian fluvial sedimentation. *Quaternaire*, **15**, 5–15.

BRIANT, R. M., COOPE, G. R. ET AL. 2004b. Fluvial response to Late Devensian aridity, Baston, Lincolnshire, England. *Journal of Quaternary Science*, **19**, 479–495.

BRIANT, R. M., BATEMAN, M. D., COOPE, G. R. & GIBBARD, P. L. 2005. Climatic control on Quaternary fluvial sedimentology: a Fenland Basin river in England. *Sedimentology*, **52**, 1397–1424.

BRIANT, R. M., BATES, M. R., SCHWENNINGER, J.-L. & WENBAN-SMITH, F. F. 2006. A long optically-stimulated luminescence dated Middle to Late Pleistocene fluvial sequence from the western Solent Basin, southern England. *Journal of Quaternary Science*, **21**, 507–523.

BRIGGS, D. J., COOPE, G. R. & GILBERTSON, D. D. 1985. The chronology and environmental framework of early man in the Upper Thames Valley. *British Archaeological Reports, British Series*, **13**, 1–176.

BROWN, A. G., KEOGH, M. & RICE, R. J. 1994. Floodplain evolution in the East Midlands, U. K.: the Lateglacial and Holocene alluvial records from the Soar and Nene valleys. *Philosophical Transactions of the Royal Society of London*, **A348**, 261–293.

CHURCH, M. 1974. *Hydrology and Permafrost with Respect to North America. Permafrost Hydrology: Proceedings of Workshop Seminar*. Canadian National Committee, International Hydrological Decade, Environment Canada, Ottawa, 7–20.

CLAPHAM, A. R., TUTIN, T. G. & WARBURG, E. F. 1981. *Excursion Flora of the British Isles*, 3rd edn. Cambridge University Press, Cambridge.

COLLINS, P. E. F., FENWICK, I.-M., KEITH-LUCAS, D. M. & WORSLEY, P. 1996. Late Devensian river and flood plain dynamics and related environmental change in North West Europe, with particular reference to a site at Woolhampton, Berkshire, England. *Journal of Quaternary Science*, **11**, 357–375.

COOPE, G. R. 1968. An insect fauna from Mid-Weichselian deposits at Brandon, Warwickshire. *Philosophical Transactions of the Royal Society of London*, **254B**, 425–456.

COOPE, G. R. 1986. Coleopteran analysis. *In*: BERGLUND, B. E. (ed.) *Handbook of Holocene Palaeoecology and Palaeohydrology*. Wiley, Chichester, 703–711.

COOPE, G. R. 2000. Middle Devensian (Weichselian) coleopteran assemblages from Earith, Cambridgeshire (UK) and their bearing on the interpretation of 'Full glacial' floras and faunas. *Journal of Quaternary Science*, **15**, 779–788.

COOPE, G. R. & PENNINGTON, W. 1977. The Windermere Interstadial of the Late Devensian. *In*: COOPE, G. R. *Fossil Coleopteran Assembing the Devensian (last) Cold Stage*. Philosophical Transactions of the Royal Society of London, **B280**, 337–339.

COOPE, G. R., GIBBARD, P. L., HALL, A. R., PREECE, R. C., ROBINSON, J. E. & SUTCLIFFE, A. J. 1997. Climatic and environmental reconstructions based on fossil assemblages from Middle Devensian (Weichselian) deposits of the River Thames at South

Kensington, Central London. *Quaternary Science Reviews*, **16**, 1163–1195.

DAY, P. 1996. Devensian Late-glacial and early Holocene environmental history of the Vale of Pickering, Yorkshire, England. *Journal of Quaternary Science*, **11**, 9–24.

FRENCH, C. A. I., MACKLIN, M. G. & PASSMORE, D. G. 1992. Archaeology and palaeochannels in the Lower Welland and Nene valleys: alluvial archaeology at the fen edge, Eastern England. *In*: NEEDHAM, S. & MACKLIN, M. G. (eds) *Alluvial Archaeology in Britain*. Oxbow Monograph, **27**. Oxbow Press, Oxford, 169–176.

GIBBARD, P. L. 1995. The formation of the Strait of Dover. *In*: PREECE, R. C. (ed.) *Island Britain: a Quaternary Perspective*. Geological Society Special Publication, **96**, 15–26.

GIBBARD, P. L. & LEWIN, J. 2002. Climate and related controls on interglacial fluvial sedimentation in lowland Britain. *Sedimentary Geology*, **151**, 187–210.

GODWIN, H. 1975. *The History of the British Flora: a Factual Basis for Phytogeography*, 2nd edn. Cambridge University Press, Cambridge.

HOLYOAK, D. T. 1982. Non-marine mollusca of the Last Glacial Period (Devensian) in Britain. *Malacologia*, **22**, 727–730.

HOLYOAK, D. T. 1983. The colonization of Berkshire, England, by land & freshwater Mollusca since the Late Devensian. *Journal of Biogeography*, **10**, 483–498.

HOLYOAK, D. T. 1984. Taphonomy of prospective plant macrofossils in river catchments on Spitsbergen. *New Phytologist*, **98**, 405–423.

HOLYOAK, D. T. & SEDDON, M. B. 1984. Devensian and Holocene fossiliferous deposits in the Nene Valley, Central England. *Mercian Geologist*, **9**, 127–150.

HORTON, A. 1989. *Geology of the Peterborough District*. Memoirs of the British Geological Society Sheet, **158**, England and Wales. HMSO, London.

HUGHES, P. D. M., MAQUOY, D., BARBER, K. E. & LANGDON, P. G. 2000. Mire-development pathways and palaeoclimate records from a full Holocene peat archive at Walton Moss, Cumbria, England. *The Holocene*, **10**, 465–479.

IVERSEN, J. 1944. Viscum, Hedera and Ilex as climatic indicators. A contribution to the study of the postglacial temperature climate. *Geologiska Föreningeus i Stockholm Förhandlingar*, **66**, 463–483.

JOHNSEN, S. J., CLAUSEN, H. B. *ET AL.* 1997. The delta O-18 record along the Greenland Ice Core Project deep ice core and the problem of possible Eemian climatic instability. *Journal of Geophysical Research – Oceans*, **102**, 26397–26410.

KASSE, C. 1999. Late Pleniglacial and Late Glacial aeolian phases in the Netherlands. *GeoArchaeoRhein*, **3**, 61–82.

KASSE, C., BOHNCKE, S. J. P. & VANDENBERGHE, J. 1995. Fluvial periglacial environments, climate and vegetation during the Middle Weichselian in the Northern Netherlands with special reference the Hengelo Interstadial. *Mededelingen Rijks Geologische Dieust*, **52**, 387–413.

KERNEY, M. P., GIBBARD, P. L., HALL, A. R. & ROBINSON, J. E. 1982. Middle Devensian river deposits beneath the 'Upper Flood Plain' terrace of the River Thames at Isleworth, West London. *Proceedings of the Geologists Association*, **93**, 385–393.

KNOX, J. C. 1993. Large increases in flood magnitude in response to modest changes in climate. *Nature*, **361**, 430–432.

KOCH, K. 1992. *Die Käfer Mitteleuropas: Ökologie. 3*. Goecke & Evers, Krefeld, 1–389.

KOCHEL, R. C. & BAKER, V. R. 1988. Palaeoflood analysis using slackwater deposits. *In*: BAKER, V. R., KOCHEL, R. C. & PATTON, P. C. (eds) *Flood Geomorphology*. Wiley, New York, 357–376.

KOZARSKI, S. 1983. River channel adjustment to climate change in west central Poland. *In*: GREGORY, K. J. (ed.) *Background to Palaeohydrology*, Wiley, Chichester, 355–374.

LEWIS, S. G., MADDY, D. & SCAIFE, R. G. 2001. The fluvial system response to abrupt climate change during the last cold stage: the upper Pleistocene River Thames fluvial succession at Ashton Keynes, UK. *Global and Planetary Change*, **28**, 341–359.

LOWE, J. J. & WALKER, M. J. C. 2000. Radiocarbon dating the last glacial-interglacial transition (c14–9 14C ka BP) in terrestrial and marine records: the need for new quality assurance protocols. *Radiocarbon*, **42**, 53–68.

MACKLIN, M. G. 1999. Holocene river environments in prehistoric Britain: human interaction and impact. *In*: EDWARDS & SADLER (eds) *Holocene Environments of Prehistoric Britain*. Quaternary Proceedings, **7**. Wiley, Chichester, 521–530.

MACKLIN, M. G. & LEWIN, J. 2003. River sediments, great floods and centennial-scale Holocene climate change. *Journal of Quaternary Science*, **18**, 101–105.

MADDY, D. 1999. English Midlands. *In*: BOWEN, D. Q. (ed.) *A Revised Correlation of Quaternary Deposits in the British Isles*. Geological Society Special Report, **23**. Geological Society, London.

MARTINSON, D. G., PISIAS, N. G., HAYS, J. D., IMBRIE, J., MOORE, T. C. & SHACKLETON, N. J. 1987. Age dating and the orbital theory of the ice ages: development of a high resolution 0–300,000 year chronostratigraphy. *Quaternary Research*, **27**, 1–29.

MIALL, A. D. 1996. *The Geology of Fluvial Deposits: Sedimentary Facies, Basin Analysis and Petroleum Geology*. Springer, Berlin.

MORGAN, A. 1969. A Pleistocene fauna and flora from Great Billing, Northamptonshire, England. *Opuscula Entomologica, Lund*, **34**, 109–129.

MURRAY, A. S. & WINTLE, A. G. 2000. Luminescence dating of quartz using an improved single aliquot regenerative-dose protocol. *Radiation Measurements*, **32**, 57–73.

OLSSON, U. 1986. Radiocarbon dating. *In*: BERGLUND, B. E. (ed.) *Handbook of Holocene Palaeoecology and Palaeohydrology*, Wiley, Chichester, 275–296.

PREECE, R. C. & ROBINSON, J. E. 1982. Mollusc, ostracod and plant remains from early postglacial deposits near Staines. *The London Naturalist*, **61**, 6–15.

PRESCOTT, J. R. & HUTTON, J. T. 1994. Cosmic ray contributions to dose rates for luminescence and ESR dating: large depths and long term time variations. *Radiation Measurements*, **23**, 497–500.

ROBERTS, R. G., JONES, R. & SMITH, M. A. 1994. Beyond the radiocarbon barrier in Australian prehistory. *Antiquity*, **68**, 611–616.

ROSE, J., TURNER, C., COOPE, G. R. & BRYAN, M. D. 1980. Channel changes in a lowland river catchment over the last 13,000 years. *In*: CULLINGFORD, R. A., DAVIDSON, D. A. & LEWIN, J. (eds) *Timescales in Geomorphology*. Wiley, Chichester, 159–175.

SCHIRMER, W. 1995. Valley bottoms in the late Quaternary. *Zeitschrift für Geomorphologie Suppl.*, **100**, 27–51.

SHOTTON, F. W. 1968. The Pleistocene succession around Brandon, Warwickshire. *Philosophical Transactions of the Royal Society of London*, **254B**, 387–400.

SHOTTON, F. W., BLUNDELL, D. J. & WILLIAMS, R. E. G. 1969. Birmingham University Radiocarbon Dates III. *Radiocarbon*, **11**, 263–270.

SHOTTON, F. W., BLUNDELL, D. J. & WILLIAMS, R. E. G. 1970. Birmingham University Radiocarbon Dates IV. *Radiocarbon* **12**, 385–399.

SMITH, A. J. 1985. A catastrophic origin for the palaeovalley system of the eastern English Channel. *Marine Geology*, **64**, 65–75.

SMITH, A. J. 1989. The English Channel – by geological design or catastrophic accident? *Proceedings of the Geologists' Association*, **100**, 325–337.

STACE, C. 1991. *New Flora of the British Isles*. Cambridge University Press, Cambridge.

STRAND, A. 1946. *Nord-Norges Coleoptera*. Tromsø Museums Årshefter Nnaturhistorisk AVD. NR34, Vol. 67, no. 1, 1–625.

STUIVER, M., REIMER, P. J. ET AL. 1998. INTCAL98 radiocarbon age calibration, 24,000–0 cal BP. *Radiocarbon*, **40**, 1041–1084.

TAYLOR, J. R. 1997. *An Introduction to Error Analysis*, 2nd edn. University Science Books, Sausalito, CA.

THORN, C. E. 1988. *Introduction to Theoretical Geomorphology*. Unwin Hyman, London.

TØRNQVIST, T. E., DEJONG, A. F. M., OOSTERBAAN, W. A. & VAN DER BORG, K. 1992. Accurate dating of organic deposits by AMS C-14 measurement of macrofossils. *Radiocarbon*, **34**, 566–577.

TØRNQVIST, T. E., WALLINGA, J., MURPHY, A. S., DEWOLF, H., CLEVERINGA, P. & DEGANS, W. 2000. Response of the Rhine-Meuse system (west-central Netherlands) to the last Quaternary glacio-eustatic cycles: a first assessment. *Global & Planetary Change*, **27**, 89–111.

TURNEY, C. S. M., COOPE, G. R., HARKNESS, D. D., LOWE, J. J. & WALKER, M. J. C. 2000. Implications for the dating of Wisconsinan (Weichselian) late-glacial events of systematic radiocarbon age differences between terrace plant macrofossils from a site in South West Ireland. *Quaternary Research*, **53**, 114–121.

VAN DER PLICHT, J. 2000. Introduction: The 2000 Radiocarbon Varve/Comparison Issue. *Radiocarbon*, **42**, 313–322.

VAN HUISSTEDEN, KO (J.) & KASSE, C. 2001. Detection of rapid climate change in last glacial fluvial successions in the Netherlands. *Global and Planetary Change*, **28**, 319–339.

VAN HUISSTEDEN, J. (KO), GIBBARD, P. L. & BRIANT, R. M. 2001. Periglacial fluvial systems in northwest Europe during marine isotope stages 4 and 3. *Quaternary International*, **79**, 75–88.

VANDENBERGHE, J., KASSE, C., BOHNCKE, S. & KOZARSKI, S. 1994. Climate related river activity at the Weichselian–Holocene transition: a comparative study of the Warta and Maas rivers. *Terra Nova*, **6**, 476–485.

VELEGRAKIS, A. F., DIX, J. K. & COLLINS, M. B. 1999. Late Quaternary evolution of the upper reaches of the Solent River, southern England, based upon marine geophysical evidence. *Journal of the Geological Society of London*, **156**, 73–87.

WALKER, M. J. C., COOPE, G. R. & LOWE, J. J. 1993. The Devensian (Weichselian) Lateglacial Palaeoenvironmental record from Gransmoor, East Yorkshire, England. *Quaternary Science Reviews*, **12**, 659–680.

WALKER, M. J. C., BJÖRCK, S., LOWE, J. J., CWYNAR, L. C., JOHNSEN, S., KNUDSEN, K.-L., WOHLFARTH, B. & INTIMATE Group. 1999. Isotopic 'events' in the GRIP ice core: a stratotype for the Late Pleistocene. *Quaternary Science Reviews*, **18**, 1143–1150.

WALKER, M. J. C., COOPE, G. R., SHELDRICK, C., TURNEY, C. S. M., LOWE, J. J., BLOCKLEY, S. P. E. & HARKNESS, D. D. 2003. Devensian Late-glacial environmental changes in Britain: a multi-proxy record from Llanilid, South Wales, UK. *Quaternary Science Reviews*, **22**, 475–520.

WEST, R. G. 1980. Pleistocene Forest History in East Anglia. *New Phytologist*, **85**, 571–622.

WEST, R. G. 1991. On the origin of Grunty Fen and other landforms in southern Fenland, Cambridgeshire. *Geological Magazine*, **128**, 257–262.

WEST, R. G. 2000. *Plant Life of the Quaternary Cold Stages: Evidence from the British Isles*. Cambridge University Press, Cambridge.

WEST, R. G., ANDREW, R. & PETTIT, M. 1993. Taphonomy of plant remains on floodplains of tundra rivers, present and Pleistocene. *New Phytologist*, **123**, 203–221.

WOHLFARTH, B., SKOG, G., POSSNERT, G. & HOLMQUIST, B. 1998. Pitfalls in the AMS radiocarbon dating of terrestrial macrofossils. *Journal of Quaternary Science*, **13**, 137–145.

WOO, M.-K. 1990. Permafrost hydrology. *In*: PROWSE, T. D. & OMMANNEY, C. S. L. (eds) *Northern Hydrology: Canadian Perspectives*. National Hydrological Research Institute Science Report, **1**. Environment Canada, Saskatoon, 63–75.

Fluvial solar signals

C. VITA-FINZI

Department of Mineralogy, Natural History Museum, Cromwell Rd, London SW7 5BD, UK
(e-mail: cvitafinzi@aol.com)

Abstract: The fluvial history of the Mediterranean basin and the Near East includes depositional evidence for a latitudinally diachronous, locally bipartite, episode of aggradation by equable streams during the period AD 500–1900. According to the Leopold gullying model, the key requirement would have been an increase in the proportion of small, non-erosive rains. Theoretical considerations supported by general climate models suggest that a decrease in solar radiation at the UV wavelengths would lead to equatorward displacement of the subtropical jet streams and the associated mid-latitude depressions. Atmospheric $\Delta^{14}C$ values show a gradual decline from c. 7000 BP followed by a temporary resurgence after AD 500 which includes peaks corresponding to the Oort, Wolf, Spörer, Maunder, Dalton and other solar minima. Reduced irradiance could account for channel aggradation by favouring cyclonic at the expense of convectional precipitation. Confirmation of a solar–fluvial link would benefit both solar history and flood forecasting.

A long and comprehensive history is essential for the interpretation of the Sun's present behaviour. Direct measurement of solar energy output has been possible only since the launch of the first dedicated satellites in 1978. For earlier times reliance is placed on the indirect evidence of sunspots and the abundance of ^{14}C and ^{10}Be recovered from tree rings and ice cores, supplemented by palaeoclimatic data, tracks and cosmogenic isotopes in meteorites and lunar samples, the results of numerical modelling, and comparison with other stars (Vita-Finzi 2002).

This chapter tries to show that in certain favoured locations river history is demonstrably linked to the Sun's activity and that its reconstruction can usefully complement existing sources because it depends on tangible evidence, potentially spans the period since the earliest fluviatile deposits were laid down (at least 3.2×10^9 years ago; Hessler *et al.* 2004), and bears on a much wider range of subaerial environments than those suited to growing trees or accumulating ice.

Moreover, fluvial studies are local enough to reveal environmental gradients and to reflect narrow time bands. Although some progress has been made in finding associations between solar radiation and sea-surface temperature or cloud cover, there is still little agreement over the interaction between incoming radiation and climate (see Hoyt & Schatten 1997). The fluvial stratagem would sidestep the problem by focusing on specific weather systems. Progress on this front would be of benefit not only to astrophysical modelling but also to the forecasting of river behaviour because any periodicities that might emerge from solar data would impinge on the associated hydrological processes, just as position on the 11-year sunspot cycle is already taken into account in planning the launch of artificial satellites and assessing the impact of space weather on telecommunications.

Solar variability at the $10^2 - 10^3$ a scale

Recent years have seen the publication of extensive surveys of variation in total solar irradiance (e.g. Pap & Frölich 1999). The changes include both episodic events such as flares and coronal mass emissions lasting a few seconds or minutes, and periodic or quasi-periodic events with repeat times measured in minutes, days or years, the most familiar being the c. 11 a sunspot or Schwabe cycle. The evidence derives both from direct observation and from indirect indicators of solar activity such as aurorae and sunspots.

Besides demonstrating the great range of phenomena that belie the term 'solar constant', these accounts draw attention to the variability of different parts of the solar spectrum. Satellite measurement during solar cycles 21 and 22 (June 1976 to September 1986 and September 1986 to May 1996) has revealed changes in total irradiance at the top of the atmosphere of c. 0.1% over the solar cycle (Pap & Frölich 1999); the ultraviolet (UV; <400 nm), however, may fluctuate by about 1% at 300 nm, 10% at 200 nm, and as much as 200% at the shortest wavelengths (Rottman 1999).

For the last four centuries, eyewitness record of sunspot numbers provide a reasonable measure of variation in solar irradiance but they say nothing about spectral changes. Earlier times are documented by the ^{14}C (radiocarbon) record of tree rings (Fig. 1), which is inversely related to solar activity, as it is produced by galactic cosmic rays (GCR) from which the atmosphere is shielded by the fluctuating solar wind; any increase in the production of solar cosmic rays when the sun is at its most active will not materially counteract this effect (Bonino et al. 2001).

Agreement with the sunspot evidence is good, whence the identification from tree-ring ^{14}C of periods of depressed solar activity other than the Maunder (1645–1715), which was first revealed by analysis of sunspot records. The minima include the Oort (AD 1020–1065), the Wolf (AD 1275–1350) and the Spörer (AD 1395–1530). However, the radiocarbon signal is also influenced by the geomagnetic field and by the many biogeochemical processes that involve atmospheric carbon (Stuiver et al. 1991). The consensus is that, apart from minor climatic effects, modulation of the GCR flux by the solar wind accounts for Δ^{14}C fluctuations with a period of c. 200 a or less, whereas further shielding by the Earth's dipole moment explains the long-term fluctuations on which the solar record is superimposed (Sternberg 1992). Once the geomagnetic signal has been allowed for, the isotopic time series can therefore be converted into changes in irradiance by linear scaling based on historical records and on studies of sun-like stars (Bard et al. 2000).

The ^{10}Be solar narrative derived from cores drilled into ice caps and glaciers (Beer et al. 1990) is also influenced by climatic and geomagnetic change (Field et al. 2006). The short atmospheric residence time ^{10}Be of 1–2 years compared with c. 20 years for ^{14}C results in better time-resolution at the price of greater variability from place to place. A number of discrepancies emerge when the results are compared with those for ^{14}C, however, including the failure of ^{10}Be to detect the Maunder Minimum (Hoyt & Schatten 1997). The need for complementary sources is further illustrated by the evidence of aurora sightings, which point to a solar minimum in about 1765 as well as the Spörer, Maunder and Dalton (Silverman 1992).

In view of these disagreements it is no surprise to find that estimates of the net increase in solar irradiance since the Maunder Minimum of AD 1645–1715 vary between 0.2 and 0.3–0.5%. Yet even the lower value, which tallies with observations of the interplanetary magnetic field extrapolated to its presumed solar sources (Lockwood & Stamper 1999), has as a corollary an increase of 0.7% in the broad UV (Lean 2000). The figures calculated by Fligge & Solanki (2000) are more

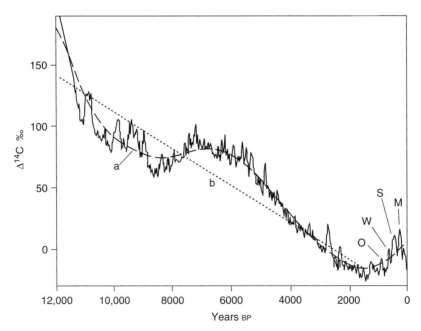

Fig. 1. Fluctuations in atmospheric Δ^{14}C for the last 10,000 years based on INTCAL87 tree-ring data and 8000 spline (**a**) after Stuiver & Braziunas (1992). Four sunspot minima are labelled: M, Maunder; S, Spörer; W, Wolf; O, Oort. A linear trend (**b**) has been added.

substantial: about 3% for the UV (<300 nm) and 1.3% for the near-ultra violet (NUV, 300–400 nm).

Heliohydrology

There are two main approaches to the study of the solar factor in climatic history. In one, fluctuations in the Sun's activity are correlated with meteorological variables, such as the 10.7 cm solar flux with the quasi-biennial cycle or $\Delta^{14}C$ with the $\delta^{18}O$ of the Camp Century Greenland ice core (Damon & Jirikowic 1992). In the other, circulation models are adjusted to incorporate changes in the radiation reaching the ground, reradiated from the ground or absorbed by atmospheric gases and by stratospheric ozone, or in the flux of GCR mediated by net rainfall or cloud cover.

The approach in this paper is to discover whether geological evidence can be used to detect anomalous solar activity. Any such evidence may reveal changes which are progressive in time or in location, and will hinge on weather systems rather than individual variables recorded at a station.

An association between solar activity and the global circulation pattern has long been suspected. By the 1920s a southward shift of the North Atlantic depression tracks had been shown to occur during sunspot maxima; the average displacement was subsequently put at 3° (Herman & Goldberg 1985). The solar cycle period emerges from records of winter storm tracks in the North Atlantic especially clearly once they are sorted according to the phase of the Quasi-Biennial Equatorial Wind Oscillation (QBO; Tinsley 1988). The trend tallies with the suggestion by historians of climate that during the Little Ice Age of 1550–1700 depressions travelled on tracks 5–10° further south than nowadays, leading to an increase in rainfall at latitudes 50–60° N (Lamb 1964). A possible mechanism for the effect was suggested by Willett (1949), who reported indirect evidence for significant variation in the UV part of the spectrum during the sunspot cycle, and inferred that the resulting circulation changes would lead to an increase by 39% in winter precipitation at sunspot maximum over much of the Mediterranean and NW Africa.

Later studies of the solar factor pointed to shifts of the weather systems in the opposite sense. Landscheidt (1987) found that increases in solar UV output would result in heating of stratospheric ozone and in an associated poleward shift in the latitude of the jet streams. The reverse effect would presumably accompany the low production of stratospheric ozone during solar (and thus UV) minima. Computer modelling by Haigh (1996) of the response of the atmosphere to the 11-a solar activity cycle using a general circulation model also found that, at solar maximum, a small increase in UV radiation in the 175–320 nm range caused substantial stratospheric heating, with some positive feedback because additional O_3 absorbs longer wavelength UV energy (Haigh 1999; Shindell et al. 1999). In both hemispheres the stratospheric winds were thereby strengthened and the tropospheric subtropical jet streams were displaced poleward. As the location of the westerly jets governs the latitudinal extent of the Hadley cells, this poleward shift resulted in a similar displacement of the descending limbs of the cells and thus of the mid-latitude storm tracks. The amount of poleward shift was in the region of 70 km (Haigh 1996) or less than 1° of latitude.

Frequency maps for winter extratropical storms during 1961–1998 show that cyclonic activity at solar maximum tends to be concentrated in the extreme north of the Atlantic (Fig. 2a), whereas at solar minimum it is more diffuse and affects large parts of the Mediterranean and the Near East (Fig. 2b). The question remains whether pronounced and prolonged changes in solar activity lead to more substantial shifts and distortion of the cyclonic belts. Evidence (Kirov & Georgieva 2002) of an inverse relationship between solar activity on a centennial timescale and both the North Atlantic Oscillation (NAO) and the El Niño-Southern Oscillation (ENSO) demonstrates that the association between solar UV and the terrestrial circulation is a complex one; yet that would seem all the more reason to identify the major strands of the knotted web.

Palaeometeorology

The recovery of changes in circulation from pre-instrumental times is making steady progress. As regards wind direction, the requisite resolution will doubtless soon be made possible at sea by isotopic analysis of corals (Shen et al. 1992) and on land by advances in the dating of aeolian deposits (Preusser et al. 2002). In favoured locations it can already be derived from the pollen evidence, especially where it is buttressed by meteorological observations.

At Honolulu, in Hawai'i, cyclic changes in wind direction help to identify the location of the Pacific anticyclone (Leopold, pers. comm. 2005). Instrumental data show that the northeast trade wind changed during the period of record, which began in 1907, to a more easterly sector, arguably because the average track of the anticyclones in the northeast Pacific had shifted south in the winter season until 1936 (Beale 1927; Wentworth 1949). It reached its most easterly direction about 1939 (when the international sunspot index shows

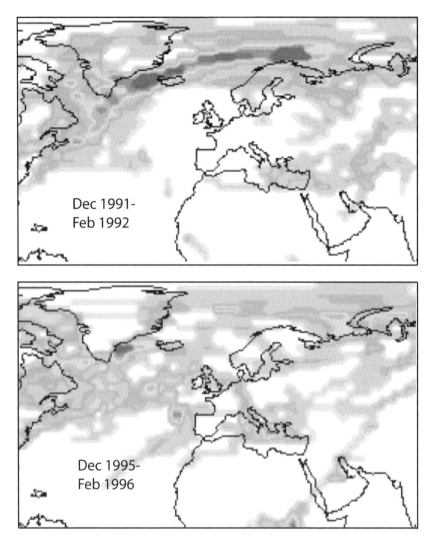

Fig. 2. North Atlantic depression frequency plots for three winter months (December, January and February) near sunspot maximum (1991, second peak) and minimum (1996) for Solar Cycle 22 (http://data.giss.nasa.gov/stormtracks). Courtesy of the Goddard Institute for Space Studies.

a maximum) and then sharply reversed to become as northwesterly as it had been in 1907. Corresponding changes in rainfall appeared to come from shifts in the isohyetal pattern best explained by a movement of the Pacific high or a secular change in the height of the trade wind subsidence inversion (Leopold 1949).

Pollen studies at a number of sites reflected earlier such precipitation changes resulting from vertical displacement of the upper boundary of the mid-Pacific Trade Wind Inversion (TWI) cloud layer (Selling 1945; Hotchkiss & Juvik 1999; Burney *et al.* 1995). The anticyclone moves east and south in winter (Fig. 3); according to Leopold (pers. comm. 2005), if these winter conditions were to become more frequent or more pronounced (with the Pacific High occupying a position that in today's climate is confined to winter), the wind over East Maui would not be drained of moisture as it moved up over the mountain, the rain forest would gradually move to a higher elevation, and bogs could then form near the mountaintop.

Few other locations combine palaeobotanical data with appropriate meteorological observations over several decades. Fluvial activity, unlike wind, is an indirect guide to the atmospheric circulation, but in any derived alluvial deposits it offers

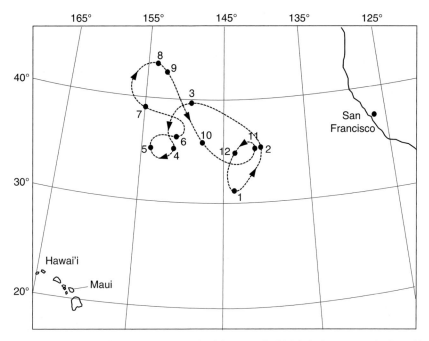

Fig. 3. Mean position of the Pacific High at each month of the year. The high is in the most southerly position in January (numeral 1) and in the most northerly in August (numeral 8). After Beale (1927).

the potential for stratigraphic, datable reconstruction which refers to specific latitudes and altitudes and can therefore be used to trace regional variations and gradients.

The latest valley fill above the modern floodplain in numerous river basins in Europe and North Africa (Fig. 4a) yields ages of about AD 500–1900. A broadly contemporaneous deposit accumulated in North America (Leopold & Vita-Finzi 1998). Early work on the subject depended for its chronology mainly on archaeological and historical dating, which only provided limiting ages. Recent years have witnessed great progress in the dating and interpretation of Mediterranean alluvial sequences, and this is reflected in Table 1 and Figure 4b. Regrettably a number of ^{14}C dates for Spanish rivers could not be included because they refer to features described as slackwater deposits or palaeofloods rather than a single fill which could be interpreted in the context of the other sites listed here.

Table 1 shows 24 more ^{14}C ages than in an earlier compilation (Vita-Finzi 1995a) and it spans some 17.5 degrees of latitude. The dates refer to a variety of depths in the fill and of locations within the drainage basins, and no correction for elevation has been attempted. As in earlier plots, aggradation appears to have been time-transgressive with respect to latitude. At one location (Hasan Langhi, Iran, in Fig. 4b) three successive measurements show that 2 m of alluvium accumulated in about 700 years.

The deposits listed in Table 1 are sometimes bipartite. They invariably display a high degree of sorting, with coarse silt and fine sand often dominant, well-developed stratification and drab colours (often in contrast with the red haematitic pigment of its source deposits), all indicative of syn- or post-depositional reducing conditions and sustained flow in channels now or previously characterized by ephemeral or strongly seasonal discharge. The resulting silt-clay depletion (Vita-Finzi 1971) would have given rise offshore to material comparable to the clay-rich flood deposits that have been identified in 2000 years of deposition off California (Schimmelmann et al. 2002). The associated regime was equable, a term used here to indicate reduced extremes of discharge.

The latitudinal diachronism indicated by the ^{14}C ages (Fig. 4b) was initially taken to reflect a progressive but temporary southward shift in the cyclonic belts of the Northern Hemisphere (Vita-Finzi 1995b), which had provoked an increase in the proportion of cyclonic to convection rainfall, that is to say a decrease in the importance of erosive rains (Leopold 1951). The diachronism persists in the new plot, though with reduced gradient, and the excellent match between the timing of the phase of

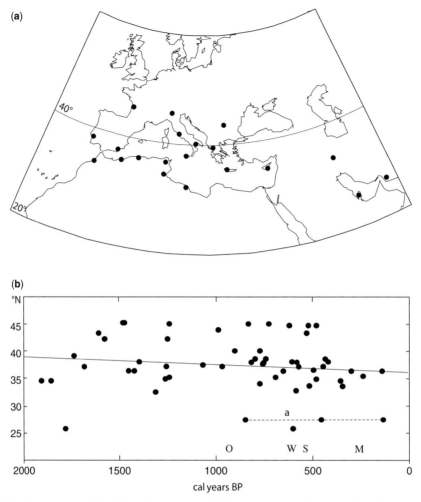

Fig. 4. (a) Regions represented in Table 1. (b) Latitude/cal ^{14}C age plot for sites listed in Table 1. Sample TX-7633 (Cacchiavia) lies close to the trend line but has been omitted to avoid unduly stretching the plot. O, W, S and M indicate approximate timing of Oort, Wolf, Spörer and Maunder minima respectively.

valley filling and the sequence of solar minima now provides a plausible mechanism. At some of the sites listed in Table 1 aggradation includes one or at most two hiatuses; the ^{14}C curve suggests that there should be three or four. However, the time-transgressive nature of the deposit could imply that the earlier solar minima were manifested as aggradation only in the north and the later ones only in the south.

Maas & Macklin (2002) have argued that there is a probable Mediterranean-wide correlation between large-scale climatic/oceanic circulation and river hydrodynamics, the link being provided by changing storm frequency and magnitude. On Crete, analysis of daily precipitation data suggests that negative phases of the winter NAO are characterized by an increase in the number of prolonged, high-intensity storms. Such storms, particularly those with five-day and greater duration, appear to be significant in triggering major floods in the Aradena Gorge. According to Trigo et al. (2000), it is a reduction not in the totality but in the intensity of cyclogenesis (associated with a northward shift in the main Atlantic storm tracks) that accounts for a decline in wet-season rainfall over the Mediterranean in 1979–1996. Modelling by Tucker & Slingerland (1997) shows how an increase in runoff intensity may lead to a rapid expansion of the channel network, with the resulting increase in sediment supply generating aggradation along the main

Table 1. ^{14}C ages for the last Mediterranean valley fill

Location	Latitude	^{14}C age years BP	cal age years BP[a]	Laboratory no.	Reference
Algeria					
El Abadia	36°17'	700 ± 80	650 ± 70	MC-2530	2
El Abadia	36°17'	120 ± 70	140 ± 110	BETA-2678	2
El Abadia	36°17'	270 ± 70	300 ± 130	MC-2531	2
Chéria-Mezera	34°52'	1350 ± 70	1260 ± 65		7
Cyprus					
Vasilikos	34°50'	470 ± 80	480 ± 90	OxA-804	2
Gialias	35°09'	735 ± 40	690 ± 120	Poz-1396	13
Gialias	35°09'	1325 ± 65	1240 ± 60	A-45457	13
France					
Manaurie	44°57'	1335 ± 90	1240 ± 85	I-16794	2
Greece					
Voidomatis	40°00'	1000 ± 50	900 ± 60	OxA-191	6
Voidomatis	40°00'	800 ± 100	770 ± 95	OxA-192	6
Omalos (Crete)	35°19'		240 ± 25[b]	BETA-80842	9
Iran					
Khorramabad	33°30'	330 ± 105	345 ± 130	I-3214	2
H. Langhi	27°21'	110 ± 90	135 ± 115	HAR-1114	2
H. Langhi	27°21'	460 ± 170	455 ± 155	HAR-1707	2
H. Langhi	27°21'	930 ± 80	845 ± 75	HAR-1706	2
Italy					
Fiano Romano	42°12'	1670 ± 95	1575 ± 115	I-4801	2
Fiano Romano	42°12'	1350 ± 100	1250 ± 100	I-4802	2
Valchetta	37°25'	1140 ± 160	1065 ± 160	I-881	2
Crescenza	37°25'	1400 ± 100	1310 ± 95	I-882	2
Moncalieri	44°59'	790 ± 50	725 ± 35	R-618A	2
Moncalieri	44°59'	900 ± 50	830 ± 65	R-618	2
C. Monferrato	45°10'	1595 ± 50	1480 ± 55	R-622A	2
C. Monferrato	45°10'	1580 ± 50	1470 ± 50	R-622α	2
Treia	43°18'	1675 ± 45	1605 ± 60	I-6109	2
Treia	43°18'	490 ± 30	530 ± 10	I-6110	2
Tresinaro	44°41'		520 ± 70[b]		8
Tresinaro	44°41'		620 ± 50[b]		8
Tresinaro	44°41'		480 ± 60[b]		8
Caccchiavia	39°05'	1810 ± 70	1735 ± 90	TX-7632	5
Caccchiavia	39°05'	2800 ± 60	2915 ± 75	TX-7633	5
Libya					
Ganima	32°42'	610 ± 110	585 ± 85	Q-656	2
Morocco					
Mellah	33°34'	490 ± 90	515 ± 95	I-2693	2
Bou Regreg	33°58'	800 ± 200	770 ± 170	L-398A	2
Beth	34°30'	280 ± 60	355 ± 80	GrN-2198	2
Sidi Kacem II	34°30'	1900 ± 30	1855 ± 30	GrN-5571	2
Sidi Kacem I	34°30'	1950 ± 30	1905 ± 30	GrN-5572	2
Portugal					
Boi	37°50'	780 ± 50	750 ± 135	BETA-2909	3
Estoi	37°05'	520 ± 60	570 ± 50	UM-1332	3
Porches	37°06'	1750 ± 90	1680 ± 110	ANU-2815	3
Qatar					
Ras Abaruk	25°43'	600 ± 90	600 ± 55	HAR-639	2
Bir Markhiyah	25°43'	1850 ± 90	1780 ± 110	HAR-638	2
Rumania					
Teleorman	43°52'	1050 ± 60	985 ± 60	BETA-147288	12

(Continued)

Table 1. Continued

Location	Latitude	^{14}C age years BP	cal age years BP^a	Laboratory no.	Reference
Spain					
Ayna	38°31′	820 ± 35	740 ± 30	SSR-726	2
Ayna	38°31′	850 ± 50	795 ± 70	SRR-727	2
Ayna	38°36′	780 ± 110	755 ± 100	SRR-728	2
Ayna	38°29′	420 ± 65	435 ± 80	SRR-729	2
Aguas	37°08′	1040 ± 40	965 ± 35	Cam-59178	4
Aguas	37°08′	430 ± 50	445 ± 70	BETA-100599	4
Rambla Ancha	37°08′	1340 ± 50	1255 ± 50	BETA-100600	4
Carboneras	36°59′	390 ± 60	420 ± 75	BETA-84855	10
Librilla	37°53′	550 ± 75	580 ± 50	OxA-5372	14
Librilla	37°53′	880 ± 65	815 ± 75	AA-22067	14
Tunisia					
R'mel	36°21′		1420 ± 25^b		1
R'mel	36°21′		1450 ± 30^b		1
el Akarit	37°58′	1470 ± 190	1395 ± 200		7
el Akarit	37°58′	610 ± 110	605 ± 65		7
Medjerda	36°27′	440 ± 25	495 ± 30	BETA-135719	11

^{a}Calibrated using CalPal2005_SFCP (courtesy of Universität zu Köln, September 2005) for consistency except when only the calibrated age was available. ^{b}The mean of the 1 SD calibrated range is used here as the BP central value. Sites for which only limiting maxima are cited (e.g. Cavone in ref. 5) are omitted. As authors do not always specify the sample material or whether the age has been normalized, these matters are not tabulated. Some laboratory numbers could not be traced.
References: 1, Mørch 1995; 2, Vita-Finzi 1995a; 3, Devereux 1983; 4, Schulte 2002; 5, Abbott & Valastro 1995; 6, Lewin et al. 1991; 7, Ballais 1995; 8, Veggiani 1983; 9, Maas et al. 1998; 10, Bell et al. 1997; 11, Faust et al. 2004; 12, Howard et al. 2003; 13, Devilliers 2005; 14, Calmel-Avila 2002.

network followed by downcutting as the sediment supply tapers off.

In the southern Mediterranean, lightning-producing storms are responsible for the bulk of the rainfall (Adamo et al. 2003); in the central and eastern Mediterranean, space-based instruments showed that in 1998–2003 the monthly and seasonal correlation coefficients between rainfall and lightning varied between 0.81 and 0.98 (Price & Ferdermesser 2006). The solar wind inhibits the flux of GCR. Although several studies have demonstrated a link between cosmic ray flux and cloud cover, the underlying process remains elusive (see e.g., Kristjánsson et al. 2004; Harrison & Stephenson 2005). However there is general acceptance that charged particles are effective cloud condensation nuclei (Pruppacher & Klett 1997; Tinsley & Heelis 1993). As the flux of GCR varies with the 11-a cycle (Reid 1986), one might expect their role to be enhanced during prolonged intervals of depressed solar activity.

Discussion

The dipole is favoured as the main control of long-term ^{14}C fluctuations to some extent because they tally with those derived from ^{10}Be in ice cores and with the outcome of box-core models (e.g. Stuiver et al. 1991). Yet agreement with ^{10}Be could arise just as well from a common heliomagnetic origin; there is no theoretical reason why a sine-type curve should characterize fluctuations in the dipole, and the results of modelling the dipole from ^{14}C data (Sternberg 1992) are particularly unimpressive for the last 2000 years. Moreover there are substantial inconsistencies in the dipole data for Europe, Asia and other parts of the world (Yang et al. 2000). In short, the secular decline on which the sinusoidal spline is impaled (Fig. 1b) could just as well reflect a corresponding increase in solar activity which culminated in about AD 500 and gave way to a spell of solar minima.

The data would then point to a solar oscillation (it would be premature to term it a periodicity) with a $\lambda/2$ of about 5000 years. Some such phenomenon was suspected by Russell (1975), when he noted that the geomagnetic aa index showed a 63% increase from solar cycle 13 to solar cycle 18. The decline in GCR flux implied by an increased solar magnetic flux at the close of the post-AD 500 minimum is corroborated by the ^{44}Ti of H chondrites that fell between 1810 and 1997 and by ^{10}Be data from the Dye3 core in Greenland (Cini Castagnoli et al. 2004).

Confirmation of a link between fluvial regimes and GCR would encourage the search for other kinds of proxy solar history in the geological record. The implications for climatology and environmental analysis are equally clear, as the solar component of meteorological change – and consequently of river behaviour and soil degradation – would be clarified, and any regularities in its fluctuation harnessed to forecast flood and drought.

I thank Adrian Harvey for helpful criticisms, and Mark Chandler and Jeff Jonas, Columbia University, Goddard Institute for Space Studies, for permission to reproduce two of their storm track plots. This paper is dedicated to the memory of Luna B. Leopold.

References

ABBOTT, J. T. & VALASTRO, S. JR. 1995. The Holocene alluvial records of the chorai of Metapontum, Basilicata, and Croton, Calabria, Italy. *In*: LEWIN, J., MACKLIN, M. G. & WOODWARD, J. C. (eds) *Mediterranean Quaternary River Environments*. Balkema, Rotterdam, 195–205.

ADAMO, C., SOLOMON, R., GOODMAN, S., CECIL, D., DIETRICH, S. & MUGNAI, A. 2003. Lightning and precipitation. *Proceedings of 3rd Plinius Conference on Mediterranean Storms*, Ajaccio, France.

BALLAIS, J.-L. 1995. Alluvial Holocene terraces in eastern Maghreb: Climate and anthropogenic controls. *In*: LEWIN, J., MACKLIN, M. G. & WOODWARD, J. C. (eds) *Mediterranean Quaternary River Environments*. Balkema, Rotterdam, 183–194.

BARD, E., RAISBECK, G., YIOU, F. & JOUZEL, J. 2000. Solar irradiance during the last 1200 years based on cosmogenic nuclides. *Tellus*, **52**, 985–992.

BEALE, E. A. 1927. The northeast trade winds of the North Pacific. *Monthly Weather Review*, **55**, 211–221.

BEER, J. ET AL. 1990. Use of ^{10}Be in polar ice to trace the 11-year cycle of solar activity. *Nature*, **347**, 164–166.

BELL, J. W. S., AMELUNG, F. & KING, G. C. P. 1997. Preliminary late Quaternary slip history of the Carboneras fault, southeastern Spain. *Journal of Geodynamics*, **24**, 51–66.

BONINO, G., CINI CASTAGNOLI, G., CANE, D., TARICCO, C. & BHANDARI, N. 2001. Solar modulation of the galactic cosmic ray spectra since the Maunder minimum. *In*: SIMON, M., LORENZ, E. & POHL, M. (eds) *Proceedings of the 27th International Cosmic Ray Conference*, Hamburg, Copernicus Gesellschaft, Katlenburg-Lindau, Germany, 3769–3772.

BURNEY, R. V., DECANDIDO, L. P., BURNEY, F. N., KOST EL-HUGHES, T. W., STAFFORD, T. W. JR & JAMES, H. F. 1995. A Holocene record of climatic change, fire ecology, and human activity from montane Flat Top Bog, Maui. *Journal of Paleolimnology*, **13**, 209–217.

CALMEL-AVILA, M. 2002. The Librilla 'rambla', an example of morphogenetic crisis in the Holocene (Murcia, Spain). *Quaternary International*, **93–94**, 101–108.

CINI CASTAGNOLI, G., CANE, D., TARICCO, C. & BHANDARI, N. 2004. GCR flux decline during the last three centuries: extraterrestrial and terrestrial evidences. *In*: KAJITA, T., ASAOKA, Y., KAWACHI, A., MATSUBARA, Y. & SASAKI, M. (eds) *Proceedings of the 28th International Cosmic Ray Conference*, Tsukuba, Japan. Universal Academy Press, Tokyo, 4045–4048.

DAMON, P. E. & JIRIKOWIC, J. L. 1992. Solar forcing of climatic change? *In*: TAYLOR, R. E., LONG, A. & KRA, R. S. (eds) *Radiocarbon after Four Decades*. Springer, New York, 117–129.

DEVEREUX, C. M. 1983. Recent erosion and sedimentation in southern Portugal. PhD thesis, University of London.

DEVILLERS, B. 2005. Morphogenèse et anthropisation holocènes d'un bassin versant semi-aride: le Gialias, Chypre. Unpublished PhD thesis, University Aix-Marseille I, Marseille.

FAUST, D., ZIELHOFER, C., BAENA, E. R. & DIAZ DEL OLMO, F. 2004. High-resolution fluvial record of late Holocene geomorphic change in northern Tunisia: climatic or human impact? *Quaternary Science Reviews*, **23**, 1757–1775.

FIELD, C. V., SCHMIDT, G. A., KOCH, D. & SALYK, C. 2006. Modeling production and climate-related impacts on ^{10}Be concentration in ice cores. *Journal of Geophysical Research*, **111**, D15107, doi:10.1029/2005JD006410.

FLIGGE, M. & SOLANKI, S. K. 2000. The solar spectral irradiance since 1700. *Geophysical Research Letters*, **27**, 2157–2160.

HAIGH, J. D. 1996. The impact of solar variability on climate. *Science*, **272**, 981–984.

HAIGH, J. D. 1999. Modelling the impact of solar variability on climate. *Journal of Atmospheric and Solar-Terrestrial Physics*, **61**, 63–72.

HARRISON, R. G. & STEPHENSON, D. B. 2006. Empirical evidence for a nonlinear effect of galactic cosmic rays on clouds. Proceedings of the Royal Society A, doi:10.1098/rspa.2005.1628.

HERMAN, J. R. & GOLDBERG, R. A. 1985. *Sun, Weather and Climate* (first published by NASA, Washington, 1978). Dover, New York.

HESSLER, A. M., LOWE, D. R., JONES, R. O. L. & BIRD, D. K. 2004. A lower limit for atmospheric carbon dioxide levels 3.2 billion years ago. *Nature*, **428**, 736–738.

HOTCHKISS, S. & JUVIK, J. O. 1999. A late-Quaternary pollen record from Ka'au Crater, O'ahu, Hawaii. *Quaternary Research*, **52**, 115–128.

HOWARD, A. J., MACKLIN, M. G., BAILEY, D. W., MILLS, S. & ANDREESCU, R. 2004. Late-glacial and Holocene river development in the Teleorman Valley on the southern Romanian Plain. *Journal of Quaternary Science*, **19**, 271–280.

HOYT, D. V. & SCHATTEN, K. H. 1997. *The Role of the Sun in Climate Change*. Oxford University Press, New York.

KIROV, B. & GEORGIEVA, K. 2002. Long-term variations and interrelations of ENSO, NAO and solar activity. *Physics and Chemistry of the Earth*, **27**, 441–448.

KRISTJÁNSSON, J. E., KRISTIANSEN, J. & KAAS, E. 2004. Solar activity, cosmic rays, clouds and climate – an update. *Advances in Space Research*, **34**, 407–415.

LAMB, H. H. 1964. Climatic changes and variations in the atmospheric and ocean circulations. *Geologische Rundschau*, **54**, 486–504.

LANDSCHEIDT, T. 1987. Long-range forecasts of solar cycles and climatic change. *In*: RAMPINO, M. R., SANDERS, J. E., NEWMAN, W. S. & KÖNIGSSON, L. K. (eds) *Climate History, Periodicity and Predictability*. Van Nostrand Reinhold, New York, 421–445.

LEAN, J. 2000. Evolution of the Sun's spectral irradiance since the Maunder Minimum. *Geophysical Research Letters*, **27**, 2425–2428.

LEOPOLD, L. B. 1949. The annual rainfall of East Maui. *The Hawaiian Planters' Record*, **53**, 47–63.

LEOPOLD, L. B. 1951. Rainfall frequency: an aspect of climatic variation. *Transactions of the American Geophysical Union*, **32**, 347–357.

LEOPOLD, L. B. & VITA-FINZI, C. 1998. Valley changes in the Mediterranean and America and their effects on humans. *Proceedings of the American Philosophical Society*, **142**, 1–17.

LEWIN, J., MACKLIN, M. G. & WOODWARD, J. C. 1991. Late Quaternary fluvial sedimentation in the Voidomatis basin, Epirus, Northwest Greece. *Quaternary Research*, **35**, 103–115.

LOCKWOOD, M. & STAMPER, R. 1999. Long-term drift in the coronal source magnetic flux and the total solar irradiance. *Geophysical Research Letters*, **26**, 2461–2464.

MAAS, G. S. & MACKLIN, M. G. 2002. The impact of recent climate change on flooding and sediment supply within a Mediterranean mountain catchment, southwestern Crete, Greece. *Earth Surface Processes and Landforms*, **27**, 1087–1105.

MAAS, G. S., MACKLIN, M. G. & KIRKBY, M. J. 1998. Late Pleistocene and Holocene river development in Mediterranean steepland environments, southwest Crete, Greece. *In*: BENITO, G., BAKER, V. R. & GREGORY, K. J. (eds) *Palaeohydrology and Environmental Change*. Wiley, Chichester, 153–165.

MØRCH, H. F. C. 1995. A dating of a valley fill in NE-Tunisia – an analysis of radio carbon from the Oued R'mel basin. *In*: DIETZ, S., SEBAÏ, L. L. & BEN HASSEN, H. (eds) *Africa Proconsularis I*, Carlsberg Foundation and Danish Research Council for the Humanities, Copenhagen, 51–55.

PAP, J. M. & FRÖLICH, C. 1999. Total solar irradiance variations. *Journal of Atmospheric and Solar-Terrestrial Physics*, **61**, 15–24.

PREUSSER, F., RADIES, D. & MATTER, A. 2002. A 160,000-year record of dune development and atmospheric circulation in southern Arabia. *Science*, **296**, 2018–2020.

PRICE, C. & FEDERMESSER, B. 2006. Lightning-rainfall relationships in Mediterranean winter thunderstorms. *Geophysical Research Letters*, **33**, L07813, doi:10.1029/2005GL024794.

PRUPPACHER, H. R. & KLETT, J. D. 1997. *Microphysics of Clouds and Precipitation*, 2nd edn. Kluwer, Dordrecht.

REID, G. C. 1986. Electrical structure of the middle atmosphere. *In*: *The Earth's Electrical Environment*, The National Academies Press, Washington, DC, 183–194.

ROTTMAN, G. 1999. Solar ultraviolet irradiance and its temporal variation. *Journal of Atmospheric and Solar-Terrestrial Physics*, **61**, 37–44.

RUSSELL, C. T. 1975. On the possibility of deducing interplanetary and solar parameters from geomagnetic records. *Solar Physics*, **42**, 259–269.

SCHIMMELMANN, A., LANGE, C. B. & LI, H.-C. 2002. Major flood events of the past 2,000 years recorded in Santa Barbara Basin sediment. *In*: WEST, G. J. & BUFFALOE, L. D. (eds) *Proceedings of the Eighteenth Annual Pacific Climate (PACLIM) Workshop*, California Department of Water Resources, Technical Report 68 of the Interagency Ecological Program for the San Francisco Estuary, 83–103.

SCHULTE, L. 2002. Climatic and human influence on river systems and glacier fluctuations in southeast Spain since the Last Glacial Maximum. *Quaternary International*, **93–94**, 85–100.

SELLING, O. H. 1945, *Studies in Hawaiian Pollen Statistics, Part III*. Bernice P. Bishop Museum, Special Publication **39**.

SHEN, G. T., LINN, L. L., CAMPBELL, T. M. & FAIRBANKS, R. G. 1992. A chemical indicator of westerly winds in corals from the central tropical Pacific. *Journal of Geophysical Research*, **97**, 12689–12697.

SHINDELL, D., RIND, D., BALACHANDRAN, N., LEAN, J. & LONERGAN, P. 1999. Solar cycle variability, ozone, and climate. *Science*, **284**, 305–308.

SILVERMAN, S. M. 1992. Secular variation of the aurora for the past 500 years. *Reviews of Geophysics*, **30**, 333–351.

STERNBERG, R. S. 1992. Radiocarbon fluctuations and the geomagnetic field. *In*: TAYLOR, R. E., LONG, A. & KRA, R. S. (eds) *Radiocarbon after Four Decades*. Springer, New York, 93–116.

STUIVER, M. & BRAZIUNAS, T. F. 1992. Evidence of solar activitry variations. *In*: BRADLEY, R. S. & JONES, P. D. (eds) *Climate Since AD 1500*. Routledge, London, 593–605.

STUIVER, M., BRAZIUNAS, T. F., BECKER, B. & KROMER, B. 1991. Late-glacial and Holocene atmospheric $^{13}C/^{12}C$ change: climate, solar, oceanic and geomagnetic influences. *Quaternary Research*, **35**, 1–24.

TINSLEY, B. A. 1988. The solar cycle and the QBO influences on the latitude of storm tracks in the North Atlantic. *Geophysical Research Letters*, **15**, 409–412.

TINSLEY, B. A. & HEELIS, R. A. 1993. Correlations of atmospheric dynamics with solar activity: evidence for a connection via the solar wind, atmospheric electricity, and cloud microphysics. *Journal of Geophysical Research*, **98**, 10375–10384.

TRIGO, I. F., DAVIES, T. D. & BIGG, G. R. 2000. Decline in Mediterranean rainfall caused by weakening of Mediterranean cyclones. *Geophysical Research Letters*, **27**, 2913–2916.

TUCKER, G. E. & SLINGERLAND, R. 1997. Drainage basin responses to climate change. *Water Resources Research*, **33**, 2031–2047.

VEGGIANI, A. 1983. Degrado ambientale e dissesti idrogeologici indotti dal deterioramento climatico nell'alto medioevo in Italia. I casi riminesi. *Studi Romagnoli*, **34**, 123–146.

VITA-FINZI, C. 1971. Heredity and environment in clastic sediments: silt/clay depletion. *Bulletin of the Geological Society of America*, **82**, 187–190.

VITA-FINZI, C. 1995a. Medieval mud and the Maunder minimum. *Paläoklimaforschung*, **16**, 95–103.

VITA-FINZI, C. 1995b. Solar history and paleohydrology during the last two millennia. *Geophysical Research Letters*, **22**, 699–702.

VITA-FINZI, C. 2002. *Monitoring the Earth*. Terra, Harpenden.

WENTWORTH, C. K. 1949. Directional shift of trade winds at Honolulu. *Pacific Science*, **3**, 86–92.

WILLETT, H. C. 1949. Solar variability as a factor in the fluctuations of climate during geological time. *Geografiska Annaler*, **31**, 295–315.

YANG, S., ODAH, H. & SHAW, J. 2000. Variations in the geomagnetic dipole moment over the last 12 000 years. *Geophysical Journal International*, **140**, 158–162.

Exploring the links between sediment character, bed material erosion and landscape: implications from a laboratory study of gravels and sand–gravel mixtures

LYNNE FROSTICK, BRENDAN MURPHY & RICHARD MIDDLETON

Hull Environment Research Institute (HERI) and Department of Geography, University of Hull, Hull HU6 7RX, UK (e-mail: l.e.frostick@hull.ac.uk)

Abstract: Rates of landscape evolution and landform development depend on the capacity of the main transporting medium, predominantly water and the river system, to move material away from the site of production in the upland slopes. In upland areas the main river bed material is coarse gravels, with various admixtures of finer sand and silt. This paper reports a series of flume experiments to investigate the impact of admixtures of finer material on the entrainment of coarse particles. In all experiments the main framework of the bed was made up of quartz-density gravels with a mean particle size of 8 mm. In some of the experiments unimodal, 0.09 mm mean particle size, quartz sands were introduced upstream of the experimental section and transported into place in order to simulate a common condition in natural river beds, of sand migrating over a static gravel bed. New image analysis techniques were developed to extract data from video recordings of the experimental runs. These revealed important differences in entrainment processes among the experiments with a distinctive contrast between the clean gravel and sand–gravel runs. Observations suggest that the presence of sand increases the rates of gravel entrainment and leads to a distinctive patchiness in bed break-up which will encourage bed form development. In the mixtures, sand removal prior to gravel entrainment destabilizes the bed and allows large areas to become entrained. This contrasts with clean gravels where grains tend to entrain individually. These observations show the importance of bed material character in controlling river form and process and point to its role in controlling sediment flux through the landscape.

Catastrophic, infrequent and large-scale events, for example major fault movements and large volcanic eruptions (see e.g. Frostick & Reid 1989, 1990; Clavero *et al.* 2004; Chorowicz *et al.* 2005), have an immediate visual impact on a landscape, creating new topography and fresh rock surfaces. These change the energy balance in the landscape and result in new patterns of weathering, erosion and deposition which begin a new cycle of landscape development. Although low-frequency large-scale events exert overall control on surface topography, at medium to small scales both the character of the land surface and the location and character of sedimentary deposits are direct consequences of the spatial patterns of sediment production and transfer. These are the timescales which are generally significant in the lives of human beings, since events with recurrence intervals in excess of 10^2 may change a landscape dramatically but have a low probability of occurring in a single life span.

In particular the movement of coarse sediment as bed load through a river system and, by implication, the storage and preservation of the same sediment as fluvial deposits control the character of the local river landscape and the morphology of the alluvial valley. Understanding the variability in these processes is core to the predictive modelling of landscape evolution, as well as to the interpretation of ancient river deposits. However, relatively little is known about the character of, and controls on, the spatial distribution of bed load transport, largely because it is a process which is difficult to measure under field conditions (Reid & Frostick 1984; Reid *et al.* 1985; Rennie & Millar 2004). This is because it occurs when the river is in flood and the water is not only opaque, but also deep, fast flowing and dangerous.

One alternative is to infer bed load fluxes from morphological change (e.g. Lane *et al.* 1995), but this can only give time-integrated 'minimum' values that are not appropriate for all applications. Where field measurements of bed load transport are made, they are generally made using point samplers, either portable (e.g. Helley-Smith; Emmett 1980; Habersack & Laronne 2002) or installed in the bed of the stream (e.g. Birkbeck bed load samplers; Reid *et al.* 1980) with the results extrapolated stream-wide. These have revealed strong temporal variability (Einstein 1942; Reid *et al.* 1985; Reid & Frostick 1985;

Gomez 1991; Frostick & Jones 2002), but have done little to elucidate stream-wide spatial patterns of entrainment. Yet we know there is a predictable relationship between topography on the bed of the stream, near-bed turbulence (Nelson et al. 1995) and three-dimensional flow patterns (Lane 1998), which should translate into spatially variable bed load transport rates.

The work of Rennie & Millar (2004) is almost unique in having explored this problem in both sand and gravel bed reaches of the Fraser river. Their results show a good relationship between bed load velocity and both near-bed and depth-averaged water velocity. However they concentrate on a single river where the D_{50} is assumed to be representative of the grain size of the bed and also where water velocity is uncontrolled and will inevitably change throughout the data collection period.

The work of Wilcock & McArdle (1997) and Wilcock et al. (2001) has shown that bed load transport rates increase as gravel and sand sized sediments are mixed, although the reason for this remains obscure. If the effect of grain size on transport rates is to be incorporated into predictive models, there is a need for experimentation to investigate why sand–gravel mixtures behave differently from unimodal gravels. Some progress has been made in determining the impact on near-bed flow of adding sand to gravel. Of particular interest is the work by Sambrook Smith & Nicholas (2005), who used particle imaging velocimetry to quantify the two-dimensional flow field of a gravel bed with and without sand. Their experiments showed that shear stress and turbulent kinetic energy decreased whilst near-bed velocity increased in the presence of sand. They also noted the suppression of turbulence well away from the bed. However, they did not examine the impact of these changes on entrainment processes, sediment flux and reach scale bed characteristics. This paper describes a series of experiments designed to visualize the processes of entrainment for gravels and sand–gravel mixtures in order to identify differences which might help account for enhanced rates of bed load transport in bimodal sediments. It also examines the many implications of this work for landscape evolution at the reach to drainage basin scales.

Methods

Experimental programme

Experiments were carried out in a standard Armfield S6 glass-sided tilting flume (12 × 0.3 × 0.45 m) with a 3 m working section sufficiently distant from inlet and outlet to ensure approximately uniform flow conditions. The programme of experiments was designed to reproduce three commonly occurring perennial river scenarios (Table 1):

1. Inter-flood low flows – these flows were used to condition the bed to simulate more closely the bed of a river at the end of a flood. The sand matrix material was also introduced during these flows but upstream of the experimental section. It was then allowed to migrate into the experimental reach until it filled all of the surface pore spaces but did not cover the framework gravels. Feeding the sand in from upstream gave a fair representation of conditions towards the end of a flood and the bed was than considered ready to undertake the remainder of the experiments. Throughout this phase flows were maintained at levels well below the theoretical critical value of shear velocity (u_c^*) for the gravel material (0.45 m s^{-1}) but above that for the movement of sand particles. The mean velocity at 0.6 of the flow depth was less than 0.4 m s^{-1} at all times. This simulates between-flood flow conditions in a flashy perennial river where the bed does not undergo major disturbance.
2. Small floods – during these experiments, conditions were close to the threshold for gravel movement and well above the threshold for sand transport. The peak shear velocities were just in excess of 0.45 m s^{-1}, close to u_c^* for the gravel. This simulates flows where the gravel particles are disturbed to a limited extent, causing 'fluttering', but not displacement. These flows were used to condition the gravel and sand–gravel beds to approximate more closely to those of natural rivers prior to the entrainment experiments
3. Large floods – during these experiments both sand and gravel were entrained and the bed underwent extensive erosion. Flow conditions were maintained well above the threshold for gravel transport and shear velocities reached values around 0.6 m s^{-1}.

All experiments were carried out under uniform and sub-critical flow conditions with Froude numbers ranging between 0.3 and 0.9. The higher values were reached only in the third set of experiments when the gravels were entrained and values for the other two sets of experiments ranged between 0.3 and 0.5.

A total of 33 experiments were carried out during the two periods of experimentation, eight of which focused directly on the contrast between unimodal and bimodal sediments and are reported here.

Table 1. *Experimental conditions*

Experiment	Maximum shear velocity, m s^{-1}	Mean bed slope at the start of experiments	Average water depth (m)	Froude range
(1) Inter flood	0.39	0.017	0.15	0.3–0.5
(2) Small flood	0.47	0.017	0.19	0.3–0.5
(3) Large flood	0.62	0.017	0.25	0.8–0.9

Bed material

In the unimodal experiments the bed was made up of quartz-density gravels with a mean grain size of 8 mm. This was introduced into the flume by hand and then water worked at velocities close to the entrainment threshold (experiment 2 above) prior to carrying out the framework entrainment experiments (experiment 3). The material included white-painted clasts marked with black dots (Brasington *et al.* 2000; Middleton *et al.* 2000, and Fig. 1) to facilitate digital video recording and tracking of individual particle entrainment through the side wall of the flume. In the sand and gravel mixture experiments a unimodal, 0.09 mm quartz sand was introduced upstream of the experimental section and transported into place under flows below the entrainment threshold for the larger gravel material (experiment 1 above). This simulated a common condition in natural river beds, that of a finer material migrating over a static gravel bed on the recession limb of a flood hydrograph or during a small flood.

During the early experiments, which were partially reported in Allan & Frostick (1999), the bed was frozen and extracted in sections to assess the impact of side wall interference on sand movement. The results showed no significant difference in the quantity, distribution and grain size of the both sands and gravels between the side and the centre of the bed by the end of the experiments. The conclusion from this work was that side wall effects on the sedimentary processes being studied were small and movements recorded through the side wall of the flume can be considered to be reasonably representative of the bed as a whole.

Image analysis techniques

Several image analysis techniques were developed to process the digital video data collected during these and other experiments. The first is a development of earlier software (Middleton *et al.* 2000) for tracking the movement of marked particles through progressive video frames. The output from the program is a file of co-ordinates, indicating the position of the target black dots on individual particles throughout the experiment.

A technique for identifying significant locations of particle movement using between-frame differences in pixel values has already been demonstrated under the same conditions as those used in these experiments (Brasington *et al.* 2000). This compares identical pixels in sequential images from a video recording and then designates each pixel as white, if static, or black if there has been movement beyond a pre-set threshold. This generates a sequence of images highlighting the areas of change with black fringes and allowing the spatial patterns of movement to be picked out and analysed for any period of time in the experiment.

Fig. 1. Image (extracted from digital video through glass sidewall of flume) of the bed material used in the experiments showing the white particles with black dots used to track and analyse the character of individual grain movements.

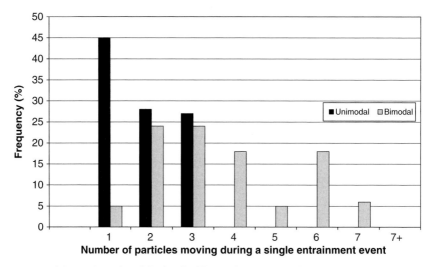

Fig. 2. Frequency of the numbers of particles involved in entrainment events for unimodal gravels (black) and sand-gravel (bimodal) mixtures (grey). Note the dominance of single particle events in the gravel experiments and the dominance of multi-particle events in the sand–gravel mixture experiments.

Results

Analyses of the video recordings have revealed distinctive differences in entrainment between the experiments involving unimodal gravels and those where the gravels are mixed with sands which are not directly linked with changes in bed shear or roughness. There were differences in the spatial patterns of entrainment, the timing of significant entrainment events and the processes which initiate these events.

Unimodal gravel experiments

Spatial patterns of entrainment. In the unimodal gravel experiments, examination of the individual (25 frames per second) digital video images of the bed showed that gravel particles were entrained from the surface either individually or in very small groups of two or three. It was very rare to see more than two particles dislodge from the same area of the bed within 1 s of each other and even over longer periods it was rare to see more than three particles moving out of any area of the bed. There was no evidence of patchy entrainment, particles coming instead from all areas of the bed with no preference for particular sections (Fig. 2).

When the image difference plots were examined it was evident that particle movement was confined to the very surface of the bed (one grain diameter) with no evidence of readjustment below this depth. As the surface was disturbed, the spaces between the grains enlarged slightly (Fig. 3a), but this dilation did not penetrate into sub-surface layers.

Temporal patterns of entrainment. The time intervals between significant entrainment events is very variable (Fig. 4). In most cases these intervals range from 1 to 3 s, with few intervals greater than 4 s.

Entrainment processes and bed readjustment. Careful examinations of video clips of the entrainment processes during the experiments have revealed that a variety of initiating events can cause particle movement. These could not be quantified but were categorized according to the way in which each entraining particle moved during the first few milliseconds and by observed interactions with incoming particles transported from upstream. The five distinctive categories used here are: lift, this was assumed only where the initial movement on all of the registration dots was vertical; drag, where there is evidence of rotation and the initial movement of the registration dots had a strong downstream component; impact, where a particle was bounced out of the bed by an incoming saltating grain from upstream; upstream domino, were a particle moves as a result of other particles pushing it from upstream; and downstream domino, where the removal of particles downstream removed obstacles to movement and destabilized the bed structure. The frequency of each type of initiating event is plotted in Figure 5. This shows that the dominant events that bring about entrainment in unimodal gravels are upstream domino, impacts and lift.

The organization of particles on the bed was examined both during each experiment and at its

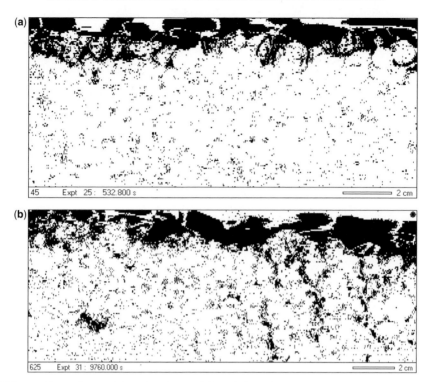

Fig. 3. Differenced image for gravel (**a**) and sand–gravel (**b**) beds during bed entrainment events. Note that movement is restricted to the surface layers in the gravel bed but penetrates to depths of several grains in the sand–gravel mixture.

end (Fig. 6). The gravels used in this experiment comprised flint and chert pebbles, most of which approximated ovate spheroids, with two equal axes and one long one. It was therefore possible to determine the positioning of the long axis for many of the particles on a series of still photographs taken through the sidewall of the flume. Where measurement was possible, the angle made by the long axis with a line drawn parallel with the bed surface was used as a measure of the orientation and organization of the bed in general. This showed that unimodal gravels are relatively poorly organized, and the orientation of the long axis of surface grains ranges widely, with a modal value of 23% between 120 and 140° (i.e. inclined upstream), but with almost 20% of the observations inclined at between 0 and 20° (a slight downstream inclination). The remainder of the observed orientations were predominantly upstream, but at shallow angles close to the general bed slope.

Sand–gravel experiments

Spatial patterns of entrainment. In the bimodal sand–gravel experiments the individual digital video images of the bed showed that the gravel particles tended to entrain in clumps, often with seven or more particles either moving at once or in very close succession (Fig. 2). Particles were never seen moving singly, and often one area of the bed was the focus of erosion over periods of 5 s or more. Entrainment was always patchy and some areas of the bed were eroded repeatedly, with the bed regaining stability by particle sliding from upstream into the depression.

The image difference plots show that movement and readjustment amongst particles at depth in the bed is a feature of these sand–gravel mixtures. As the bed surface was disturbed, pore spaces between the particles enlarged and the bed dilated. However, this dilation was not confined to the near surface but was propagated down through the bed along defined pathways which can be seen clearly in Figure 3b. This dilation sometimes penetrated as far as five grain diameters for several seconds, but across the majority of the bed and for most of the time the bed dilated to a depth of no more than two grain diameters.

Temporal patterns of entrainment. The time intervals between entrainment events were short, generally less than 10 s (almost 70% of the observations falling into this category, Fig. 4). However, more than 20% of the observed intervals between

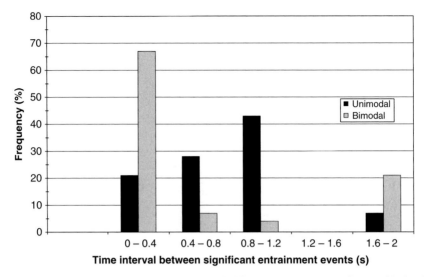

Fig. 4. Frequency distribution of the time intervals between significant entrainment events for gravel (unimodal) and sand–gravel (bimodal) experimental sections of the bed.

entrainment events were significantly longer, ranging between 40 and 50 s. Only approximately 10% of the observations fell outside these two categories.

Entrainment processes and bed readjustment. Examination of the video recordings of sand–gravel experiments have shown a very interesting range of processes involved in entrainment (Fig. 5), with long sequences of events causing the entrainment of groups of particles. The sequences began with an individual gravel particle starting to move, either as a result of impacting particles from upstream or as a direct result of fluid movement, which exposed the sand in the pores to rapid erosion. This destabilized other gravel particles which slid and adjusted, some of them also entraining. This left a depression in the bed which exposed the grains immediately downstream to erosion and so on. By this mechanism

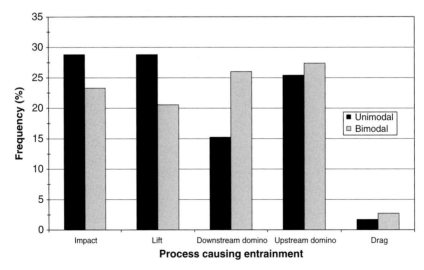

Fig. 5. Frequency distribution of the different processes initiating entrainment for individual particles during gravel (unimodal) and sand–gravel (bimodal) experiments.

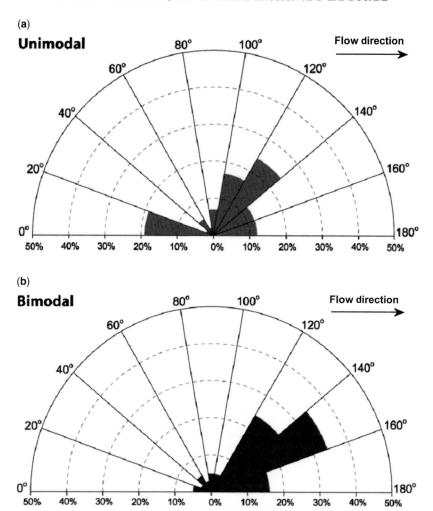

Fig. 6. Orientation of the long axis of surface clasts as seen through the side wall of the flume at the end of the entrainment experiments for (**a**) unimodal and (**b**) bimodal materials. The measurements plotted are the angle made by surface particle long axes with a line drawn parallel to the mean bed surface.

the focus of erosion migrated progressively downstream. Entrainment only ceased once both the sands and gravels were stabilized through removal of some particles and the readjustment of others. Dominant processes of entrainment were therefore interpreted as due to the domino effects, both up and downstream (over 50% of the observed movements).

The removal of sand and readjustment of particles led to a much more organized bed than in unimodal gravels. The gravel bed particles had dominant long axis orientations of between 140 and 160° (Fig. 6). Three-quarters of the surface pebbles measured had long axes inclined in an upstream direction.

Discussion

Comparison of clean gravel and sand–gravel results

There are distinctive differences in both entrainment processes and bed character between the gravel and the sand–gravel mixtures used in these experiments which have implications for the rates at which these types of material will either accumulate within, or move through, the landscape. These variations are both temporal and spatial, and there is evidence that they may result in bed surface characteristics that have knock-on effects for the susceptibility of the bed to erosion.

During entrainment in the unimodal gravel, particles acted more or less singly, each individual particle moving out of the bed in response to impacts from other moving grains and fluid movement. The periodicity and location of these movements appeared to be controlled by the distribution of these forces in time and space, with domino effects only small scale and therefore less important. There was little evidence of large-scale destabilization of the bed, although on some occasions one particle moving pushed and dislodged the one or two downstream of it.

This contrasts with the sand-laden beds where entrainment was much more patchy. In these experiments the bed destabilized in response to the rapid erosion of the sand, leading to entrainment of many of the surrounding particles, as well as readjustment and reorientation of the remaining ones. Upstream and downstream domino effects were the dominant processes, reflecting the concentration of erosion in areas where the bed was already destabilized.

There were also marked differences in the temporal patterns of entrainment between the two bed material types. Intervals between entrainment events for the gravel-only experiments were very variable, whereas in the sand-filled experiments the intervals were either very short or quite long, with few observations falling in between. This reflects the differences in process between the experiments, and points to a fundamental difference in the way in which gravel and sand–gravel mixtures might behave in the natural environment. These differences are probably a consequence of changes in near-bed flow structure produced by adding sand to a gravel bed (see Sambrook Smith & Nicholas 2005) and changes in surface bed permeability which must alter interactions between flows above and below the bed surface but which have yet to be studied in detail.

The character of the bed surface reflects the differences in process, with the extensive sliding and readjustment of the sand filled bed resulting in better developed organization of the bed surface. With the material selected for the experiments reported here there was no possibility for the development of an armour layer. However, if a wider range of grain sizes had been included in the gravel sediment, it might be expected that the many bed readjustments which occurred in the sand-filled bed would favour the selective entrainment of smaller particles and the retention of coarser ones to form an armour layer. Both organization of surface particles and armouring protect the bed from erosion at shear values close to the entrainment threshold (see e.g. Karim & Holly 1986; Duizendstra 2001). This suggests that sand-packed gravel beds might develop surfaces that can resist erosion during small to moderate floods events and during the early stages of big floods, delaying the commencement of bed load transport. The consequence of this is that, although sand enhances bed load transport rates for gravels once the bed is mobilized, it also delays its onset.

Implications for sediment flux

It is evident from this work that the flux of gravels through the alluvial landscape will reflect, to a greater or lesser extent, the range of grain sizes available for transport. This has been reported by a number of previous authors (e.g. Deitrich et al. 1989; Ferguson et al. 1996; Buffington & Montgomery 1999) and explained in terms of various contributing processes including changes in hydraulic roughness and slope as well as selective entrainment and deposition. The results of the two sets of experiments reported here contribute some process observations concerning the differences between gravels and sand–gravel mixtures. These add weight to existing explanations of how the range of grain sizes influences the temporal and spatial patterns of gravel transport in both laboratory flumes and in the field. For example, the work of Rennie & Millar (2004) on alluvial bed load transport velocities in the sand and gravel bed sections of the Fraser river demonstrated both strong spatial and temporal variations in transport velocities for the gravel bed reach. This may be a direct reflection of the mean grain size of the gravel, but it could be linked with the patchy distribution of fine pore-filling sediments which could alter the susceptibility of the gravel to entrainment.

Interestingly, research reported by Wilcock et al. (2001) on the transport of gravels and sand–gravel mixtures in a laboratory flume showed that mixtures of sand in gravel enhanced the mobility of the gravel. The conclusions reported here are consistent with their observations, and explained by the large-scale but patchy erosion of the gravel beds which is induced by adding sand. Results reported elsewhere show that on average sand-filled gravels produce bed load transport 20% higher than clean gravel equivalents (Frostick et al. 2006).

Observations of bedload transport rates in gravel bed rivers have resulted in widespread reports of pulses of bed load transport which do not reflect similar changes in bed shear (Reid & Frostick 1985; Reid et al. 1985). These occur at a range of intervals, from fractions of an hour to several hours. The results reported here show that, on any given area of the bed, very short time interval variations are a feature of sand–gravel mixtures but do not occur in gravels alone. When this is scaled up to river-wide bed load fluxes the patchy and sporadic nature of entrainment is likely to result in pulses of sediment when several areas of the bed destabilize at one time. If the results of these experiments can be generalized it might be suggested that the pulsing will be more regular and more pronounced in gravel bed rivers where pore spaces are partly filled with fine sediment. In addition, the enhanced bed organization in sand-filled gravels requires higher flow velocities to begin bed load transport, delaying its onset. This is consistent with field observations that bed load transport is delayed following long periods of low flows when matrix fines accumulate in the bed surface (Reid & Frostick 1985).

Implications for bed form development

Bed forms in gravel bed rivers are dominated by large-scale bars, particularly in braided reaches and at meander bends (Fredsoe 1982; Reid & Frostick 1994). Cross- and downstream sorting of grain sizes is a feature of the these bars, with fine bar tails reflecting lateral sorting on the distal flanks of bars (Ferguson & Ashworth 1992). Coarser grains tend to follow the river talweg whilst finer material climbs up onto the bar as the flow diverges. This spatial pattern of grain size distribution is augmented by the bar exposure during low flows, when fine material accumulates in the talweg channel gravels.

The differences in entrainment between gravel and sand–gravel mixtures will therefore have implications for the development and maintenance of course river channel bars. Areas of the bar where gravel is admixed with sand will be less stable than those where gravel alone comprises the bed. This will lead to changes in the bar shape and steepness as the sandy gravels move less readily but in larger quantities. There is therefore a direct link between the morphological features found in the river channel, the range of grain sizes available for transport, the distribution of these grain sizes across the bed and flood magnitude and frequency. These complex interactions have yet to be explored in detail.

Implications for landscape evolution

Although these experiments focus on details of how gravel material moves into and out of transport in a sub-aqueous environment, the results do have implications at the broader landscape scale. The landscape comprises areas of sediment production and areas of sediment deposition and the overall balance between these two dictates both the morphology and character of an area. The results of these experiments have shown that, if the sediment produced in the slopes is dominated by gravel size material, it will remain close to the slope for longer, since clean gravels appear to erode at a more consistent but slower rate. By contrast, if the slope produces a mixture of gravels and sands, this material will move faster and in a more sporadic manner. Since the wider landscape is also the sum of all its smaller scale parts, the influence of grain size mixing on the size and shape of alluvial deposits will also influence the form of the land. If these differences are translated directly into different sediment flux rates then there may be implications for the rate at which sediment is transferred into the local basin, and therefore for the character and deposits of the basin margin.

Conclusions

The laboratory experiments outlined in this paper suggest that the grain size distribution of sediment that is available for transport through the headwaters of the river systems will exert a strong influence on the rate and timing of sediment flux. This in turn will influence the character of the river system and the nature of the deposits which may be preserved in its bed and banks. The differences in the susceptibility to, and nature of, entrainment in clean gravel and sand–gravel mixtures lead to some interesting conclusions about how bed forms might develop in different ways in response to the particulate material that is available. At the bar scale areas of the bar where sand is admixed with gravel are likely to resist entrainment during small floods whilst cleaner areas of gravel are mobilized and reshaped. However, during larger floods areas of mixed grain size will erode more rapidly, causing major changes in morphology. At reach scale, the gravel–sand transition section of the river is likely to experience intense but sporadic and patchy entrainment that may help maintain and enhance bedforms and contribute to the separation of the two size grades and downstream fining through selective entrainment and transport.

There are also implications for landscape models, emphasizing the need for the inclusion of grain size measures in the flux–power relationships if such models are to capture some of the complexities of the controls on sediment transport in the natural environment. It is therefore essential that more consideration is given to the range of grain sizes available for transport, with less emphasis on the mean grain size. This has already been recognised in some reach and fewer landscape scale models (Hoey & Ferguson 1994; Gasparini *et al.* 1999). The importance of grain size as a factor will vary with the timescale being modelled. Over shorter timescales models that ignore considerations of variations in the range of bed material sizes are more likely to produce misleading results. Over longer timescales it may well be that any variations will be evened out and become relatively insignificant. If the implications of spatial and temporal changes in the patterns of sand and gravel mixing are to be fully incorporated into models, it will make modelling much more complex, but over shorter timescales and smaller spatial scales the impact on the predictive accuracy of the model may be significant. At longer time scales and over larger spatial scales such differences may not be important; however the levels of improvement that might be achieved have yet to be evaluated.

This work was funded by Leverhulme and NERC. We would like to thank James Brasington for his help during the early stages of the project and John Garner and Linda Love for their assistance during the preparation of this paper. Thanks are also due to Rob Ferguson and Brian Turner, who refereed the paper and whose comments helped improve the text considerably.

References

ALLAN, A. & FROSTICK, L. E. 1999. Framework dilation, winnowing, and matrix particle size: the behaviour of some sand–gravel mixtures in a laboratory flume. *Journal of Sedimentary Research*, **69**, 21–26.

BRASINGTON, J., FROSTICK, L. E., MIDDLETON, R. & MURPHY, B. J. 2000. Detecting significant sediment motion in a laboratory flume using digital video image analysis. *Earth Surface Processes and Landforms*, **25**, 191–196.

BUFFINGTON, J. M. & MONTGOMERY, D. R. 1999. Effects of hydraulic roughness on surface textures of gravel-bed rivers. *Water Resources Research*, **35**, 3507–3522.

CHOROWICZ, J., DHONT, D., AMMAR, O., RUKIEH, M. & BILAL, A. 2005. Tectonics of the Pliocene Homs basalts (Syria) and implications for the Dead Sea fault zone activity. *Journal of the Geological Society, London*, **162**, 259–271.

CLAVERO, J. E., SPARKS, R. S. J., PRINGLE, M. S., POLANCO, E. & GARDEWEG, M. C. 2004. Evolution and volcanic hazards of Taapaca volcanic complex, Central Andes, Northern Chile. *Journal of the Geological Society, London*, **161**, 603–618.

DEITRICH, W. E., KIRCHNER, J. W., IKEDA, H. & ISEYA, F. 1989. Sediment supply and the development of the coarse surface layer in gravel-bedded rivers. *Nature*, **340**, 215–217.

DUIZENDSTRA, H. D. 2001. Determination of the sediment transport process in a gravel-bed river. *Earth Surf. Proc. Landforms*, **26**, 1381–1393.

EINSTEIN, H. A. 1942. Formulas for the transportation of bedload. *Transactions of the ASCE*, **107**, 561–573.

EMMETT, W. W. 1980. A field calibration of the sediment trapping characteristics of the Helley-Smith bedload sampler. *US Geological Survey Professional Paper* **1139**.

FERGUSON, R. I. & ASHWORTH, P. J. 1992. Spatial patterns of bedload transport and channel change in braided and near braided rivers. *In*: BILLI, P., HEY, R. D., THORNE, C. R. & TACCONI, P. (eds) *Dynamics of Gravel Bed Rivers*. Wiley, Chichester, 477–492.

FERGUSON, R., HOEY, T., WATHEN, S. & WERRITY, A. 1996. Field evidence for rapid downstream fining of river gravels through selective transport. *Geology*, **24**, 179–182.

FREDSOE, J. 1982. Shape and dimensions of stationary dunes in rivers. *Journal of the Hydraul. Division, ASCE*, **108**, 932–947.

FROSTICK, L. E. & JONES, S. J. 2002. Impact of periodicity on sediment flux in alluvial systems: Grain to basin scale. *In*: JONES, S. J. & FROSTICK, L. E. (eds) *Sediment Flux to Basins: Causes Controls and Consequences*, Geological Society London, Special Publication, **191**, 81–96.

FROSTICK, L. E. & REID, I. 1989. Is structure the main control on sedimentation in rifts? *Journal of African Earth Sciences*, **8**, 168–182.

FROSTICK, L. E. & REID, I. 1990. Structural controls of sedimentation patterns and implications for the economic potential of the East African Rift basins. *Journal of African Earth Sciences*, **10**, 307–318.

FROSTICK, L. E., MURPHY, B. & MIDDLETON, R. 2006. Unravelling Flood History Using Matrices in Fluvial Gravel Deposits. IAHR.

GASPARINI, N. M., TUCKER, G. E. & BRAS, R. L. 1999. Downstream fining through selective particle sorting in an equilibrium drainage network. *Geology*, **27**, 1079–1082.

GOMEZ, B., 1991. Bedload transport. *Earth Science Reviews*, **31**, 89–132.

HABERSACK, H. M. & LARONNE, J. B. 2002. Evaluation and improvement of bed load discharge formulas based on Helley-Smith sampling in an alpine gravel-bed river. *Journal of Hydraulic Engineering, ASCE*, **128**, 484–499.

HOEY, T. B. & FERGUSON, R. 1994. Numerical simulation of downstream fining by selective transport in gravel bed rivers: model development and illustration. *Water Resources Research*, **30**, 2251–2260.

KARIM, M. F. & HOLLY, F. M. 1986. Armoring and sorting simulation in alluvial rivers. *Journal of Hydraulic Engineering ASCE*, **112**, 705–715.

LANE, S. N. 1998. Hydraulic modelling in hydrology and geomorphology: a review of high resolution approaches. *Hydrological Processes*, **12**, 1131–1150.

LANE, S. N., RICHARDS, K. S. & CHANDLER, J. H. 1995. Morphological estimation of the time integrated bed-load transport rate. *Water Resources Research*, **31**, 761–772.

MIDDLETON, R., BRASINGTON, J., MURPHY, B. J. & FROSTICK, L. E. 2000. Monitoring gravel framework dilation using a new digital particle tracking method. *Computers & Geosciences*, **26**, 329–340.

NELSON, J. M., SHREVE, R. L., MCLEAN, S. R. & DRAKE, T. G. 1995. Role of near-bed turbulence structure in bed load transport and bed form mechanics. *Water Resources Research*, **31**, 2071–2086.

REID, I. & FROSTICK, L. E. 1984. Particle interaction and its effect on the threshold of motion in coarse alluvial channels. *Memoir of the Canadian Society for Petroleum Geology*, **10**, 61–68.

REID, I. & FROSTICK, L. E. 1985. Dynamics of bedload transport in Turkey Brook, a course-grained alluvial channel. *Earth Surface Processes and Landforms*, **10**, 315–326.

REID, I. & FROSTICK, L. E. 1994. Fluvial transport and deposition. In: PYE, K. (ed.) Sediment Transport and Depositional Processes, Blackwell Science Ltd, 89–132.

REID, I., LAYMAN, J. & FROSTICK, L. E. 1980. The continuous measurement of bedload discharge. Journal of Hydraulic Research, 18, 243–249.

REID, I., FROSTICK, L. E. & LAYMAN, J. 1985. The incidence and nature of bedload transport during flood flows in coarse-grained alluvial channels. Earth Surface Processes and Landforms, 10, 33–44.

RENNIE, C. D. & MILLAR, R. G. 2004. Measurement of the spatial distribution of fluvial bedload transport velocity in both sand and gravel. Earth Surface, Processes and Landforms, 29, 1173–1193.

SAMBROOK SMITH, G. H. & NICHOLAS, A. P. 2005. Effect on flow structure of sand deposition on a gravel bed: results from a two-dimensional flume experiment. Water Resources Research, 41, W10405.

WILCOCK, P. R. & MCARDLE, B. W. 1997. Partial transport of a sand/gravel sediment. Water Resources Research, 33, 235–245.

WILCOCK, P. R., KENWORTHY, S. T. & CROWE, J. C. 2001. Experimental study of the transport of mixed sand and gravel. Water Resources Research, 37, 3349–3358.

Inferring bedload transport from stratigraphic successions: examples from Cenozoic and Pleistocene rivers, south central Pyrenees, Spain

STUART J. JONES[1] & LYNNE E. FROSTICK[2]

[1]Department of Earth Sciences, South Road, Durham University, Durham DH1 3LE, UK
(e-mail: stuart.jones@durham.ac.uk)

[2]Department of Geography and Hull Environment Research Institute, University of Hull, Hull HU6 7RX, UK

Abstract: Geologists and geomorphologists have long been concerned with rates of sediment transfer as bedload in gravel-bed rivers, especially as rates of sediment transfer are important factors controlling river aggradation and incision. Bedload transport equations, originally derived for Holocene streams, have been used widely in modern gravel-bed river systems. However, palaeohydraulic reconstructions have received less attention and are generally dismissed as inaccurate since most are estimated to be at least an order of magnitude out. This study focuses on deriving stream power, bedload transport rates and efficiency estimates for Oligo–Miocene and Plio–Pleistocene gravel-bed river deposits from the south central Pyrenees, Spain. The basic data used in the palaeohydraulic calculations are estimates of palaeoslope, palaeovelocity, palaeodepth and the volume of sediment accreted in yearly flood events on gravel bars. Analyses of data from these ancient river systems yield more accurate estimates of relative stream power, bedload transport rates and efficiency parameters. This study illustrates the need for understanding the palaeohydraulics of river systems in order to characterize ancient rivers. Gravel-bed rivers with low sediment supply and high bedload transport rates incise. Conversely, when sediment supply is abundant, bedload transport rates and efficiency are low and the river system aggrades.

It has long been recognized that understanding the behaviour of ancient river systems is fundamental to the successful reconstruction of palaeo-basins (e.g. Burbank 1992; Vincent & Elliott 1997; Nichols & Hirst 1998; Jones et al. 2001; Jones 2004). An important goal is to understand more fully the behaviour of ancient rivers, an objective achieved largely through the use of palaeocurrent measurements, detailed architectural studies and by comparison with modern rivers. Measurements of palaeoflow properties such as velocity, sediment flux and considerations of palaeoefficiencies are rare in the geological literature, primarily due to the fact it is intrinsically more difficult both to obtain data and to interpret them. Sediment flux is a first-order control on the pattern and distribution of sedimentary facies in depositional basins. Consequently, sediment flux provides evidence of dynamic geomorphic processes operating in sedimentary basins and any fluctuations must be recorded in the preserved sedimentary deposits (Laronne & Reid 1993; Frostick & Jones 2002).

A range of palaeohydraulic studies have appeared in the literature (e.g. Baker 1974; Gardiner 1983; Steer & Abbott 1984; Williams 1984; Ryder & Church 1986; Church et al. 1990), but most are concerned more with the stratigraphy and general fluvial morphology than fluxes, owing to the inherent difficulty in obtaining appropriate field data. However, recent research into sediment transport and depositional processes in modern rivers has significantly improved the range of potential methods available for palaeohydraulic reconstruction (Maizels 1989; Hoey 1992; Laronne et al. 1992; Reid et al. 1996, 1997; Jones 2002; Carling et al. 2003).

In this paper a critical attempt is made at estimating the bedload flux through the application of palaeohydraulic analyses from the proto-Rio Cinca of the south central Pyrenees, Spain. Equations developed by fluvial engineers and geomorphologists are used to quantify the palaeohydraulic properties of Cenozoic and Pleistocene gravel-bed river sediments. The data collection methods and the errors involved are discussed.

Pyrenean geology

The Pyrenees are a nearly linear mountain belt some 200 km wide and extending for 450 km along the border between France and Spain (Fig. 1). They are part of the Alpine chain that formed during Late Cretaceous to Miocene times due to north-directed convergence and limited underthrusting of the Iberian lower crust and lithospheric mantle

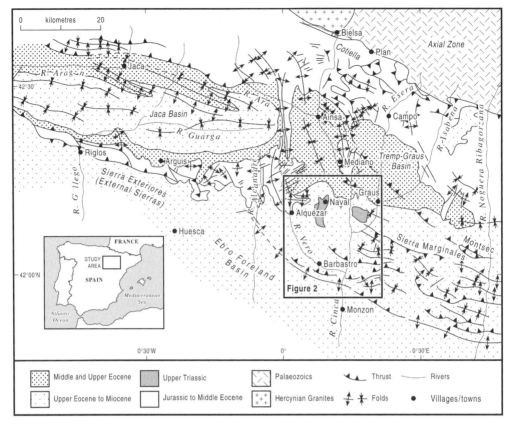

Fig. 1. Geological map of the south central Pyrenees and Ebro Foreland Basin. The study area around the town of Barbastro is shown in Figure 2. The small inset map shows the approximate position of the study area within the Pyrenean chain.

beneath the European plate (Muñoz 1992). The main part of the range within the Iberian plate consists of uplifted Hercynian Palaeozoic basement rocks, the Axial Zone, flanked to the north and south by a fold–thrust system of Mesozoic–Cenozoic sediments (Puigdefàbregas et al. 1992).

The thrust deformation of the Pyrenean orogen is strongly diachronous from east to west. In the southern Pyrenees, basins developed during Palaeocene to early Eocene times, ahead of southerly propagating thrust sheets. Collectively referred to as the South Pyrenean Basin (Puigdefàbregas 1975), the region became compartmentalized during the early Eocene epoch in response to the incorporation of the foreland into the developing thrust sheets (Fig. 1). Throughout much of this period of basin evolution, sediment was largely derived from erosion of the actively deforming Axial Zone to the north. On entering the basins, the sediments were directed westward, forming an axial depositional system that progressively becomes marine towards the Atlantic Ocean. Subsequent thrust deformation during the middle-late Eocene and into the Oligo–Miocene further changed the basin configuration and established the Ebro Basin as a closed continental, terminal basin (Hirst & Nichols 1986; Puigdefàbregas et al. 1992). Coarse sediment continued to be supplied from the north, and was directed to the S–SSW and into the Ebro Foreland Basin (Jones et al. 2001; Jones 2004).

Oligo–Miocene and Plio–Pleistocene (proto-Rio Cinca) fluvial systems

The initial development of a proto-Rio Cinca was during the late Oligocene when the first gravel-bed river deposits developed in the Naval to El Grado area (Fig. 2). The drainage network was structurally constrained by several oblique ramp anticlines, by smaller scale N–S orientated anticlines at the eastern limit of the External Sierras and by thrust fronts along the western margin of the South

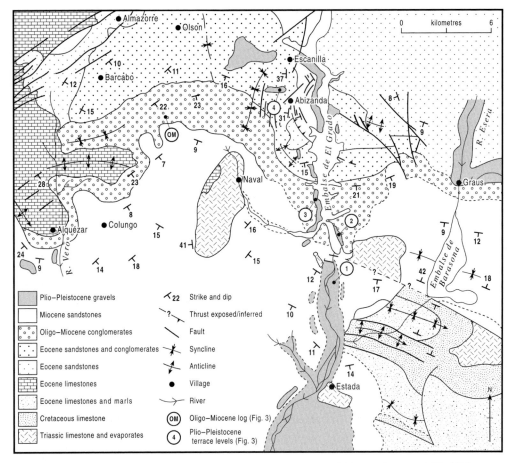

Fig. 2. Sketch geological map of the study area illustrating the distribution of Oligo–Miocene and Plio–Pleistocene gravel-bed river deposits. Important structural features, villages within the study area and location of the type logged sections are all shown.

Central Unit (e.g. Sierra Marginales; Fig. 1). This regional structural trend maintained the position of the proto-Rio Cinca during the Oligo–Miocene in the Naval–El Grado area and during the late Miocene the Rio Cinca became fixed in its present day position through incision (Jones 1997; Jones et al. 1999; Figs 1 & 2).

Extensive field work has been carried out on the Oligo–Miocene and Plio–Pleistocene gravel-bed river deposits along the middle to upper reaches of the modern Rio Cinca, south central Pyrenees, Spain (Fig. 2). The Oligo–Miocene deposits are dominated by sub-horizontal masses of pebble- and cobble-grade conglomerates, lying with an angular unconformity over the older rocks. In comparison, the Pliocene to recent history of the fluvial system is contained in a suite of terraces that flank the present day river valley (Fig. 3). These conglomerates are coarser-grained with cobble–boulder size clasts (Fig. 4).

The conglomerates in both successions commonly occur in erosionally based, broad channelized units (between 40 and 300 m wide and up to 3 m thick), interpreted as being deposited from widespread, weakly channelized flows carrying coarse bedload (Fig. 3a, b). Stratified conglomerates that fine upwards are common, and these deposits often have cobble and/or boulder basal lags. Cobble conglomerates with a diffuse stratification occur frequently, often imbricated and with abundant pebble clusters. Individual beds are defined by the alternation of open and closed framework fabrics, which may be only a few clasts thick and are particularly well preserved in the Plio–Pleistocene terraces (Fig. 3c). Imbrication and pebble clusters are particularly common throughout

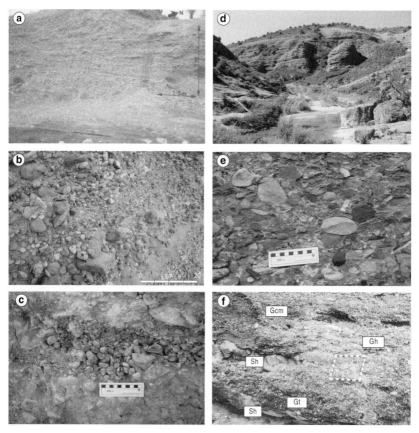

Fig. 3. Plio–Pleistocene (**a**–**c**) and Oligo–Miocene (**d**–**f**) gravels and conglomerates. (**a**) Gravel Terrace (T3) illustrating broad *c.* 50 m-wide channel-fill multistory units with gravel bars and associated accretionary foresets. (**b**) Polymict, clast supported, well rounded, pebble to cobble size gravels. Horizontally stratified with abundant imbrication and pebble clusters. (**c**) Diffusion texture. Variation from open to closed framework pebble–cobble-size horizontally stratified gravels. (**d**) View to the NE along the Rio Vero gorge directly beneath the village of Alquézar (see Figs 1 & 2). Approximately 250 m of conglomerates are preserved dipping gently to the south. (**e**) Polymict, clast-supported, well-rounded, pebble–cobble size conglomerates. Scale bar: 20 cm. (**f**) Conglomerates with horizontal stratification, trough cross-bedding and clast supported (Gh, Gt, Gcm) often associated with broad channel-fill complexes. Lenses and stringers of coarse- to medium-grained sandstones are a common occurrence (Sh). Lithofacies classification scheme after Miall (1978) and Jones *et al.* (2001).

both successions and would suggest armouring of the river-bed in a perennial river system (Reid *et al.* 1992; Laronne & Reid 1993).

Accretionary foresets of conglomerates are well preserved throughout the entire succession and may exceed 4 m thick in early terrace levels of the Plio–Pleistocene conglomerates (Fig. 5). Many of the units grade in grain size from cobble clasts to coarse sandstones and palaeocurrent data indicate that they formed at frontal or oblique margins of gravel bars. Differentiation of lateral from frontal (downstream) accretion elements of mid-channel bars is an important part of this study. Where the orientations of an accretionary unit and included cross-bedding and any other palaeocurrent indicators are within 60° of each other, it indicates that frontal accretion has occurred in a downstream direction. Where the orientations of the accretionary unit and cross-bedding are perpendicular to each other (>60°) then lateral accretion of the bar is interpreted. Both types of accretionary units would be expected to be present in different parts of the same bar and it is important to distinguish between them for any palaeohydraulic study of ancient river systems. This criterion is similar to that used by Miall (1978) and Jones *et al.* (2001).

Horizontally stratified sandstones and siltstones associated with the accretionary surfaces are interpreted as having been deposited from a waning flood event and representing supra-bar deposits.

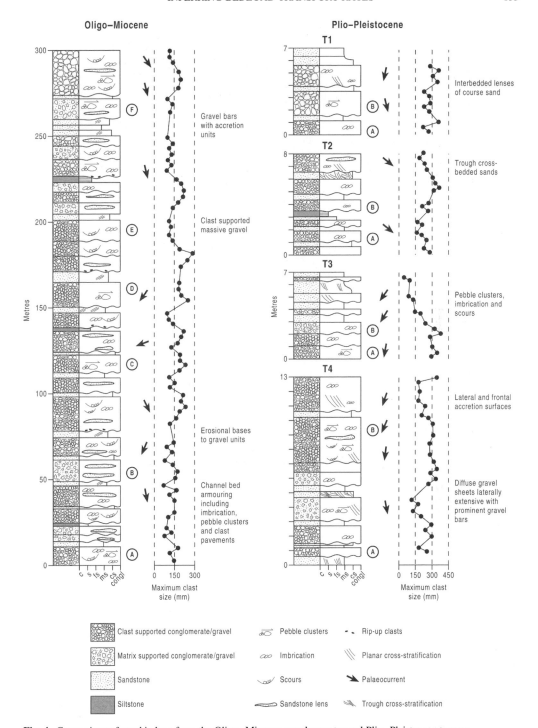

Fig. 4. Comparison of graphic logs from the Oligo–Miocene conglomerates and Plio–Pleistocene terrace gravels. All palaeocurrent directions are predominantly to the south, with each arrow representing at least 30 readings taken from cross bedding, pebble imbrication and clusters. The logs for each of the Plio–Pleistocene terrace levels correspond with those shown in Figure 2. Letters refer to sample sites as in Tables 1 & 3.

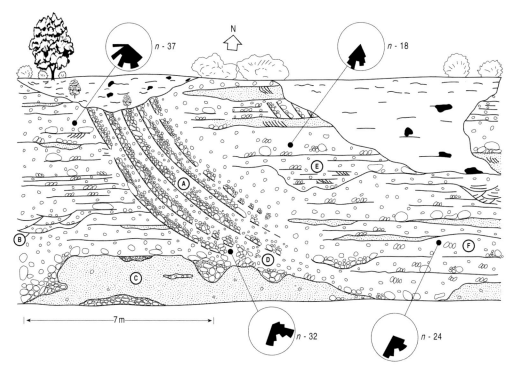

A Well-developed lateral-frontal accretion conglomerates (Gl)
B Channel-fill complex with trough-cross bedding (Gt & Gh)
C Massive coarse sandstone (minor Sh & Sp)
D Deep scours with cobble–boulder fills and smaller scours (Cse)
E Channel-fill complex with stratified conglomerates (Gh, Gcm)
F Sheet conglomerate with horizontal stratification (Gcm, Gh)

Fig. 5. Field sketch of Plio–Pleistocene terrace level (T4) to the west of Abizanda (Figs 2 & 3a–c). Large-scale frontal accretion foresets dominate the exposure. Accretionary units of ~4 m high and individual sets between 0.5 and 1 m thick with internal grading are common. The lithofacies classification used is after Miall (1978). Figures next to palaeocurrent roses refer to the total number of measurements taken at a particular location.

Planar cross-stratified conglomerates and coarse-grained pebbly sandstones are present in all sequences (Fig. 6). Set heights are commonly between 20 and 40 cm, and the sets are arranged in 1–2 m scale stacked units. They are usually associated with channel-fill complexes and interpreted as bar-top or seasonal channels only active during flood events. Trough cross-bedded sets of gravel are less common and are interpreted to represent the migration of dunes downstream within the channels (Figs 3e, f, 5 & 6). Many of the channels and channel-fill complexes are typically strongly erosional with deep scours, lined with cobble clasts and sandstone intraclasts up to 0.4 m long. Sandstone rip-up clasts and cave-in deposits from bank undercutting occur at the lateral limits of many of the channel-fills.

The stratigraphic range from Oligocene to Recent and the abundance of conglomerates provide evidence for the longevity of the Rio Cinca as a major transverse gravel-bed river supplying sediment to the Ebro foreland basin (Jones et al. 1999, 2001; Jones 2004).

Measurement strategy

To assemble the palaeohydraulic data necessary to fully understand the hydraulic character of the coarse-grained braided proto-Rio Cinca, several parameters need to be inferred.

Stream power

Specific stream power (W m^{-2}) is the potential rate of work done by a flow on a unit area of a channel bed (Bagnold 1966), such that

$$W = \rho g h S u \qquad (1)$$

where ρ is the density of water (1000 kg m^{-3}), g is the gravitational constant (9.81 m s^{-2}), h is the

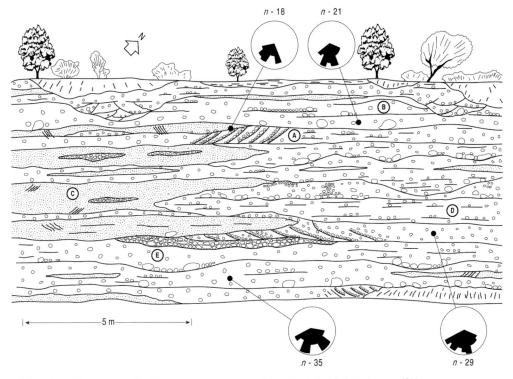

A Lateral accretion conglomerates (**Gl**)
B Sheet conglomerate, convex top (**Gcm**)
C Massive sandstone (minor **Sh**, **Sp** & **Csh**)
D Horizontally stratified conglomerate (**Gh**)
E Massive clast-supported conglomerate sheet (**Gh**, **Gcm**)

Fig. 6. Field sketch of the Oligo–Miocene conglomerates at Alquézar (see Figs 2 & 3d–f). The gravel-bed river deposits are dominated by gravel sheets, gravel bars and well-preserved lateral and frontal accretion units, complex channel-fills and laterally restricted coarse- to medium-grained sandstones.

water depth (m), S is the channel gradient and u is the mean flow velocity (m s^{-1}).

In this study this parameter is termed 'relative stream power' due to the data being an estimation of the palaeoriver specific stream power. Theoretical considerations show that sediment transport is related to the power expended upon the bed by the flow (stream power). Data from flume studies and short-term field measurements support a link between stream power and bedload transport in channels (e.g. Bagnold 1980), hillslope erosion by overland flow and incision in bedrock channels (e.g. Wohl 1993; Seidl & Kirchner 1994). The relationship between stream power and bedload transport rate is particularly important for long-term patterns of channel-bed incision and deposition. However, before stream power can be estimated, several variables need to be determined from the ancient gravel-bed river deposits, including water depth, channel gradient and flow velocity (Table 1). None of these can be measured directly and may be estimated using the following methods that provide the underpinning for palaeohydraulic analysis of these systems.

Palaeodepth estimation

The calculation requires an estimate of average depth, which approximates to that of a section of equivalent modern river. This can be achieved through measuring as many depth indicators as possible over a chosen outcrop area (e.g. large-scale planar cross-stratification, lenticular channel-fills, fining upward successions). These depth indicators were chosen carefully to avoid examples with extensive erosion or where deep erosion scours occur, since these can be misinterpreted as a channel cross-section. Based on multiple measurements and comparison with channel depths in modern gravel-bed rivers, the proto-Rio Cinca had a flow depth in the range of 0.5–2.5 m for the Oligo–Miocene and 0.8–4.5 m for the Plio–Pleistocene (Table 1).

Table 1. *Summary of palaeohydraulic data for the Oligo–Miocene and Plio–Pleistocene gravel-bed river systems*

	T1		T2		T3		T4	
	a^a	b	a	b	a	b	a	b
			Plio–Pleistocene Terrace levels					
S		0.0038		0.0044		0.0043		0.0024
D_{50} (mm)	147	105	86	66	44	56	135	77
D_{84} (mm)	270	207	168	123	164	126	291	133
D_{90} (mm)	300	235	183	139	181	144	327	164
D_{90}/D_{50}	2.04	2.23	2.12	2.1	4.11	2.57	2.42	2.12
Y (m)	1.8	2.6	3.1	1.4	2.2	1.2	1.7	2.9
n: Limerinos	0.029	0.040	0.036	0.034	0.036	0.037	0.042	0.042
n: Strickler	0.023	0.031	0.029	0.026	0.029	0.029	0.032	0.032
u: Limerinos (m s^{-1})	1.6	4.8	3.1	3.3	3.1	2.7	4.8	3.8
u: Strickler (m s^{-1})	2.1	6.2	3.8	4.2	3.8	3.5	6.3	5
Stream power (W m^{-2})	214	465	414	201	287	136	192	260
			Oligo–Miocene					
	a^a	b	c	d	e	f		
S	0.0021	0.0024	0.0067	0.0038	0.014	0.0028		
D_{50} (mm)	21	36	115	69	103	24		
D_{84} (mm)	47	98	218	130	176	69		
D_{90} (mm)	54	110	235	157	197	86		
D_{90}/D_{50}	2.6	3.05	2.04	2.12	1.91	3.58		
Y (m)	0.95	1.4	1.6	1.7	0.7	0.8		
n: Limerinos	0.028	0.025	0.023	0.021	0.030	0.033		
n: Strickler	0.022	0.019	0.017	0.016	0.024	0.025		
u: Limerinos (m s^{-1})	1.6	1.3	1.1	0.95	1.7	1.8		
u: Strickler (m s^{-1})	2.0	1.7	1.5	1.2	2.2	2.4		
Stream power (W m^{-2})	268	194	115	93	163	113		

^{a}Location of data collection as in Figure 4.

Palaeoslope/gradient

Two methods were used to infer palaeoslope of the proto-Rio Cinca river systems and the results were compared with gradients of modern rivers in similar tectonic settings. Firstly, through detailed mapping combined with air photograph recognition, it is possible to recognize four sets of fluvial terraces (Figs 4 & 7), many of them paired at similar topographic heights above the present Rio Cinca (Jones et al. 1999). Long profiles for each of the terrace levels have been reconstructed from which a palaeoslope for the Plio–Pleistocene gravel-bed river has been estimated (Fig. 7). Secondly, the palaeoslope for the Oligo–Miocene gravel-bed river systems was estimated using the equation derived by Paola & Mohrig (1996).

$$S_{est} = 0.094(D_{50})/h \qquad (2)$$

The median size or D_{50} from modern rivers is calculated by taking a large number of samples and then pooling the results (Table 1) and h is the depth (Ferguson & Paola 1997). The basic sampling criterion is that the largest clast in the sample accounts for no more than 0.1–1% of the total sample volume (Church et al. 1987; Wolcott & Church 1991; Milan 1999). The technique cannot be exactly replicated in indurated ancient sediments. Measurements of the conglomerates grain sizes were obtained by direct measurement of the b-axis of at least 100 clasts per sample population. This can be viewed as the application of the Wolman-count method (Wolman 1955). More recent applications have been devised using empirical demonstrations of the approximate similarity of surface and subsurface grain-size distributions when compared over a common range of sizes (Rice & Haschenburger 2004). This technique has not been applied in this study but could provide a more expedient alternative to standard methods for large, perennial gravel-bed rivers.

The sediments of the proto-Rio Cinca comply well with the sedimentological criteria that

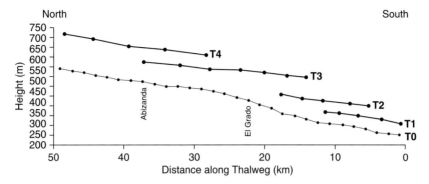

Fig. 7. Long profile of the Rio Cinca, illustrating the present day river profile and the four terrace levels. Adapted from Jones *et al.* (1999). T4, First river terrace; T3, second terrace; T2, third terrace; T1, fourth terrace; T0, present day river profile.

Paola & Mohrig (1996) suggest as being important for obtaining the best results using equation (2) (i.e. noncohesive gravel banks; the absence of rapid, flash flood deposits). Comparison of the results using the direct measurement and calculations using the Paola & Mohrig (1996) algorithm with data from modern rivers in similar tectonic and climatic settings indicates that, for gravel-bed river deposits, the palaeoslope is within the range 0.001–0.01 (Table 1).

Palaeoflow/velocity

Velocity is a factor determining a stream's competence, which is defined as the largest particle a stream can move as bedload. Competence is a measure of the stream's transporting ability and can vary along the course of the river (Komar 1987; Komar & Carling 1991). In the gravel-bed river deposits of the Oligo–Miocene and Plio–Pleistocene, where the sediments are dominated by cobble size conglomerates and cobble–boulder gravels respectively, flow competence can be estimated with a considerable degree of accuracy (Ferguson & Paola 1997). Errors arise from problems of supply limitation that may be reflected in grain size variations, but these can at least be roughly appraised by examining sedimentary structure development on lag palaeosurfaces (e.g. Dietrich *et al.* 1989).

The clast sizes were recorded for the Oligo–Miocene conglomerates and the Plio–Pleistocene gravels. Four random grid counts of 100 clasts each were made at each location. The *b*-axis of clasts larger than 10 mm were measured or, where no *b*-axis was evident, an average of the *a*- and *c*-axes was used. From these data, the 90th percentile of grain size (D_{90}) was calculated (Table 1). True maximum clast size was not used because, by definition, it represents the extreme coarse end of size distribution with little statistical significance. Several techniques can be used to equate grain size to palaeovelocity. They are mainly based on empirical relationships derived from flume and river data. To estimate velocity, a flow resistance relation and a method to estimate the resistance coefficient are needed (Ryder & Church 1986). At high flows over gravels, the various logarithmic and power law flow resistance formulae converge to the particle Manning equation, so-called when the Manning resistance number is calculated from bed grain size. An early empirical formulation of a roughness coefficient that can be applied to the Manning formula, from Strickler (see Henderson 1966; Ryder & Church 1986), is

$$n = 0.038 D_{90}^{1/6} \text{ for } D_{90} > 10 \text{ mm} \quad (3)$$

Limerinos (1970) modified the roughness relationship such that

$$n/R^{1/6} = 0.113/1.16 + 2.0 \log(R/D_{90}) \quad (4)$$

where $R \approx h$ is the hydraulic radius/[depth]. When velocities are estimated using Strickler's *n*, values are variable, but generally in excess of 5 m s^{-1} and are rather high. For Limerinos's *n*, values of between 0.95 and 4.8 m s^{-1} were calculated (Table 1). Comparison of the data with velocity criteria for designed cobble–gravel channels suggests that the Limerinos results provide the most realistic result for palaeovelocity, for all but apparently the shallowest of channels. The Strickler approach yields velocity estimates that appear less sensitive to errors in the input paramters, but are relevant only to the shallowest flows (Neill 1973). Several other equations, are discussed by Steer & Abbott (1984) and when applied to these data the results

Table 2. *Comparison of channel flow velocities from several examples of gravel-bed rivers*

Location of river	Reference	Channel velocities (m s^{-1})
Proto-Rio Cinca, Spain	This study	0.95–1.8^a
		1.6–4.8^b
Piccaninny Creek, Australia	Wohl (1993)	2.01–3.04
Eocene Ballena Gravel, California	Steer & Abbott (1984)	2.5–4
Brandywine Creek, Pennsylvania	Wolman (1955)	0.5–3
Fraser River, Canada	Ryder & Church (1986)	1.8–6

^{a}Estimated range of palaeovelocity of the Oligo–Miocene (Limerinos, m s^{-1}).
^{b}Estimated range of palaeovelocity for the Plio–Pleistocene (Limerinos, m s^{-1}).

are highly comparable with the modern Rio Cinca and modern gravel-bed rivers generally (Table 2).

Bedload transport

Bedload transport formulae are founded upon the premise that a specific relationship exists between hydraulic variables, sedimentological parameters and the rate at which the bedload is transported. Many bedload transport equations are available for generating an estimate of sediment transport rates. However, although many of the equations used arise from different deterministic approaches, the fact that the problem they try to solve is reasonably well defined inevitably means that there is considerable generic similarity. This means that the choice of one or other equation may be dictated by the availability within the database of the requisite input parameters. The review of the predictive power of bedload transport equations by Gomez & Church (1989) concluded that many of the formulae used to estimate bedload transport in gravel-bed rivers were of limited use and none were universally applicable. Despite this conclusion and the many widely used bedload transport equations (e.g. Gomez & Church 1989; Parker 1990), it is Bagnold's (1980) equation based upon stream power that provides the best fit for both flume and field measurements and is particularly useful where there are limited hydraulic data available, as in ancient gravel-bed rivers (Gomez & Church 1989; Reid & Frostick 1993; Reid et al. 1996; Garcia et al. 2000). Bagnold's (1980) equation is used in this study.

As part of this study a field method has been developed that can estimate bedload transport in ancient fluvial sediments. This involves measuring the cross-sectional area and the lateral extent of accretionary foresets of a gravel bar (Fig. 8). The gravel-bed river deposits of the proto-Rio Cinca exhibit well preserved and exposed accretionary foresets of gravel bars. These provide an ideal measure and easily comparable, well-defined units between separate sections and terrace levels. Frontal accretionary units are preferable, but units lateral and oblique to the palaeoflow direction can provide very valuable palaeohydraulic data. Relative bedload transport rate is defined as

$$i_b \Delta t = 0.25 \rho_s (bhl) \quad (5)$$

where i_b is the relative bedload transport rate, Δt is the time interval (1 year), ρ_s is the bulk density of gravel (2550 kg m^{-3}), and b, h and l are the breadth, height and length respectively of the accretionary unit (m) and the porosity of the conglomerate/gravel when initially deposited is taken as 25%. Before these raw data can be utilized to generate an estimate of a bedload transport rate, the data have to represent a rate of accretion. Several studies of modern gravel-bed river systems have implied yearly flood events and associated rates of accretion of gravel bars reflecting a strong seasonality (Steer & Abbott 1984; Ryder & Church 1986; Laronne & Duncan 1992). However, to estimate bedload transport (i_b) for the palaeoriver systems expressed as kg a^{-1} m^{-1}, it is necessary to assume that the accretion rate equates to a transport rate and ignores the bypassing of sediments, providing a minimum estimate of annual bedload transport (Table 3).

Efficiency (%)

The efficiency of a river relates the power available to that used in transporting sediment downstream (Bagnold 1980, 1986). Efficiency is defined as

$$\% = 100 i_b / (\omega / \tan \alpha) \quad (6)$$

where i_b is the relative bedload transport, ω is relative stream power and α is the internal angle of friction for cohesionless granular material (33° is widely used as suggested by Bagnold 1966). Again, for this parameter the term relative efficiency is used, especially as the accuracy of the

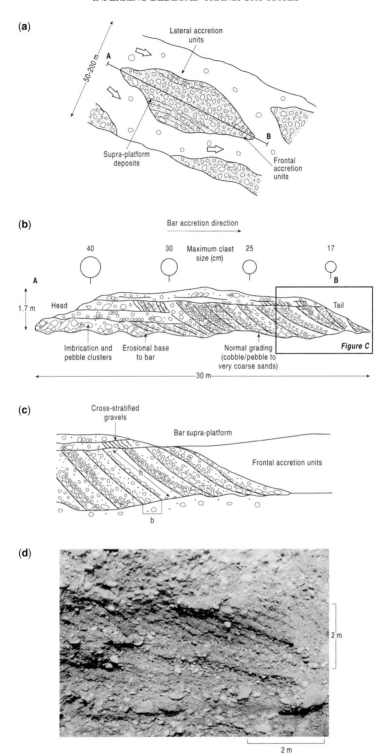

Fig. 8. Determination of a bedload transport rate from accretionary foresets of gravel bars. Such a technique is applicable to both gravel-bed and sandy braided river systems, where accretionary foresets are easily recognizable.

Table 3. *Minimum estimates of bedload transport rates and efficiency as representative examples for the Oligo–Miocene and Plio–Pleistocene fluvial sediments*

	T1[a]	T2	T3	T4
		Plio–Pleistocene		
	1	2	3	4
h (m)	1.2	0.59	1.43	1.73
b (m)	1.3	0.73	1.15	1.52
l (m)	3.1	2.61	3.4	3.5
Volume (m^3)	4.84	1.12	5.59	9.20
Bedload transport rate (kg a^{-1} m^{-1})	1.28	0.29	1.48	2.43
Efficiency (%)	0.38	0.045	0.33	0.82
		Oligo–Miocene		
	a[b]	b	c	d
h (m)	0.5	0.3	0.45	0.37
b (m)	0.3	0.25	0.38	0.26
l (m)	1.2	0.9	1.05	0.84
Volume (m^3)	0.18	0.068	0.018	0.008
Bedload transport rate (kg a^{-1} m^{-1})	0.06	0.045	0.012	0.005
Efficiency (%)	0.034	0.073	0.016	0.002

[a] Plio–Pleistocene terrace levels. Data collected from location (a) for each terrace (Fig. 4).
[b] Location of data collection as in Figure 4.

result is dependent upon both the quality of data collected and assumptions made in calculating relative stream power and bedload transport (Tables 1 & 3).

Error analysis

As with all palaeohydraulic studies, many of the variables have to be estimated using empirical relationships between modern river processes, behaviour and ancient deposits. As a result several assumptions are implicit to the process and must be considered when assessing the results.

Stream power

The two main sources of error involved in estimating stream power for ancient deposits are possible bias in estimating palaeodepth and palaeovelocity (Table 1). The estimated palaeodepths tend to be an underestimate of the actual depth. In these coarse-grained fluvial deposits, the most common preserved features affording estimates of palaeodepth are lenticular channel fills, and fining-upward successions. A fundamental problem with either of these is that the preservation is nearly always incomplete in the sense that the upper part of the feature has been removed by subsequent erosion – usually by a younger channel. Although the missing information cannot be restored, Paola and Borgman (1991) demonstrate that the error in using preserved unit thickness to estimate palaeodepth can be constrained by statistical arguments. The basis of the argument is that the bias of the depositional record towards extreme scours compensates in part for the effect of only partial preservation. It is likely that a maximum underestimate of actual depth is around a factor of 0.6 (Paola & Mohrig 1996; Paola & Borgman 1991). Thus the use of palaeodepths based on preserved thickness does not lead to severe underestimation of the true depth. In estimating a palaeovelocity the actual data are again biased towards an underestimate and the empirical relationship used is dependent on the input parameters.

The method of Paola & Mohrig (1996) used to estimate palaeoslope has an implicit grain size bias as, in general, the surface size distribution appears to be a better predictor of palaeoslope than the subsurface size distribution. Using subsurface deposits will cause palaeoslope to be underestimated. Fortunately the tendency to underestimate grain size is offset by the tendency to underestimate palaeodepth. The net result is estimated slopes (equation (3)) that are within a factor of 2 of actual slopes (Paola & Mohrig 1996).

Bedload transport

For estimates of bedload transport using equation (5), uncertainty arises from bias in estimating the porosity, the volume of bedload per accretionary unit, and the assumption that accretionary units are the product of yearly events (Fig. 8; Table 3). The porosity bias is such as to provide an underestimate. Statistically the maximum underestimate of the actual porosity is about a factor of 0.3.

Bedload transport estimated from accretionary foresets will always be an underestimate. This is because not all sediment is deposited as part of bar frontal accretion units; a proportion will be transported through the reach, and/or removed by subsequent erosional events (see Ashmore & Church 1998). All sediment that has been lost cannot be restored and, to minimize these errors, gravel bars should be carefully chosen in the field and those with extensive erosional contacts with overlying beds, usually a channel base, should be avoided. However, the use of preserved unit thickness to estimate a bedload transport rate can be constrained in a similar manner as for palaeodepth explained above (Paola & Borgman 1991; Paola & Mohrig 1996). The general bias of the depositional record towards extreme erosional scoured contacts compensates in part for the limited overall preservation of a unit. Comparison with modern braided streams the expected minimum preservation fraction of bar accretionary units is in the range of 50–60%. This does not lead to excessive underestimation of bedload transport rates.

An analysis of flows in modern rivers suggests that bankfull discharge has a recurrence interval of about 1.5 years (Leopold & Maddock 1953), but there is also a strong annual flood series that can be extracted from such data. However, the timing of bankfull discharge events varies and in several palaeohydraulic studies a 2-year or even 2.3-year flood discharge has been used (e.g. Steer & Abbott 1984). Although a yearly flood occurrence has been used in this study, when a 2-year or 2.3-year flood rate is applied, a very similar distribution of data is observed for the Oligo–Miocene and Plio–Pleistocene gravel-bed palaeo-river systems as presented below.

Suspended sediment

Suspended sediment in modern rivers will, like bedload, also be subject to non-capacity behaviour, but it is, nevertheless, more amenable to estimation using theoretical procedures (Williams 1989). Evidence suggests that suspended sediment can account for as much as 90% of the total load in perennial streams and there is some validity in the presumption that the flux of suspended sediment can be a measure of sediment yield. However, although the relationship between water and suspended sediment discharge is better than for bedload, problems of supply variations from tributaries and side slopes as well as temporary storage within, and release from, the bed and banks make prediction difficult (Frostick & Jones 2002). In this study only the bedload transport rate is calculated and as a result should not be equated to sediment yield or efflux of the Oligo–Miocene or Plio–Pleistocene river systems.

Results

The palaeohydraulic results from the Oligo–Miocene and Plio–Pleistocene gravel-bed rivers are plotted graphically to illustrate the differences between the two systems (Figs 9 & 10). Figure 9 gives relative bedload transport rate as a function of relative stream power. The trends in both sets of data are relatively well defined considering what is known about sediment transport in gravel-bed river channels (Reid & Frostick 1986). Such well-defined trends reflect a significant difference between the Oligo–Miocene and Plio–Pleistocene fluvial systems.

At lower values of relative stream power, implied bedload transport of the Plio–Pleistocene proto-Rio Cinca are as much as 400 times greater than those for the Oligo–Miocene system. The slopes of the least-squares log–log relationship between relative stream power and relative bedload transport are variable due to the spread of each data set, though it can be shown to be 2.27 and 3.79 for the Plio–Pleistocene and Oligo–Miocene systems, respectively. The extrapolation of the two data sets suggests that the two river systems would behave similarly only if the relative stream power were to approach 1200 W m^{-2}.

As the two separate data sets are interpreted to represent perennial gravel-bed rivers that were supplied with coarse erosional sediment from the uplifted Axial Zone and along the southern flank of the Pyrenees, an explanation of the differing relative bedload transport is not immediately apparent.

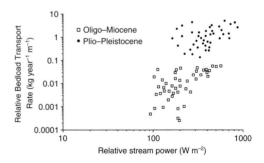

Fig. 9. Logarithmic plot of relative bedload transport rate (i_b) and relative stream power (ω) for the Oligo–Miocene and Plio–Pleistocene gravel-bed rivers (see Frostick & Jones 2002, Fig. 4).

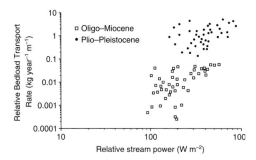

Fig. 10. Plot of relative efficiency (%) in transporting sediment as a function of relative stream power (ω) for data from the Oligo–Miocene and Plio–Pleistocene gravel bed rivers of the south central Pyrenees, Spain. This figure illustrates the effects of sediment availability on the behaviour of a river. The Plio–Pleistocene perennial river (proto-Rio Cinca) is at least an order of magnitude more efficient in carrying bedload downstream.

Differences in sediment supply, the general waning of thrust tectonic activity in the area and climatic change between the two systems are all possible controls on variations in bedload (Jones et al. 1999; Jones 2002; Frostick & Jones 2002).

Figure 10 shows the relative efficiency of each river as relative stream power varies. Not only is the Plio–Pleistocene proto-Rio Cinca much more efficient at transporting cobble to boulder size bedload (mean efficiency is 1.63% against a mean of 0.025% for the Oligo–Miocene system over approximately the same hydraulic range), but its efficiency does not vary significantly with relative stream power (Tables 1 & 3). This suggests that sediment supply and the ability of a river to transport sediment are very important parameters. It also explains the difference between the more incisive Plio–Pleistocene phase of the proto-Rio Cinca development and the aggradational Oligo–Miocene times. This relationship has not been previously identified, and is an important conclusion of this study.

Discussion

A final aim of this paper is to discuss the plausible causes for the results presented in direct comparison with recent studies of modern gravel-bed rivers and the implications for river behaviour in their middle to upper reaches.

The hydraulics of modern gravel-bed rivers and the effects of bed structure upon the timing and amounts of bedload transport have been studied in detail by many researchers (e.g. Reid & Frostick 1986, 1987; Dietrich et al. 1989; De Jong 1991; Reid et al. 1992; Hoey 1992). Understanding the detailed structure of beds in gravel-bed rivers is important, especially as armouring and development of pebble clusters will delay entrainment, as found in the proto-Rio Cinca sediments, and reduce bedload transport (Best 1996; Reid et al. 1997).

It has been suggested that similar differences in bedload transport behaviour can be explained by the different vertical structures of gravel-bed rivers. Laronne & Reid (1993) show that the difference in bedload transport and efficiency of gravel-bed rivers is primarily due to the presence of surface armour layers in perennial rivers compared with little to no armouring in ephemeral and seasonal rivers. It has been argued (Laronne et al. 1994) that poor layer development or non-layering in seasonal and ephemeral rivers is a function of rapid flood rise and fall characteristic of flash floods and that this leads to higher bedload transport rates and overall efficiency.

At first appearance the two data sets presented (Tables 1 & 3) suggest that the Plio–Pleistocene (proto-Rio Cinca) could be an ephemeral to seasonal river system compared with a perennial Oligo–Miocene system (Laronne & Reid 1993; Reid et al. 1997). However, detailed field studies show that both palaeoriver systems were perennial with well-developed armour, preserved as distinct coarse pebble layers with abundant clusters and imbrication. The similarity between ephemeral rivers (e.g. Nahal Yatir; Laronne & Reid 1993) and the proto-Rio Cinca needs to be explained.

Part of the explanation for the differences in bedload transport rates may lie with the sediment supply to each river bed. During Oligo–Miocene times the gravel-bed rivers were continually supplied with coarse erosional sediment from the Axial Zone and along its southern flank, allowing the accumulation of several hundreds of metres of pebble–cobble size conglomerates (Fig. 3d). However, during Plio–Pleistocene times the aggradational episodes identified by terrace formation, intercalated by periods of incision, represent a fluctuating sediment supply to the proto-Rio Cinca fluvial system (Fig. 4). This may be the result of waning thrust tectonics and uplift, and a change in climate from semi-arid to temperate-humid conditions at the end of the Miocene (Calvo et al. 1993; Jones et al. 1999).

Bagnold (1966) incorporated an efficiency term into the early version of his bedload equation, but few have utilized such a parameter and it is frequently neglected in hydraulic studies. If the limitations of the palaeohydraulic data set are fully appreciated then it is acceptable to calculate a relative efficiency. The Plio–Pleistocene rivers were more effective sediment carriers, than their Oligo–Miocene counterparts (Fig. 10). In addition direct observations and inference suggest that these differences in efficiency are related to the difference in sediment supply. It is possible that the greater efficiency is related to a partial breakup of the armour layer and the exploitation of the finer substrate under high-stage flood events, with restricted sediment supply. This has been illustrated recently through a series of flume experiments to investigate the behaviour of sand–gravel mixtures (Allan & Frostick 1999). These experiments reveal that the interplay between sediment winnowing, sediment removal and infiltration during gravel framework entrainment means that each deposit is the product of river discharge history. However, during periods when sediment supply was more abundant, gravel terraces formed with armouring and bedload deposits. Additionally, the notably higher efficiency of the proto-Rio Cinca during the Plio–Pleistocene is further verified by the river's improved carrying capacity for predominantly larger clast sizes (Fig. 4).

Conclusions

Ryder & Church (1986) proposed that their palaeohydraulic analyses of the Fraser River are only approximate results and that 'palaeohydraulic exercises remain essentially qualitative'. However, the data in this study illustrate that the palaeohydraulic reconstruction of the Oligo–Miocene and Plio–Pleistocene gravel-bed rivers can be estimated with a reasonable amount of confidence. The results suggest there may be palaeohydraulic criteria for distinguishing between rivers that were incising and those that were aggrading and these are summarized below.

1. Comparison of the Oligo–Miocene and Plio–Pleistocene gravel-bed river systems shows, on average, that the Plio–Pleistocene ones are up to 200–300 times more efficient at transporting coarse bedload of cobble to boulder size gravel downstream than their Oligo–Miocene counterparts.
2. This study illustrates the need to try to understand the palaeohydraulics of river systems, as it allows the character of palaeorivers to be inferred, particularly as to whether they were incising or aggrading. For a gravel-bed river with low sediment supply, as during the Plio–Pleistocene, the bedload transport rate and efficiency are high and the river incises. Alternatively, when sediment supply is abundant, the bedload transport rates and efficiency are low and the river system aggrades.

The data generated by this study are minimum estimates of bedload transport and efficiency. Their value lies in the ability to compare different phases of a single long-lived river system, but further caution is needed when using these methods to compare data from different river systems and contrasting tectonic settings.

Several attempts have been made to integrate valley formation and mountain building into models of landscape development (e.g. Beaumont *et al.* 1992; Chase 1992). The incorporation of the palaeohydraulic data of the river systems into the interpretation of tectonically active orogens is of crucial importance to the quality of these geomorphological models. The ability to infer the palaeohydraulics of ancient drainage systems, allowing comparison between rivers and their behaviour, has important consequences for understanding sediment supply, the size of material carried by rivers, the erosion of uplifting orogens, and the infill of adjacent sedimentary basins.

We thank J. Allen and T. Astin for early discussions and advice at the University of Reading. We also thank C. Paola, J. Baas, J. Best, D. Mohrig, J. Howell and A. Russell for their critical comments on earlier versions of the manuscript. B. Turner and J. Wainwright are thanked for thoughtful comments that improved the manuscript and for editorial assistance.

References

ALLAN, A. F. & FROSTICK, L. E. 1999. Framework dilation, winnowing and matrix particle size: the behaviour of some sand–gravel mixtures in a laboratory flume. *Journal of Sedimentary Research*, **69**, 21–26.

ASHMORE, P. E. & CHURCH, M. A. 1998. Sediment transport and river morphology: a paradigm for study. *In*: KLINGEMAN, P. C., BESCHTA, R. L., KOMAR, P. D. & BRADLEY, J. B. (eds) *Gravel-bed Rivers in the Environment International Gravel-bed Rivers Workshop 4*. Water Resources Publications, United States, 115–148.

BAGNOLD, R. A. 1966. *An Approach to the Sediment Transport Problem from General Physics*. United States Geological Survey Professional Paper, **422-I**.

BAGNOLD, R. A. 1980. An empirical correlation of bedload transport rates in flumes and natural rivers. *Proceedings of the Royal Society London*, **372A**, 453–473.

BAGNOLD, R. A. 1986. Transport of solids by natural water flow: evidence for a world wide correlation. *Proceedings of the Royal Society London*, **405A**, 369–374.

BAKER, V. R. 1974. Paleohydraulic interpretation of Quaternary alluvium near Golden, Colorado. *Quaternary Research*, **4**, 94–112.

BEAUMONT, C., FULLSACK, P. & HAMILTON, J. 1992. Erosional control of active compressional orogens. *In*: MCCLAY, K. R. (ed.) *Thrust Tectonics*. Chapman & Hall, London, 1–18.

BEST, J. 1996. The fluid dynamics of small-scale alluvial bedforms. *In*: CARLING, P. A. & DAWSON, M. R.

(eds) *Advances in Fluvial Dynamics and Stratigraphy.* Wiley, Chichester, 67–125.

BURBANK, D. W. 1992. Causes of recent Himalayan uplift deduced from deposited patterns in the Ganges basin. *Nature*, **357**, 680–683.

CALVO, J. P., DAAMS, R. ET AL. 1993. Up-to-date Spanish continental Neogene synthesis and paleoclimatic interpretation. *Revista Sociedad Geológica España*, **6**, 29–40.

CARLING, P., ZHIXIAN, C., KIDSON, R. & HERGET, J. 2003. Palaeohydraulics of extreme flood events: reality and myth. *In*: GREGORY, K. J. & BENITO, G. (eds) *Palaeohydrology: Understanding Global Change.* Wiley, Chichester, 325–336.

CHASE, C. G., 1992. Fluvial landsculpting and the fractal dimension of topography. *Geomorphology*, **5**, 39–57.

CHURCH, M., MCLEAN, D. G. & WOLCOTT, J. F. 1987. River-bed gravels: sampling and analysis. *In*: THORNE, C. R., BATHURST, J. C. & HEY, R. D. (eds) *Sediment Transport in Gravel-bed Rivers.* Wiley, Chichester, 43–87.

CHURCH, M., WOLCOTT, J. & MAIZELS, J. 1990. Paleovelocity: a parsimonious proposal. *Earth Surface Processes and Landforms*, **15**, 475–480.

DE JONG, C. 1991. A reappraisal of the significance of obstacle clasts in cluster bedform dispersal. *Earth Surface Processes and Landforms*, **16**, 727–744.

DIETRICH, W. E., KIRCHNER, J. W., IKEDA, H. & ISEYA, F. 1989. Sediment supply and the development of the coarse surface layer in gravel-bedded rivers. *Nature*, **340**, 215–217.

FERGUSON, R. & PAOLA, C. 1997. Bias and precison of percentiles of bulk grain size distributions. *Earth Surface Processes and Landforms*, **22**, 1061–1077.

FROSTICK, L. E. & JONES, S. J. 2002. Impact of periodicity on sediment flux in alluvial systems: grain to basin scale. *In*: JONES, S. J. & FROSTICK, L. E. (eds) *Sediment Flux to Basins: Causes, Controls and Consequences.* Geological Society, London, Special Publication, **191**, 81–95.

GARCIA, C., LARONNE, J. B. & SALA, M. 2000. Continuous monitoring of bedload flux in a mountain gravel-bed river. *Geomorphology*, **34**, 23–31.

GARDINER, T. W. 1983. Paleohydrology and paleomorphology of a Carboniferous meandering, fluvial sandstone. *Journal of Sedimentary Petrology*, **53**, 991–1005.

GOMEZ, B. & CHURCH, M. 1989. An assessment of bed load sediment transport formulae for gravel bed rivers. *Water Resources Research*, **25**, 1161–1186.

HENDERSON, F. M. 1966. *Open Channel Flow.* Macmillan, New York.

HIRST, J. P.P. & NICHOLS, G. J. 1986. Thrust tectonic controls on alluvial sedimentation patterns, southern Pyrenees. *In*: ALLEN, P. A. & HOMEWOOD, P. (eds) *Foreland Basins.* International Association of Sedimentologists Special Publication, **8**, 247–258.

HOEY, T. 1992. Temporal variations in bedload transport rates and sediment storage in gravel-bed rivers. *Progress in Physical Geography*, **16**, 319–338.

JONES, S. J. 1997. The evolution of alluvial systems in the south central Pyrenees, Spain. PhD thesis, University of Reading.

JONES, S. J. 2002. Transverse rivers draining the Spanish Pyrenees: large scale patterns of sediment erosion and deposition. *In*: JONES, S. J. & FROSTICK, L. E. (eds), *Sediment Flux to Basins: Causes, Controls and Consequences.* Geological Society, London, Special Publication, **191**, 171–185.

JONES, S. J. 2004. Tectonic controls on drainage evolution and development of terminal alluvial fans, southern Pyrenees, Spain. *Terra Nova*, **16**, 121–127.

JONES, S. J., FROSTICK, L. E. & ASTIN, T. R. 1999. Climatic and tectonic controls on fluvial incision and aggradation in the Spanish Pyrenees. *Journal of the Geological Society London*, **156**, 761–769.

JONES, S. J., FROSTICK, L. E. & ASTIN, T. R., 2001. Braided stream and flood plain architecture: The Río Vero Formation, Spanish Pyrenees. *Sedimentary Geology*, **139**, 229–260.

KOMAR, P. D. 1987. Selective gravel entrainment and the empirical evaluation of flow competence. *Sedimentology*, **34**, 1165–1176.

KOMAR, P. D. & CARLING, P. A. 1991. Grain sorting in gravel bed streams and the choice of particle sizes for flow competence evaluations. *Sedimentology*, **38**, 489–502.

LARONNE, J. B. & DUNCAN, M. J. 1992 Bedload transport paths and gravel bar formation. *In*: BILLI, P., HEY, R. D., THORNE, C. R. & TACCONI, P. (eds) *Dynamics of Gravel Bed Rivers.* Wiley, Chichester, 177–202

LARONNE, J. B. & REID, I. 1993. Very high rates of bedload sediment transport in desert ephemeral rivers. *Nature*, **366**, 148–150.

LARONNE, J. B., REID, I., YITSHAK, Y. & FROSTICK, L. E. 1992. Recording bedload discharge in a semiarid channel, Nahal Yatir, Israel. *Erosion and Sediment Transport Monitoring Programmes in River Basins, Proceedings of the Oslo Symposium.* IAHS Publication, **210**.

LARONNE, J. B., REID, I., YITSHAK, Y. & FROSTICK, L. E. 1994. The non-layering of gravel stream beds under ephemeral flood regimes. *Journal of Hydrology*, **159**, 353–363.

LEOPOLD, L. B. & MADDOCK, T. 1953. *The Hydraulic Geometry of Stream Channels and some Physiographic Implications.* United States Geological Survey Professional Paper, **252**.

LIMERINOS, J. T. 1970. *Determination of the Manning Coefficient from Measured Bed Roughness in Natural Channels.* United States Geological Survey Water-Supply Paper, **1898B**.

MAIZELS, J. 1989. Sedimentology, paleoflow dynamics and flood history of jokulhlaup deposits; paleohydrology of Holocene sediment sequences in southern Iceland sandur deposits. *Journal of Sedimentary Research*, **59**, 204–223.

MIALL, A. D. 1978. Facies types and vertical profile models in braided river deposits: a summary. *In*: MIALL, A. D. (ed.) *Fluvial Sedimentology.* Canadian Society of Petroleum Geologists Memoir, **5**, 597–604.

MUÑOZ, J. A. 1992. Evolution of a continental collision belt: ECORS-Pyrenees crustal balanced cross-section. *In*: MCCLAY, K. R. (ed.) *Thrust Tectonics.* Chapman & Hall, London, 235–356.

MILAN, D. 1999. Influence of particle shape and sorting upon sample size estimates for a coarse-grained upland stream. *Sedimentary Geology*, **129**, 85–100.

NEILL, C. R. (ed.) 1973. Guide to bridge hydraulics. *Roads and Transport Association of Canada*. The University of Toronto Press, Toronto.

NICHOLS, G. J. & HIRST, J. P. 1998. Alluvial fans and fluvial distributary systems, Oligo–Miocene, northern Spain: contrasting processes and products. *Journal of Sedimentary Research*, **68**, 879–889.

PAOLA, C. & BORGMAN, L. 1991. Reconstructing random topography from preserved stratification. *Sedimentology*, **38**, 553–565.

PAOLA, C. & MOHRIG, D. 1996. Palaeohydraulics revisited: palaeoslope estimation in coarse-grained braided rivers. *Basin Research*, **8**, 243–254.

PARKER, G. 1990. Surface-based transport relation for gravel rivers. *Journal of Hydraulic Research*, **28**, 417–436.

PUIGDEFÀBREGAS, C. 1975. La sedimentacion molasica en la cuenca de Jaca. *Pirineos*, **104**, 1–118.

PUIGDEFÀBREGAS, C., MUÑOZ, J. A. & VERGÉS, J. 1992. Thrusting and foreland basin evolution in the southern Pyrenees. *In*: MCCLAY, K. R. (ed.) *Thrust Tectonics*. Chapman & Hall, London, 247–254.

REID, I. & FROSTICK, L. E. 1986. Dynamics of bedload transport in Turkey Brook, a coarse-grained alluvial channel. *Earth Surface Processes and Landforms*, **11**, 143–155.

REID, I. & FROSTICK, L. E. 1987. *Toward a Better Understanding of Bedload Transport*. Society of Economic Paleontologists and Mineralogists Special Publication, **39**, 13–19.

REID, I. & FROSTICK, L. E. 1993. Fluvial sediment transport and deposition. *In*: PYE, K. (ed.) *Sediment Transport and Depositional Processes*. Blackwell, Oxford, 89–155.

REID, I., FROSTICK, L. E. & BRAYSHAW, A. C. 1992. Microform roughness elements and the selective entrainment and entrapment of particles in gravel-bed rivers. *In*: BILLI, P., HEY, R. D., THORNE, C. R. & TACCONI, P. (eds) *Dynamics of Gravel-bed Rivers*. Wiley, Chichester, 253–276.

REID, I., POWELL, D. M. & LARONNE, J. B. 1996. Prediction of bedload transport by desert flash-floods. *Journal of Hydraulic Engineering ASCE*, **122**, 170–173.

REID, I., BATHURST, J. C., CARLING, P. A., WALLING, D. E. & WEBB, B. W. 1997. Sediment erosion, transport and deposition. *In*: THORNE, C. R., HEY, R. D. & NEWSON, M. D. (eds) *Applied Fluvial Geomorphology for River Engineering and Management*. Wiley, Chichester, 95–135.

RICE, S. P. & HASCHENBURGER, J. K. 2004. A hybrid method for size characterization of coarse subsurface fluvial sediments. *Earth Surface Processes and Landforms*, **29**, 373–389.

RYDER, J. M. & CHURCH, M. 1986. The Lillooet terraces of Fraser River: a palaeoenvironmental enquiry. *Canadian Journal of Earth Sciences*, **23**, 869–884.

SEIDL, M. A. & KIRCHNER, J. W. 1994. Longitudinal profile development into bedrock: an analysis of Hawaiian channels. *Journal of Geology*, **102**, 457–474.

STEER, B. L. & ABBOTT, P. L. 1984. Paleohydrology of the Eocene Ballena gravels San Diego County, California. *Sedimentary Geology*, **38**, 181–216.

VINCENT, S. J. & ELLIOTT, T. 1997. Long-lived transfer-zone paleovalleys in mountain belts: an example from the Tertiary of the Spanish Pyrenees. *Journal of Sedimentary Research*, **67**, 303–310.

WILLIAMS, G. P. 1984. Paleohydraulic equations for rivers. *In*: COSTA, J. E. & FLEISHER, P. J. (eds) *Developments and Applications of Geomorphology*. Springer, Berlin, 343–367.

WILLIAMS, G. P. 1989. Sediment concentration versus water discharge during single hydrologic events in rivers. *Journal of Hydrology*, **111**, 89–106.

WOHL, E. E. 1993. Bedrock channel incision along Piccaninny Creek, Australia. *Journal of Geology*, **101**, 749–761.

WOLCOTT, J. & CHURCH, M. 1991. Strategies for sampling spatially heterogenous phenomena: the example of river gravels. *Journal of Sedimentary Petrology*, **61**, 534–543.

WOLMAN, M. G. 1955. *The Natural Channel of Brandywine Creek, Pennsylvania*. United States Geological Survey Professional Paper, **271**.

Planar landforms as markers of denudation chronology: an inversion of East Pyrenean tectonics based on landscape and sedimentary basin analysis

MARC CALVET[1] & YANNI GUNNELL[2]

[1]*Department of Geography, Médi-Terra, Université de Perpignan, France*
[2]*Department of Geography, UUMR 8591 CNRS, France*
(e-mail: gunnell@univ-paris-diderot.fr)

Abstract: For over half a century the Pyrenees were considered to be a mountain range in which compressional structures were ancient (pre-Oligocene) but topography was young due to late Neogene tectonic uplift. Sufficient time had been afforded for a 'peneplain' to form at low elevations, undergo vertical uplift and remain partially preserved at high elevations until present times. This model of topographic growth has since been challenged by alternative theories. One of these postulates that topography in active orogens is in a steady-state, hence mountain ranges must be monocyclic and their 'peneplains' must have formed at high altitudes during continental convergence as a result of raised foreland base levels. Here we investigate Pyrenean denudation chronology using a range of evidence including provenance stratigraphy, the cross-cutting relations between topographic and tectonic features, and the age of regolith based on fossil faunas and floras. We find that the Eastern Pyrenees underwent a punctuated topographic evolution until recent times driven primarily by tectonic forcing, including kilometre-scale rock and surface uplift after 12 Ma. Climatic and eustatic inputs were subsidiary driving mechanisms.

In addition to sheet-like geological formations such as lava flows or alluvial terraces, planar landforms can be used as markers to quantify topographic uplift. This makes them special among the diverse forms of topography at the Earth's surface. The accuracy of vertical displacement values relies on evidence concerning predeformational geometry and age of the preserved feature. Owing to the limited life span of planar landforms <1 km in length scale (e.g. fluvial or marine rock terraces) in high-energy environments, due also to limitations on the time depth of dating techniques, few such landforms allow displacements older than the Pleistocene to be estimated. Equally significant, however, are discrete remnants of Cenozoic erosion surfaces because these allow inferences to be made on topographic change reaching further back in time. Topographic evolution in mountain belts is commonly reconstructed by assuming long-term steady-state conditions, or by inverting apatite fission-track datasets that lack independent constraints from the field (e.g. Morris *et al.* 1998). However, in complex areas such as the Mediterranean collision zone, non-equilibrium conditions in late Mesozoic to Cenozoic mountain belts are likely to prevail for longer periods of time than steady-state topography can ever be realistically maintained by unsteady tectonic stress fields. Although the intuitive recognition of non-equilibrium in most of the Earth's relief systems is largely rooted in Davisian ideas on landscape evolution (Davis 1899), other approaches such as thermochronometric methods have only recently begun to be used to elucidate non-equilibrium situations in mountain belts. This raises challenging geodynamic questions because the geomorphic evidence for flat topography preserved at high elevations requires either limited long-term denudation since the erosional plain was formed, or endogenous mechanisms to have caused recent, and often rapid, surface uplift of the relict landscapes (e.g. Clark *et al.* 2005).

In this paper, we explore the implications of erosion surface occurrence in the Eastern Pyrenees as a basis for making statements about the history and evolution of mountain topography. We systematically link the erosion surfaces to other related geological archives in the landscape, namely the sedimentary and palaeontological record. We therefore propose to invert the tectonics and climate from a reading of the landscape and stratigraphy. As such, the study can stand alone as a conventional, multidisciplinary field investigation, but we feel it can also serve as a basis for promoting sampling strategies for low-temperature thermochronology. Given the sensitivity to parameter choices often encountered in fission-track analysis (e.g. Ketcham 2005), greater precision is gained from so-called forward approaches, in which thermal histories are fitted empirically to predefined, independent geomorphological and stratigraphic constraints. A forward approach predicts what combination of measured fission-track or (U–Th)/He parameters would be expected from a sample that has undergone a predefined t–T history. Such independently established histories benefit from constraints and data such as presented in this paper.

Erosion surfaces in mountain belts

The existence of elevated erosion surfaces in mountain ranges has been a subject of much debate and speculation for over a century. They have long been interpreted as uplifted peneplains, implying that the orogen was eroded down to base level and subsequently uplifted to its current elevation without the mature surfaces being entirely erased from the scenery by fluvial dissection. This view was held, for instance, for the Laramide Rockies by Davis (1911). Birot (1937) and de Sitter (1952) also reached similar conclusions for the Pyrenean Axial Zone, which corresponds to the area of pre-Mesozoic outcrops in the most uplifted and deeply eroded part of the Pyrenees (Fig. 1).

Erosion surfaces are key landforms in most landscapes because they express an attenuation of pre-existing relief, in which elevation differences between valley floors and interfluve summits have been reduced even though the final result is never as perfectly flat as young depositional or structural surfaces. Whereas structural benches in the topography are shaped by differences in rock resistance, and are therefore essentially an expression of rock mechanics, erosional bevels cut rocks of varying resistance as well as geological dips and structures. Such low-gradient topography is therefore not just a geological surface: it is the expression of a geomorphic history governed by erosional processes.

In mountain belts, erosion surfaces can be used as geomorphic gauges of the intensity of forcing mechanisms, particularly tectonics, but also climate or drainage incision, during a particular time interval. Incised erosion surfaces express a resumption of unequal erosion between valleys and interfluves, usually due to an intensification of those forcing mechanisms. Contrary to oversimplifications of Davisian doctrine, erosion surfaces may form, at least locally, at a variety of elevations above sea level providing raised base levels exist to achieve this. It can therefore not always be ruled out that scattered local erosion surfaces in a mountain range were formed locally and independently of one another instead of representing a single, regional erosional plain that has been subsequently dissected and/or fragmented by faulting.

However, the fact that this possibility constitutes a theoretical alternative to the raised peneplain model need not rule out the latter *a priori* simply because it connects conceptually with disregarded Davisian theory (see, e.g. Phillips 2002 for a discussion). Furthermore, also contrary to widespread belief, erosion surfaces are not horizontal: just as marine abrasion platforms exhibit natural seaward slopes of up to 1° in order to allow evacuation of debris, erosion surfaces grade to a local or regional base level. Because they are not initially horizontal, their local slope is therefore not an easily established function of deformational tilt by subsequent tectonic movements.

It has long been observed that, compared with the Central Pyrenees and contrary to the expected steep topography of many mountain ranges, the Eastern Pyrenees are relatively flat-topped. In order to readdress this problem after ca. 70 years (Birot 1937), we initially establish where the erosion surfaces are situated within the orogen and how many generations can be distinguished. This exhaustive survey forms the basis for subsequently investigating how old the erosion surfaces are, and what they tell us about topographic change in the orogen: did they form at their currently observed altitudes, or must we accept that they formed at lower elevations and were more recently uplifted?

Methods

For the purpose of this synthesis, we processed the STRM (2004) digital elevation data base for the Pyrenean region in order to extract a regional distribution of low-relief topography across the area corresponding to the orogenic double wedge. The procedure consisted of generating a slope map derived from a nine-pixel moving window and extracting topographic areas with mean slope angles <8°. We preferred this to an alternative approach, which would have involved mapping local relief instead of slope (e.g. Babault *et al.* 2005), because mapping local relief using coarse-pixelled DEMs (i) is apt to drown out smaller, locally discrete but qualitatively important surface remnants, and (ii) implicitly relies on a subjective evaluation of how smooth an erosion surface should be in order to fit the definition. Defining the acceptable upper relief threshold of an erosion surface is a contentious aspect of erosion surface recognition that brought some disrepute to Davis's flexible definitions of peneplain morphology (Phillips 2002).

Clearly, the slope-based approach is equally subject to large geomorphological error given that DEM analysis cannot distinguish low-angle depositional or structural slope systems from low-angle

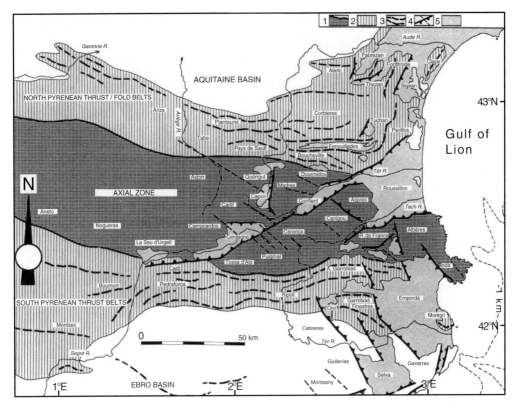

Fig. 1. Eastern Pyrenees: structural and tectonic setting. 1, Dissected high ranges in Palaeozoic rocks, abrupt mountain front; 2, outer Pyrenean fold and thrust belts with outer boundary of orogenic wedge; 3, trace of main Pyrenean hog- and razorbacks; 4, major extensional fault scarps of the eastern domain; 5, main topographic basins of eastern domain.

erosional topography. First-hand knowledge of Pyrenean topography, geological structure, and glacial forms and history is therefore essential in order to filter out all pixels that might represent non-erosional low-gradient topography, such as low-angle thrust sheets, recumbent fold limbs, planar bedding, glacial cirque and trough floors, alluvial fan and floodplain surfaces, cap-rock mesas, etc. Based on a systematic analysis of 1:25,000 topographic maps, geological maps, aerial photographs, extensive field mapping and overviews using an aircraft, we completed this task by manually eliminating all the <8° surfaces that corresponded to non-erosional topography. The resulting map is provided in Figure 2.

Based on this ground work, we then proceeded to investigate all the possible ways available in the field to age-bracket the defined erosion surfaces individually or collectively. Owing to the absence of datable volcanic deposits, age-bracketing hinged on stratigraphic correlations with marine and continental sedimentary sequences in adjacent basins based on provenance stratigraphy, and relative chronologies based on cross-cutting relations between tectonic discontinuities and topographic bevels. On limestone surfaces, a biostratigraphy of karstic fissure fillings allowed high-precision dating of the topography bevelling the outer thrust sheets forming the Corbières.

Results

At the orogen scale, erosion surfaces occur in the eastern Axial Zone, where they cut Palaeozoic igneous and metamorphic rocks as well as some Palaeozoic limestones, but are quite extensive in the outer fold belts of both the north (Corbières, Plantaurel) and south (Catalan thrust sheets and piggyback basins) where they bevel folded Mesozoic and Cenozoic limestone and sandstone conglomerate. The Pyrenees is a small orogen: c. 400 km in length and 80–100 km wide, i.e. only a little larger than the Colorado Front Range, which is also renowned for its erosion surfaces (Bradley 1987). The hypothesis that such a regionally extensive occurrence of erosion surfaces could represent a coherent, regional history rather than a haphazard collection of locally conditioned bevels cannot

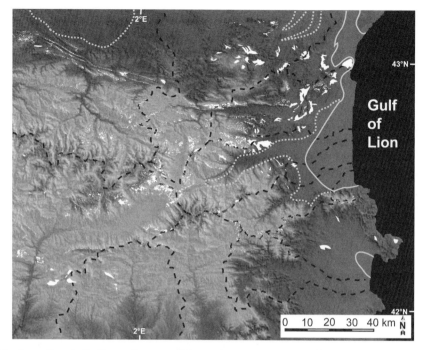

Fig. 2. Erosion surfaces in the Eastern Pyrenees. White patches represent topographic areas with gradients <8° bevelling all lithologies, structural dips and tectonic discontinuities. At this scale, the map does not differentiate between the two main generations identified: summit surfaces and pediments (see text). Solid grey line: Middle Miocene marine palaeoshoreline. Dotted grey line: inner edge of nonmarine Middle Miocene clastic sediment deposits. Black dashed line: current drainage divides. Digital elevation model derived from SRTM (2004).

therefore be ruled out. In tentative support of a regional hypothesis, we emphasize that, during the Miocene, the palaeo-Pyrenean mountain belt was surrounded on three sides by extensive marine and lacustrine bodies (Fig. 2). Only one of them, the semi-arid Ebro basin, was internally drained between Priabonian and late Pliocene times (Riba *et al.* 1983). The mountain fronts were connected to these ancient base levels to the north and south by clastic piedmonts, the northern side having been on occasions compared palaeoecologically to the modern Himalayan *terai* on the basis of fossil Miocene faunas and floras. In the east, the Pyrenees connected to the Mediterranean basin through a series of grabens that rapidly filled with clastic sediments between Chattian and Burdigalian times. The Miocene palaeoshoreline during the Langhian–Serravallian high stand occurred slightly inland of the current coastline. All these base levels exerted major controls on the erosion of the palaeo-Pyrenees.

We now assess the legacy of this erosional history through two combined approaches: the identification of two key generations of erosion surfaces in the landscape based on their relative elevations, and the Neogene erosional signal detected in the clastic basins and karstic sediment traps in the eastern portion of the Axial and outer Pyrenean zones.

Two generations of planar landforms

Any given massif in the broad area where erosion surfaces occur usually exhibits a two-tiered landscape: an upper summit surface capping the more elevated topography, and a lower pediment embayed in the residual topography that carries the upper surface. This tiered configuration is particularly striking in the scenery of the Carlit and Aston massifs, and in the eastern Corbières. The consistency of this configuration at the regional scale suggests two generations, and hence two stages, of erosion surface development (Fig. 3).

S, the residual summit surface

The most spectacular set of planar remnants in the Eastern and outer Pyrenees caps the highest topography irrespective of local lithology and structure (Fig. 3). Occurrence of residual relief on this surface is exceptional (a singular exception is seen in Fig. 3d), suggesting that this planar topography

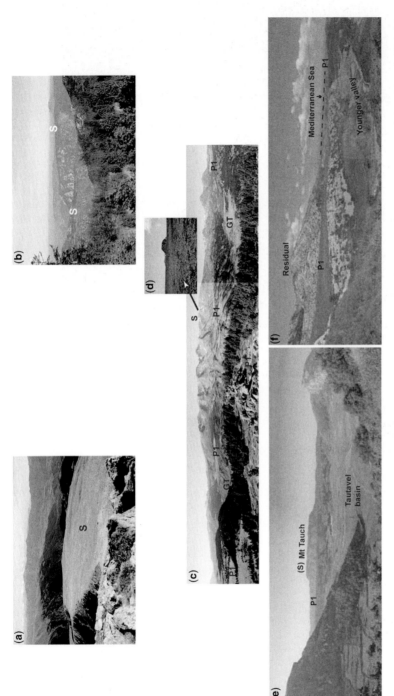

Fig. 3. Aspects of the East-Pyrenean erosion surfaces S and P1. (**a**) S in the Axial Zone cutting basement rocks (Tossa Plana, 2.9 km a.s.l., looking SE over the Cerdagne basin). (**b**) S cutting folded limestone (forêt des Fanges, 1 km a.s.l., looking north). (**c**) Panorama of the Carlit massif, looking north. Small remains of S ((**d**) 2.8 km a.s.l.) on the skyline have survived the development of the much vaster pediment P1 (c. 2.2 km a.s.l.), grading towards the Cerdagne basin and incised only by a few glacial troughs (GT). (**e**) In the Eastern Corbières, vale of Tautavel (foreground, famous for its Neanderthal history) eroded in shale below the level of P1 (middle ground, c. 400 m a.s.l.), itself notched into the residual *massif du Tauch*. This massif, like many other residual plateaux of the Corbières, is bevelled by S (here at c. 740 m a.s.l.). (**f**) P1, here cut in limestone, sloping eastward to the Mediterranean Sea (distinguishable in the distance), grading topographically to Middle Miocene shoreline deposits, and embayed in hilly, residual basement terrain.

must have been part of a remarkably smooth, and hence unmountainous, palaeolandscape. Obviously, it could be argued that high peaks existed where today deep valleys and clastic basins occur, so that the elevated planar summits we see today could represent inverted topography. However, such a (gratuitous) hypothesis would require this to be systematically true for all valleys across the entire mountain belt and so is too improbable to be pursued beyond this point. The summit surfaces are preserved near major drainage divides between the three regional base levels (Fig. 2). As such, they show no preferential spatial connection to the Ebro, Aquitaine or Mediterranean base levels. This immediately casts doubt on attempts to interpret their existence in terms of links to only one of them, for instance the Ebro basin, which is the most tempting to focus on (Coney *et al.* 1996; Babault *et al.* 2005) because it was internally drained for 30 Ma. (between 37 and ca. 6 Ma). Furthermore, the summit surfaces bear no coherent spatial relation with current drainage arteries, which is another sign of their antiquity. For all those reasons, it seems unlikely that the summit surfaces represent a random collection of local features fortuitously sharing the same topographic status in the topographic hierarchy. A regional interpretation cannot therefore be ruled out, in which the elevation of these surfaces can be extrapolated laterally to join up and define a regionally extensive palaeo-surface of low relief deformed by more recent block faulting.

P1, a spatially extensive generation of younger pediments

The population of lower pediments differs from the 'S' population by systematically occurring at lower relative elevations and being systematically dominated by higher (or 'commanding') topography (Fig. 3c), irrespective of absolute elevation values. The mean relief between the summit surfaces and the pediments ranges between *c*. 600 m in the Axial Zone (e.g. Fig. 3c) and 200–400 m or less in the outermost fold belts, as for instance in the eastern Corbières (Fig. 3e, f), the north and west margins of the Roussillon and the Emporda basins. Inland from the Mediterranean coast, they can, furthermore, be dated as post-middle Miocene given that they grade topographically to the middle Miocene Mediterranean palaeoshoreline (Fig. 2) and bevel tilted Oligocene to Aquitanian strata in the Sigean and Narbonne basins (Figs 1 & 4; Calvet 1996). Locally (see Fig. 9), younger pediments (P2, P3, etc.) have also been mapped, but remain of little relevance to the general overview we present in this paper.

Age of the erosion surfaces

Elevated low-relief surfaces in mountain ranges can, theoretically, be either pre-orogenic, synorogenic or post-orogenic. Davis popularized the view that many mountain ranges were post-orogenic uplifted peneplains. Other models, examined below, have imagined alternatives to this viewpoint.

Where cover rocks are preserved, elevated surfaces in young mountain ranges can represent exhumed stratigraphic unconformities. An example of this is the Beartooth Plateau (Montana and Wyoming), where residual Palaeozoic sandstone buttes capping the low-relief Precambrian basement surface indicate that the entire plateau was once covered by sedimentary rocks now almost entirely stripped. Stripping may be syn- or post-orogenic, but such surfaces are of little geomorphic use to understand the history of more recent topographic growth in mountain belts. The Pyrenean erosion surfaces, however, are not ancient stratigraphic unconformities fortuitously exhumed during the deep denudation of the orogenic double wedge. Two fundamental reasons support this affirmation: (i) a wide range of structural and lithological settings is affected by the erosional bevels across the breadth and length of the mountain belt, and at a wide range of elevations. Furthermore, (ii) sedimentary sequences containing abundant clasts from Axial Zone basement lithologies were being supplied as early as Bartonian times to the northern retro-foreland (Crochet 1991) and to the southern foreland basins (Reille 1971; Burbank *et al.* 1992*a*; Mató *et al.* 1994; Muñoz *et al.* 1994; Vergès *et al.* 1994). The conglomerate sequences are 2–3 km thick, so pre-orogenic palaeotopography cutting Axial Zone basement rocks could simply not have survived such deep erosion and still be preserved in the landscape.

Another potential origin for elevated low-relief summit surfaces in mountain ranges is that they may be cryoplanation (or 'altiplanation') terraces (Bryan 1946), with two major implications: (i) they are post-orogenic in age, and specifically very recent Quaternary features; and (ii) they were formed as dispersed local features rather than a regional continuum. This view not only sidesteps the exhumed pre-orogenic unconformity theory, but firmly discounts the raised peneplain theory. However, it bestows on periglacial processes formidable powers of rock bevelling that have been vigorously disputed (Guilcher 1994; Migon 2003). More generally, the assumption that the low declivity of a land surface should be the result of the most recently active geomorphic processes and their extant debris is potentially spurious. Even though periglacial processes can generate low-gradient topography locally, in the Pyrenean setting we found no evidence of periglacial activity on the lower surfaces of the outer fold belts. At higher elevations in the Axial Zone, periglacial processes have generated little more than superficial ornaments (e.g. weakly patterned ground), and the erosion surfaces frequently retain 1–10 m thicknesses of *in situ*, deeply weathered and rubified grus (Fig. 4). The pediments have also often been reshaped by glacial processes, particularly where

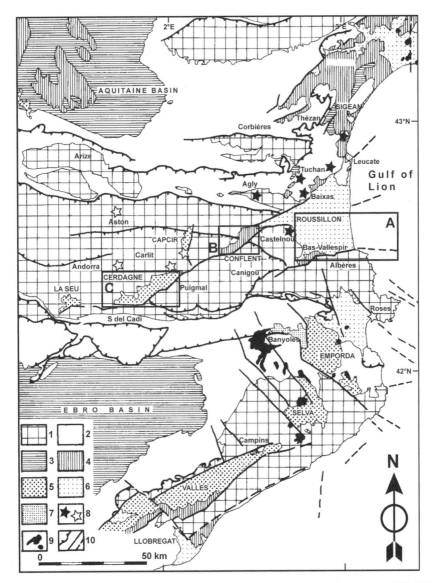

Fig. 4. Neogene deposits and structural setting in the Eastern Pyrenees. 1, Hercynian basement; 2, folded Mesozoic and Cenozoic cover sequences including foreland homoclinal strata; 3, Oligocene conglomerate and lake deposits of foreland basins and Languedoc grabens; 4, Lower and Middle Miocene sedimentary sequences (clastic and marine); 5, Upper Miocene continental sequence, and undifferentiated Miocene to Pliocene rocks (Selva); 6, marine and continental Pliocene sequence of Mediterranean seaboard; 7, Upper Pliocene and Lower Pleistocene palaeolacustrine deposits of Banyoles basin; 8, type areas where Neogene biostratigraphic ages were obtained (black stars) and key sites with preserved late Neogene palaeoweathering profiles (white stars); 9, Miocene to Pleistocene basaltic volcanism; 10, trace of main Paleogene thrusts; other faults and Neogene normal faults; blind faults. A, B and C, areas corresponding to Figs 5, 6 & 8, respectively.

major iceways have cut troughs into the planar topography (Fig. 3c), but many of those surfaces exibit grus mantles preserved beneath glacial till, suggesting that erosion by ice caps was unequally effective across the pre-glacial landscape. Finally, the periglacial theory would require that the planar surfaces were clustered within an elevation belt morphoclimatically propitious to frost riving, but this is challenged by the fact that the erosion surfaces occur from 2.9 km a.s.l. to near sea level.

In summary, the periglacial hypothesis is inconsistent because (i) it cannot explain any specifically observed local occurrence, and (ii) collectively it does not have the capacity to explain all occurrences across the region. Consequently, the low-gradient topography must have formed earlier than Quaternary times. Our investigations have revealed that the erosion surfaces all bevel Pyrenean tectonic structures (Calvet 1996), all of which recorded the latest known syntectonic compressive deformation no later than c. 32 Ma (Burbank et al. 1992a, b), though lasting longer (to 24.7 Ma) in the outermost folds despite much lower intensities of deformation after 29.5 Ma (Meigs et al. 1996). We therefore conclude that the erosion surfaces are post-orogenic but older than the Quaternary, and therefore necessarily middle to late Cenozoic.

Finer constraints on planar surface ages in the outer fold belts

Means of investigating the erosional history of the Eastern Pyrenees are unequally distributed between the outer fold belts, which are dominated by limestone lithologies and where it has been possible to date karst fissure fillings and their fossil mammalian remains; and the Axial Zone, where the scenery is dominated by basement outcrops and where the only archives of denudation (aside from low-temperature thermochronology, not addressed here) are contained in a limited set of intermontane basins such as the Conflent, Bas-Vallespir, Roussillon, Cerdagne and Seu d'Urgell (Fig. 4). Both sets of evidence, however, converge on a consistent denudation chronology for the Eastern Pyrenees.

Biochronology of karst fissure fillings (Corbières, Aspres, Roussillon)

Great stability and antiquity of the surfaces where the Pyrenees connect with the Mediterranean basin has been inferred from a widely distributed set of fossil micromammalian assemblages preserved across limestone surfaces that cut across the Devonian and Mesozoic limestone outcrops of the Corbières, Aspres and Roussillon (Aguilar et al. 1986, 1999; Faillat et al. 1990). The faunal remains are contained in hardened sandy–clayey sediment either lining small solution pits and conduits (<1 m deep), or within cracks of the limestone pavements, but also filling narrow tectonic fissures, some widened by karstic corrosion. The deepest penetrate c. 40 m below the topographic surface, but the surface occurrences are by far the most numerous and faunistically the richest. Each site offers a faunistically homogeneous assemblage, formed in situ and characteristic of a regionally established biozone. Mixed populations are exceptional and, given the excellent state of preservation of teeth (including the most brittle variety, such as bat's teeth) and the frequent occurrence of conjoinable bones, indices of reworking or disturbance are few (Aguilar & Michaux 1997). Frequently, bone-rich breccia corresponds to almost intact aggregates of Neogene raptor regurgitation pellets.

At the regional scale, these sites occur from the coast to over 40 km westward in the Corbières and Agly massif. Ages extend from 26 Ma to the Pleistocene, but the highest age concentration is in the Miocene. At the local scale, the Thuir plateau exhibits 13 known sites ranging from Chattian to early Pleistocene times concentrated over 0.8 km². The Baixas plateau, which grades to the northern edge of the Roussillon basin, numbers 46 sites dated from 21 to 1 Ma and scattered across an area of 1.5 km². Thirty-nine of those sites concentrate ages between 19 and 10 Ma. They occur in fissures, but 23 sites occur directly on the topographic surface: one Quaternary, two Upper Pliocene, one Lower Pliocene, three Tortonian, eight Serravallian and eight Burdigalian, the oldest being 18.5 Ma. Sites only a few metres apart, for instance, are 3.5, 11 and 13.5 Ma. Taken together, this evidence suggests extremely limited depths of denudation since 26 Ma, necessarily less than the depth of fissured rock within which the fossil remains are contained (i.e. c. 40 m). Corresponding denudation rates are 0.1 m/Ma, with maxima of 1.9 m/Ma. The presence of thin and fine-grained allochthonous deposits (arkose, sand, small pebbles of crystalline rocks) covering the limestone pavements after 18 Ma also indicates that these erosional limestone ramps represented smooth topography over which material from the eroding Axial Zone was being supplied to the Roussillon and adjacent basins (Calvet 1992). The small size of the clasts and their siliciclastic nature also suggests that the relief supplying them in the hinterland was subdued – a further confirmation that topography in the Agly and Madres massifs was probably no longer mountainous during middle Miocene times.

The reason why ages range from late Oligocene to Pleistocene times is because both the Thuir and

Baixas plateaux occur on the edge of the Roussillon basin, i.e. at the node where the more deeply eroding portions of the mountain belts to the west (Pyrenean Axial Zone) and north west (Corbières) grade to basins that have been depositional ever since rifting in the Mediterranean began c. 30 Ma ago. As a result, elevation and age differences between S and P1 at those sites are indistinct, although the middle Miocene modal fossil age band fits what we interpret as representing the age of P1, which is by far the most extensive occurrence throughout the Eastern Pyrenees in terms of preserved surface area. The topographic separateness between S and P1 increases northward and westward from these nodal, low-energy sites (see Fig. 3). As a result, if similar fossil sites were to be found in the future on surfaces more clearly distinguishable as S or P1, the age brackets for each would also probably be more clearly defined.

Neogene intermontane basins record a punctuated history of erosion

A systematic study of the clastic sequences present in the Bas-Vallespir and Conflent basins, with correlations eastward to the Roussillon basin, has provided some detail to the timing and rates of erosion in the Axial Zone. In the Bas-Vallespir (Fig. 5), the total thickness of barren red bed deposits reaches 2 km, and the provenance stratigraphy indicates that the debris was sourced simultaneously from the Albères horst in the south and, mostly, from the upper Tech drainage basin and the Canigou massif (Fig. 4). Correlation of eastward-dipping benchmark strata into the adjacent Roussillon basin, where on- and offshore oil wells have provided a chronologically constrained stratigraphy, suggests a late Aquitanian to early Burdigalian age for the Bas-Vallespir clastic sequence. This correlates with the nearby Conflent basin, as well as the Vallès basin in Spain, where similar Burdigalian red beds occur. In the Conflent extensional basin (Fig. 6), the clastic sequence represents sourcing of rapidly eroded weathering profiles and bedrock from the Canigou massif. Here it was possible to date the sediments *in situ* using rodent palaeofaunal remains (Baudelot & Crouzel 1974; Bandet 1975; Guitard *et al.* 1998). Results suggest a late Aquitanian to early Burdigalian episode of rapid denudation in the eastern Axial Zone.

The Bas-Vallespir and Conflent basins flank the south and north sides of the Canigou massif, respectively. Together, they therefore provide insight into the erosion history of the Canigou massif. The coherent provenance stratigraphy, biochronology and planar surface distribution suggest a punctuated history of uplift and erosion of Mt Canigou (Fig. 7). The stratified structure of the

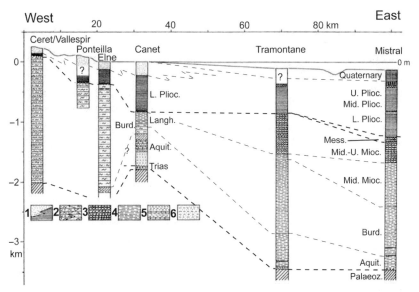

Fig. 5. Stratigraphic correlation and age of the Bas-Vallespir red beds. 1, Pliocene (marine and continental deposits); 2, terrigenous red beds (cobble and pebbles, sands and clays); 3, coastal and lagoonal deposits of regressive phase with (3a) and without (3b) gravels; 4, Miocene (marine gravel and sand or clay-sand beds); 5, continental facies – lignitic clays, marls; 6, continental facies (sandstone and grey clays). Westernmost stratigraphic log based on direct observation; other logs based on borehole data after Bourcart (1947), Gottis (1958), Cravatte *et al.* (1974) and Berger *et al.* (1988).

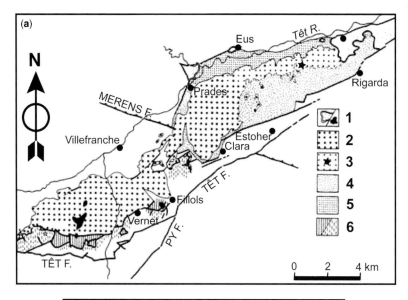
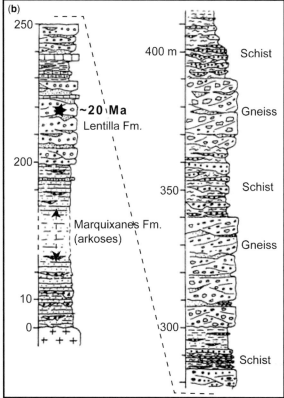

Fig. 6. The Conflent basin. (**a**) 1, Pliocene sediments; Col de Fins Formation (Upper Miocene); 2, Escaro Formation; 3, Burdigalian Lentilla Formation (star: site dated using faunal biostratigraphy); 4, Codalet Formation; 5, Marquixanes Formation. Formations 2, 3 and the top of 4 are coeval facies; 6, Gneiss and schist olistoliths supplied by the active Canigou fault palaeoscarp in Burdigalian times. (**b**) Synthetic log of Conflent basin. The star indicates dated stratum using fossil mammal biostratigraphy (Bandet 1975; Baudelot & Crouzel 1974; Guitard *et al.* 1998).

Canigou massif, in which the lithologically contrasted metamorphic nappe stack has been well characterized by detailed geological mapping (Guitard 1970; Guitard et al. 1998), allows good constraints to be placed on palaeorelief based on the basin provenance stratigraphy and sedimentology. Briefly stated, the erosional pulse recorded between c. 21 and 15 Ma resulted in extreme attenuation of relief between the palaeo-Canigou massif and its flanking basins (Fig. 7). Sediments in the Bas-Vallespir mirror the unroofing chronology of the Canigou dome, with the lower half of the series containing only micaschist and upper granitoid clasts. Massive supply of augengneiss only occurred later. The clastic fill history in the Conflent basin is identical, with micaschists at the base (Codalet Fm) followed by augengneiss clasts forming the Lower Burdigalien strata (Lentilla and Escaro formations). Pediment systems sloping towards these basins developed during this phase

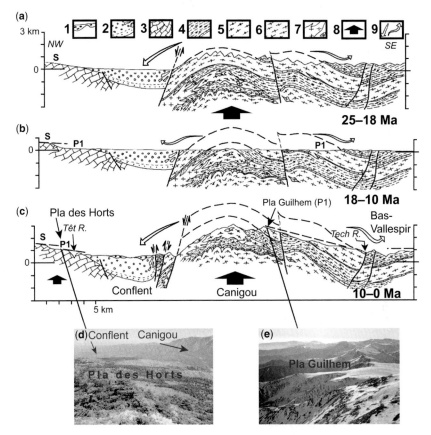

Fig. 7. Post-orogenic evolution of Mount Canigou in the Neogene. (**a**): Kilometre-scale uplift was accompanied by coarse clastic infilling of the Conflent basin between 25 and 18 Ma. Sediment input from the north (Madres massif) was subdued after Aquitanian times due to limited uplift of the northern flank (Marquixanes arkoses). Meanwhile, the southern flank recorded a basal influx of schist clasts followed by an exclusive input of olistoliths and boulders of augengneiss. (**b**) relative tectonic stability and pedimentation (18–10 Ma). Erosion only supplied fine-grained sediment to the Roussillon and Mediterranean basins (sandy and argillaceous molasse). (**c**) Latest stage of kilometre-scale uplift responsible for the present-day relief. Denudation has reached the core of the Canigou metamorphic dome structure. Micaschists, granites, migmatites and Casemi leptynites, which together form the entire exposure of the current scarp, are all represented in clastic material at the Col de Fins (Upper Miocene) and in the lower Conflent basin (Pliocene). (**d**) view from the Pla des Horts, looking SE. (**e**) The Pla Guilhem in winter, looking east from Mt Canigou. Key to ornaments: 1, Col de Fins Formation; 2, Aquitanian and lower Burdigalian; 3, Devonian marble and limestone; 4, Lower Palaeozoic micaschist and basal marble; 5, Casemi leptynites; 6, Precambrian augengneiss; 7, Canigou two-mica granite core; 8, Neogene vertical uplift; 9, clastic outflux towards the Conflent Basin, Roussillon rift and the Mediterranean passive margin. S and P1 as in text.

of subdued erosion. In that area they formed the erosion surfaces still preserved north of the Conflent (e.g. Pla des Horts, Fig. 7d) and west of Mt Canigou (Pla Guillem, Fig. 7e), respectively.

Regional uplift after *c.* 12 Ma, attaining maximum values locally in the Canigou massif, subsequently raised the low-gradient topography and initiated its dissection. Proof of this late uplift event is provided by the fact that the entire Têt Fault escarpment, with its 2 km throw, exposes the core of the metamorphic dome, i.e. the deeper micaschist, leptynite and two-mica granite lithologies of Mt Canigou, none of which were supplied to the Lentilla and Escaro formations in Burdigalian times. This vigorous uplift caused erosion that mostly destroyed the pediments in the palaeo-Canigou itself, but they remain better preserved in the north (Fig. 7). Contrary to the late Aquitanian–early Burdigalian phase of rapid uplift, the magnitude of mass flux from Mt Canigou since 12 Ma has been relatively limited, possibly due to climatic differences, but also because better integrated drainage systems have bypassed the sediment output to the outer Roussillon basin instead of the proximal Conflent and Vallespir palaeodepocentres. As a result, tectonic uplift rates have outpaced erosion rates, and neotectonic landforms, such as the northern Canigou facetted scarp front, are fairly well preserved.

Evidence of post-12 Ma regional uplift and related topographic dissection

We further substantiate that tectonics, rather than climate, was the prime factor of relief growth with an analysis linking palaeorelief evolution in the Axial Zone with the sediment stratigraphy of the Cerdagne half-graben (Figs 1, 4 & 8). The Cerdagne basin (Roca 1986; Agusti & Roca 1987; Cabrera *et al.* 1988) records 10 Ma of Tortonian and post-Tortonian geodynamic history, which has involved a subtle succession of extension, compression, erosion and uplift in the Axial Zone (Fig. 9). Firstly, the very existence of such a recent extensional basin in the Axial Zone cannot put tectonic forcing in a position subsidiary to climatic change or the Messinian salinity crisis as relief-forming factors in the Eastern Pyrenees. Secondly, its youth must lead us to accept a topographic continuum within the Axial Zone between the massifs to the north (Carlit, Andorra) and to the south of the basin (Serra del Cadí, Tossa d'Alp, Puigmal) prior to its existence. Our reconstruction (see Fig. 9 for details) is based on detailed provenance stratigraphy, lake sediments, dated faunal assemblages, a relative chronology of tectonic deformations based on the strata they affect, and geomorphic analysis using the distribution of

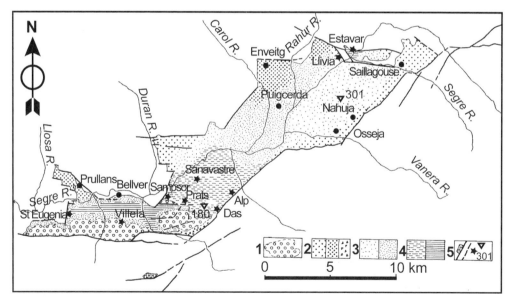

Fig. 8. The Cerdagne basin. 1, Red loams and conglomerate incorporating limestone boulders of Turolian (=Messinian) series; 2, coarse-grained fluvial and torrential facies of Vallesian (=Tortonian) series (schist, gneissic, limestone clasts); 3, fine-grained fluvial gravels, sands and lignitic loams (schist, granitic clasts); 4, deltaic (sand, clay, lignite) and lacustrine (grey clay and diatomites) facies; 5, Upper Cretaceous outcrop; Neogene faults; dated stratum based on faunal biostratigraphy (star); borehole (triangle) with depth in metres (base of Neogene sediments not reached).

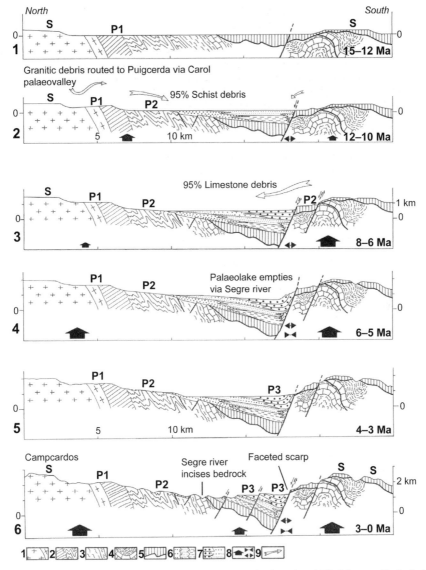

Fig. 9. Morphotectonic evolution of the Cerdagne intermontane basin since the middle Miocene. Geological ornaments: 1, Hercynian monzogranites and granodiorites; 2, Cambro-Ordovician schists with metamorphic aureole; 3, Silurian (Caradoc) schists; 4, Devonian limestone series and Carboniferous flysch; 5, sole of allochthonous Devonian limestone nappe; 6, Lower Vallesian sequence: torrential, fluvial, lacustrine and palustrine facies; 7, Upper Turolian sequence – boulder-sized debris-flow, torrential, fluvial and minor palustrine deposits; 8, vertical uplift, extension, compression; 9, clastic sourcing during late Miocene. Chronology: 1, pediment (P1) develops during middle Miocene after tectonic movements deform the summit surface (S); 2, Vallesian series is deposited during a first extensional phase on the southern boundary fault; limited uplift allows a pediment P2 to develop; 3, vigorous surface uplift of southern block (Tossa d'Alp, Sierra del Cadí, Puigmal, Canigou); 4, fault throws increase rapidly on boundary fault during late or post Turolian times; 5, pediment P3 develops cutting across upturned Turolian strata (Col de la Perche pediment); 6, Quaternary vertical uplift increases topographic elevation of Axial Zone and fragments the half graben into compartments separated by fresh faceted scarps. Note: in the main text we do not discuss pediments P2 and P3 in order to smooth out local detail of no relevance to the main model. This, however, shows (i) the importance of local variability during the recent morphotectonic history of the Pyrenees and (ii) the abundance of clues available in the field to detect relative chronologies based on topography, stratigraphy, tectonics and their cross-cutting relations.

pediments sloping towards the topographic basin (Calvet 1996). In Vallesian times (continental equivalent of the Tortonian), clastic material was mostly sourced by the northern massifs. Despite localized debris-fan facies, the alluvium is dominated by fine sand and clay containing quartz pebbles: this indicates erosion of deeply weathered saprolite. In contrast, during Turolian times (continental equivalent of the Messinian), coarse clastic fan sequences suggest a vigorous reactivation of the southern master fault, with continued fault-drag deformation during the Pliocene, as shown by the upward deflection of the entire Turolian sequence against the Serra del Cadí (Fig. 9).

Discussion: how did the surfaces form and at what palaeoelevation?

Given that 125 years of debate over stepped erosion surfaces in various parts of the Laramide Rockies have still not settled uncertainty and scepticism over their exact origin (e.g. Bradley 1987; Evanoff 1990; Gregory & Chase 1992; Leonard 2002), it would be premature to claim that the Pyrenean context is much more clear-cut. However, as already emphasized, the Pyrenees is a small and narrow orogen surrounded throughout its later history by three marine or lacustrine base levels. By comparison, the Laramide Rockies are vast and remote from marine base levels. Statistically, they are therefore likely to encompass a wider and initially more confusing range of possible scenarios. Furthermore, since the early Neogene the southern Rockies have been partly modified by the impinging Rio Grande rift system and its related volcanism.

Piedmont backfilling and the graded pediment model

Among the range of possible origins of elevated erosion surfaces in mountain belts, earlier we mentioned two major ones and explained why they were not relevant to the Pyrenean setting: exhumed pre-orogenic unconformities, and wholesale Quaternary cryoplanation. The only coherent conclusion borne out by the field evidence was that the surfaces were post-orogenic and pre-Quaternary. This age bracket, however, does not establish how the surfaces formed or at what palaeoelevation. The Davisian approach would typically postulate that the summit surface is a regional peneplain resulting from the erosion of the palaeo-Pyrenees down to its roots, followed by uplift and partial redissection. A typical anti-Davisian approach would be to suggest that, on the contrary, the erosion surfaces formed at their current elevations in a setting controlled by overfilled foreland basins where clastic backfill during orogenic convergence and uplift promoted opportunities for dwarfing relative relief. Such conditions would have readily promoted the erosion of any topography in the Axial Zone still rising above the roof of the clastic fill (Coney et al. 1996; Babault et al. 2005). This kind of conceptual model has already been applied locally in the Andes to explain erosion surfaces graded to local depositional sites generated by cut-and-fill processes (Gubbels et al. 1993). It remains compatible with the concept of steady-state orogeny because the erosion surfaces are postulated to develop during tectonic convergence instead of after it (Babault et al. 2005). Nevertheless, just like the Davisian approach, this model-dependent interpretation contains for the Pyrenees inconsistencies that we now proceed to discuss in the light of field evidence.

The key question we address first is whether the Axial Zone summit surfaces represent the upper continuation of aggradational piedmont ramps. As suggested by our reconstructions, this appears to be the case: the summit bevels (S) formed between c. 32 and c. 20 Ma in relation to aggradational surfaces forming in the three marine or lacustrine base levels. Aquitanian sediments east of the Salat river (Aquitaine piedmont) contain a few pebble beds (pebble diameter <10 cm) while debris in the overlying and conformable Burdigalian deposits does not exceed a diameter of 1–5 cm and exclusively consists of chemically resistant clasts (quartz pebbles). Such upward fining (Calvet, unpubl. data) confirms that, in the Burdigalian palaeo-Pyrenees, slope systems grading to the Aquitaine piedmont were not steep.

The fragmentation of the Axial Zone into horsts (Canigou, etc.) and grabens (Conflent, Vallespir) reduced the size of mountain blocks in the palaeo-Pyrenean mountain belt and thereby increased opportunities for reducing relief in limited geological time (<15 Ma). Given that no summit erosion surfaces have been identified in the Central Pyrenees (Fig. 10), where compression was sustained later into Neogene times and no extensional basins exist, this further supports the tentative link we establish between rapid development of planar surfaces in the crest zone of the Eastern Pyrenees on the one hand, and Mediterranean-related extensional tectonics on the other.

The clastic ramp backfilling scenario challenges the Davisian model of peneplain uplift. However, the backfilling option is also partly problematic because no tangible evidence exists to support the view that foreland clastic ramps, whether on the foreland or the retro-foreland side, backfilled into the Axial Zone. Implicitly (Coney et al. 1996), the backfilling model imagines a clastic ramp reaching

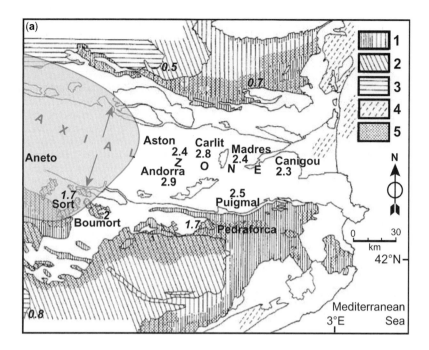

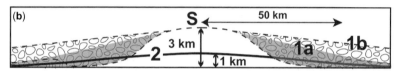

Fig. 10. Sediments in the Pyrenean foreland and retro-foreland basins and their use as tools to reconstruct palaeotopography. (**a**) Syntectonic clastic sequences. 1, Eocene (marine and continental); 2, Oligocene of Aquitaine basin, and unconformable Priabonian and Oligocene of Catalonia; 3, continental Miocene of Ebro and Aquitaine basins; the latter grade laterally to marine strata towards the NW; 4, continental Middle and Upper Oligocene of eastern basins (Miocene rocks of Mediterranean basins not shown: refer to Figs 1, 4 & 5); 5, coarse synorogenic conglomerate of Paleocene series (Palassou, Berga, Nogueras, etc.). Transparent shading: area of the Central Pyrenees devoid of summit erosion surfaces. Figures indicate elevation in kilometres (bold: Axial Zone summit surface S; italic: highest elevation of top of foreland clastic sequences). (**b**) Schematic of a two-sided steady-state mountain belt sourcing clastic sequences to its foreland basins (e.g. the palaeo-Pyrenees during the Paleocene). Declivities of piedmont ramps consisting of aggrading braided alluvial fans typically range between 1‰ and 1% (Table 1). Any palaeoslope exceeding that range (1a, grey shading) is unrealistic. In order to satisfy this constraint, the idea of erosion surfaces having formed at their current elevations and grading to piedmont ramps must satisfy one of two possibilities: either the sediments were much thicker and extensive than can currently be extrapolated from their observed elevations and spatial distributions (1b, pebble ornament), a hypothesis for which there is no proof either on the north or the south side of the Pyrenees; or, the source area (i.e. the palaeo-Pyrenees crest zone) was much lower than today and has since been uplifted (2). The most consistent solution based on field relations is hypothesis 2. Note that, in the case of the Eastern Pyrenees, three basins occur around the mountain topography, with an additional (Mediterranean) and somewhat younger basin extending out of the page towards the reader.

the crest of the Axial Zone in a manner resembling the well known Wyoming 'Gangplank' (Leonard 2002), where a clear topographic and sedimentary link can be established between the Ogalalla clastic sequence and the Tertiary pediment upon which the formation wedges out. In the Pyrenees, however, such an hypothesis remains unsupported by well-preserved geomorphic configurations analogous to the Gangplank setting. While accepting that the raised base level model seems empirically valid for local P1-type pediments surfaces in the intermontane basins (this study), was the East Pyrenean S-type summit 'peneplain' formed at its currently high elevation because of raised base levels in both the foreland basins? In the following subsection, we outline reservations against this simplistic view both in terms of absence of evidence and evidence of absence.

Palaeoelevation of the Eastern Pyrenees in middle Miocene times

The backfill hypothesis has been formulated by workers focusing exclusively on the south side of the orogen (e.g. Babault et al. 2005), lured by the appeal of a supposedly stable, internally drained base level (the palaeo-Ebro basin) to which 'bathtub' models of overfill and sudden mass release to the Mediterranean basin after the Ebro river breached the Catalan Ranges can be applied (e.g. Babault et al. 2006). This, however, is debatable in the context of a narrow and two-sided mountain belt such as the Pyrenees, where the issue of symmetry must be addressed given the distribution of erosion surfaces in all three base-level catchments of the orogen (cf. Fig. 2). Belief that the Ebro basin overfilled all the way back into the Axial Zone is refuted by the elevation and dip of the Oligocene conglomerate beds that are alleged to support this view: the roof of the outcrop of conglomerates closest to the Axial Zone (Fig. 10) reaches a maximum elevation of 1.7 km near Sort, i.e. >1 km lower than the glacially scuplted crest zone of the Central Pyrenees where, moreover, no clearly defined Cenozoic erosion surfaces exist (Fig. 10). Furthermore, the conglomerate outcrops do not penetrate into the Axial Zone. They occur only just south of the Axial Zone's mountain front. The conglomerate beds are clearly exposed in the landscape between the Noguera Pallaresa and the Noguera de Tort, and they mostly dip c. 20° towards the Axial Zone – therefore not basinward as would be expected in the case of a backfilling clastic sequence.

In order to raise the piedmont to the desired current elevation of the Pyrenean crest zone, an additional c. 1 km of missing backfill (since removed by erosion?) would have to be imagined. However, where the missing material went, what its grain size and lithology might have been, and where the palaeorelief that supplied it was situated, remain unanswered questions and highlight the gratuity of the hypothesis. In summary, backfilling of both foreland basins to the current c. 3 km a.s.l. of the Axial Zone summit surface (S) is unsubstantiated. The summit surface is more likely to have formed at a much lower palaeoelevation than today.

A further reason to challenge the view that the Axial Zone summit surface formed at its currently observed elevation is a simple geometric one: the projected slope values from the Axial Zone summit surfaces to proximal early Miocene clastic outcrops in the Aquitaine piedmont over measured distances of 50–70 km (Fig. 10) range between 3 and 4%. Such critical transport slope values are high. No alluvial fan undeformed by tectonics, and no actively aggrading megafan surface is known to attain such depositional gradients (Table 1) without involving fluvial incision and abandonment (Miall 1996). In order to restore more realistic megafan gradients of 0.1–1% for the graded palaeosurface model advocated here to be physically applicable (e.g. Horton & DeCelles 2001), we calculate that the Axial Zone summit surfaces must have formed at maximum palaeoelevations of 0.8–1 km. The early to middle Miocene palaeo-Pyrenean orogen east of the Salat river would thus have appeared as a narrow range of low hills, partly buried under its own debris with its crest zone barely rising above its aggrading piedmonts. This conclusion is broadly consistent with the views of de Sitter (1952), who postulated that 'by the end of the Miocene, the Pyrenees represented ... a gently undulating, very mature landscape, almost a peneplain with low hills, which in the center does not rise above 1000 m altitude'.

Admittedly, the mean natural slopes of fluvial-dominated Quaternary fans reported in the literature attain maximum values of c. 5% at active range fronts in northern Tibet (Meyer et al. 1998; Mériaux et al. 2005). However, such studies apply to shorter length scales (<20 km) than the ≥50 km-long Pyrenean foreland megafans portrayed by Vincent & Elliott (1997), Babault et al. (2005) and this study. Furthermore, those high-gradient fans occur in tectonically active areas reputed for high Quaternary slip rates on the Altyn Tagh fault, or where the fans ride active blind-thrust ramps with developing anticlines. Furthermore, they occur in piedmont areas backed by mountain ranges

Table 1. *Critical transport slopes on fans (after Miall 1996)*

Fan type (dominant process)	Morphometric characteristics			
	Fan slope	Fan length (head to toe) (km)	Fan width (km)	Maximum fan height (km)
Debris-flow	0.1–0.01	<5	<10	<0.25
Braided fluvial	0.001–0.0003	<40	<120	<0.15
Meandering fluvial	c. 0.0003	<150	250	<0.04

>2 km high and supplying large, poorly weathered clasts in an hyper-arid environment. This is clearly at odds with the vision of 4–5% clastic ramps reaching the bevelled summits of the Miocene Pyrenees, where the expanding 'peneplain' should have been merely supplying sand- to gravel-sized clasts to the overfilled piedmonts.

Some uncertainty therefore still exists in fan and palaeofan classification, as few studies sytematically provide useful detail on clast sizes and their spatial distribution, and how slope changes from apex to toe for a given fan length (this affects the slope-averaging procedure). Sporadic distinctions have been proposed between alluvial 'fans' generated by nodal avulsion of an unconfined single stream downstream of a bedrock-confined reach, and active alluvial 'slopes' ($2°$ or more if ≤ 15 km in length) in which flow remains channelized on the piedmont (see Smith 2000). These distinctions have hypothetically been linked to the abruptness, relief and intensity of tectonic activity of the range front. Such alluvial 'slopes', however, are reported from basin and range settings rather than collision foreland basins. Overall, it is therefore difficult to return precise palaeoelevation estimates for the formation of the Pyrenean summit peneplain. Nevertheless, given existing data on megafan slopes (cf. Horton & DeCelles 2001), a kilometre-scale component of post-orogenic uplift is a realistic proposition for the Eastern Pyrenees.

Finally, the mostly thermophilic Tortonian floras of the Cerdagne basin, currently preserved at >1 km a.s.l., share 80% of the plant species occurring in the Empordá basin, which is, and has long been, at sea level because it contains marine Tortonian sediments (Menendez Amor 1955; Alvarez Ramis & Golpe Posse 1981; Sanz de Ciria 1981). Despite the mixed fortunes, owing to large error, of using palaeofloras to determine palaeoaltitudes (Meyer 2003), this nevertheless suggests comparable initial palaeoelevations for those two basins, and hence rules out the possibility that the Cerdagne basin in Tortonian times was already an elevated (1 km a.s.l.), semi-arid basin surrounded by the high topography (>2 km a.s.l.) we observe today.

Post-orogenic crustal uplift in the Eastern Pyrenees: geodynamic implications

It now becomes clear that the model presented here of morphotectonic evolution for the Eastern Pyrenees is a hybrid between the Davisian view, which postulates peneplanation to sea level followed by vertical uplift and dissection; and the anti-Davisian view, which advocates altiplanation at ca. 3 km a.s.l. permitted by raised piedmont base levels due to foreland basin overfilling. According to the strict definition outlined by Davis (see overview in Phillips 2002), a peneplain simultaneously fulfills five conditions: (i) the topography is erosional (fluvial and subaerial), (ii) of low relief, (iii) truncates all rock types, is (iv) cut to near base level, and (v) is subcontinental in extent. We suggest that, in the Pyrenees, conditions (iv) and (v) are not met. The initial regional erosion surface (S) formed at intermediate elevations, although its exact palaeoelevation depends on the thickness of piedmont clastic fill, time and geometric constraints discussed earlier.

At sites of most vigorous uplift since c. 18 Ma, and depending on initial palaeoelevation (see above), the topography has locally gained up to 1.5 km in surface elevation. In the outer fold belts, uplift has been more limited. This composite picture suggests that, in the case of the summit surface S, we are dealing with a single regional surface differentially uplifted, dissected and fragmented after c. 12 Ma. Pediments defined as P1 all graded to separate local base levels (marine in the Corbières, clastic half-grabens in the Axial Zone, etc.) depending on location. The population of P1 pediments thus developed locally, and not necessarily strictly coevally, at different elevations (cf. Fig. 3) in response to late Neogene transtension, coastal fluctuations and fluvial incision.

Magnitudes of uplift and exhumation of blind thrust sheets from beneath the syn-orogenic conglomerate cover have been greater in the Ebro pro-wedge than in the Aquitaine retro-wedge basin, where outer Pyrenean thrusts and folds are still buried under the Cenozoic clastic wedge. This asymmetry occurs because the South-Pyrenean thrust zone has been more deeply eroded than the Aquitaine foreland. Climatic asymmetry between the more humid Aquitaine basin and the semi-arid Ebro basin would predict that the Aquitaine basin should erode more deeply than the Ebro. This is the reverse of what is observed, so we must reject a climatic hypothesis. If climatic asymmetry had been a long-term forcing factor in shaping the Pyrenees, the better supplied and externally drained Aquitaine drainage would have encroached much further back across the orogen's axis and conquered some of the Ebro drainage. Instead, drainage systems are fairly symmetrical on either side of the crest zone. This suggests that tectonic uplift has counteracted potential imbalances caused by the climatic asymmetry, and has therefore remained the main driving force behind the history of the topography until most recent times.

Another possible cause for deeper erosion of the southern thrust system is drainage integration: despite debate over the exact date and cause of the Ebro basin connecting to the

Mediterranean basin at some time after 10 Ma (Garcia-Castellanos et al. 2003; Babault et al. 2006), it is possible that this event had significant repercussions throughout the lower basin in terms of dissecting sedimentary strata, propagating knickpoints upstream towards the Axial Zone, and thus causing an isostatic response to denudation as headward erosion rejuvenated the landscape. However, models addressing this idea have reached diverging conclusions (Babault et al. 2006), so this hypothesis currently requires more evidence. Finally, and more simply, Bouguer gravity anomalies (Barnolas et al. 1996) show that crustal thickening of the Pyrenean orogen is asymmetric, the crust being noticeably thicker beneath the south-central Pyrenees than beneath the northern half of the wedge. Isostatic responses to denudation would therefore be of a greater magnitude on the Spanish than on the French side, which would in turn explain why deeper levels of denudation and exhumation of thrust systems from beneath the clastic wedges, in addition to greater mean topographic elevations, are observed on the Spanish side.

Although differences in isostatic response may explain differences in denudation depths on each side of the orogen, isostasy does not readily explain the kilometre-scale magnitude of rapid surface uplift that we inferred specifically from palaeosurface distribution in the eastern ranges. Likewise, isostasy does not explain the extensional tectonics that have generated the footwall uplands of the eastern Axial Zone. The most widely accepted preconditions for rock and/or surface uplift are crustal thickening, thinning of mantle lithosphere or dynamically maintained buoyancy. As in the Californian Sierra Nevada (Clark et al. 2005), mechanisms for explaining recent uplift in the Eastern Pyrenees are therefore still speculative. The neighbouring Catalan Ranges, for instance, attain elevations of 2 km. This has been interpreted as Neogene and Quaternary rift-related uplift involving flexural response to denudation, but nevertheless also involving an unspecified component of mantle-supported dynamic uplift given the thinness of the crust (Lewis et al. 2000). The Eastern Pyrenees share the same proximity to the Mediterranean extensional setting as the Catalan Ranges, and the distinctive geomorphic history of the Eastern as opposed to the Central or Western Pyrenees calls for an examination of subcrustal signatures by geophysical methods.

Conclusion

Based on systematic DEM, Ordnance Survey map and ground truth surveys, we have documented the distribution of erosion surfaces in the Pyrenees and debated the respective strengths of conflicting geomorphic theories in accurately characterizing Pyrenean topography. Our constraints on East Pyrenean ages and palaeoelevations during the Neogene indicate that the surfaces are all postcollisional and were formed by middle Miocene times at ≤ 1 km above palaeosea level, i.e. at elevations locally >1.5 km lower than at present. Based on this we are led to confirm that (i) denudation in this active mountain belt has been spatially heterogeneous, and that (ii) a large component of current topography in the Axial Zone is the result of post-Burdigalian uplift. At places, denudation lagged behind rock uplift, otherwise the planar markers would not have survived to this day. This suggests a long-term evolution of topography far from equilibrium involving both (i) an increase in mean elevation; and (ii) an increase in local relief occurring in the last 18 Ma due to both river entrenchment and topographic fragmentation caused by Mediterranean extensional tectonics affecting the orogen.

Our qualitative model superficially espouses several of Davis's views. The main difference is that the uplifted planar topography is not a base-level peneplain *sensu* Davis. Instead, it is a palaeosurface graded to marine or lacustrine base levels via low-gradient piedmonts, and formed at ≤ 1 km above sea-level despite having been further uplifted and dissected subsequently. Exactly what lithospheric mechanism was responsible for recent uplift and dissection of the low-gradient topography remains at this stage open to debate. Given that the morphotectonic make-up of mountain belts varies laterally along strike in far greater proportion than would suggest superficial categorizations of them as large-scale physiographic units (e.g. Clark et al. 2005, this study), erosion surfaces in mountain belts may have a range of origins depending on sub-regional and local history. The uplifted peneplain and its conceptual opposite, altiplanation, are just end members in a spectrum of more subtle causes and effects that often remain difficult to unravel objectively in terms of the three recognized forcing factors of terrestrial relief evolution: namely tectonics, climate and drainage integration. This perspective leads us to ponder the humorous words of Alexander von Humboldt (1769–1859), naturalist and contemporary of William Smith (1769–1839), whom this volume commemorates: 'There are three stages in scientific discovery: first people deny that it is true, then they deny that it is important, finally they credit the wrong person.'

We thank Eric Leonard for a very thorough review of the manuscript. Constructive comments from Michael Summerfield, Francis Lucazeau and Franck Audemard also improved a previous version of this manuscript.

References

AGUILAR, J.-P. & MICHAUX, J. 1997. Les faunes karstiques néogènes du Sud de la France et la question de leur homogénéité chronologique. Actes du Congrès Biochrom' 97, Montpellier, 14–17 avril, Ecole Pratique des Hautes Etudes, Mémoires et Travaux de l'Institut de Montpellier, **21**, 31–38.

AGUILAR, J.-P., CALVET, M. & MICHAUX, J. 1986. Découverte de faunes de micromammifères dans les Pyrénées-Orientales (France) de l'Oligocène supérieur au Miocène supérieur; espèces nouvelles et réflexion sur l'étalonnage des échelles continentales et marines. Comptes Rendus de l'Académie des Sciences Paris, série II, **303**, 755–760.

AGUILAR, J.-P., ESCARGUEL, G. & MICHAUX, J. 1999. A succession of Miocene rodent assemblages from fissure fillings in southern France: palaeoenvironmental interpretation and comparison with Spain. Palaeogeography, Palaeoclimatology, Palaeoecology, **145**, 215–230.

AGUSTI, J. & ROCA, E. 1987. Síntesis bioestratigráfica de la fosa de la Cerdanya (Pirineos orientales). Estudios Geologicos (Madrid), **43**, 521–529.

ALVAREZ RAMIS, C. & GOLPE POSSE, J. M. 1981. Sobre la paleobiología de la cuenca de Cerdanya (depresiones pirenaícas). Boletin Real Sociedad Española Historia Natural (Geologia), **79**, 31–44.

BABAULT, J., VAN DEN DRIESSCHE, J., BONNET, S., CASTELLTORT, S. & CRAVE, A. 2005. Origin of the highly elevated Pyrenean peneplain. Tectonics, **24**, TC2010, doi:10.1029/2004TC001697.

BABAULT, J., LOGET, N., VAN DEN DRIESSCHE, J., CASTELLTORT, S., BONNET, S. & DAVY, P. 2006. Did the Ebro basin connect to the Mediterranean before the Messinian salinity crisis? Geomorphology, **81**, 155–165.

BANDET, Y. 1975. Les terrains néogènes du Conflent et du Roussillon nord-occidental. Thesis, Université, P. Sabatier, Toulouse.

BARNOLAS, A., CHIRON, J.-C. & GUÉRANGÉ, B. (eds) 1996. Synthèse géologique et géophysique des Pyrénées. Vol. 1: introduction, géophysique, cycle hercynien. BRGM, Orléans.

BAUDELOT, S. & CROUZEL, F. 1974. La faune burdigalienne des gisements d'Espira de Conflent. Bulletin de la Société d'Histoire Naturelle de Toulouse, **110**, 311–326.

BERGER, G., CLAUZON, G. ET AL. 1988. Carte géologique de la France (1:50,000), feuille Perpignan (no. 1091). BRGM, Orléans.

BIROT, P. 1937. Recherches sur la morphologie des Pyrénées orientales franco-espagnoles. Baillère éditions, Paris.

BOURCART, J. 1947. Etude des sédiments pliocènes et quaternaires du Roussillon. Bulletin du Service de la Carte Géologique de la France, **218**, 395–476.

BRADLEY, W. C. 1987. Erosion surfaces of the Colorado Front Range: a review. In: GRAF, W. L. (ed.) Geomorphic Systems of North America. Geological Society of America, Centennial Special, **2**, Boulder, CO, 215–220.

BRYAN, K. 1946. Cryopedology: the study of frozen ground and intensive frost action with suggestions of nomenclature. American Journal of Science, **244**, 622–642.

BURBANK, D. W., PUIGDEFABREGAS, C. & MUÑOZ, J. A. 1992a. The chronology of the Eocene tectonic and stratigraphic development of the Eastern Pyrenean foreland basin, Norheast of Spain. Geological Society of America Bulletin, **104**, 1101–1120.

BURBANK, D. W., VERGES, J., MUÑOZ, A. & BENTHAM, P. 1992b. Coeval hindward- and forward-imbricating thrusting in the South-central Pyrenees, Spain: timing and rates of shortening and deposition. Geological Society of America Bulletin, **104**, 3–17.

CABRERA, L., ROCA, E. & SANTANACH, P. 1988. Basin formation at the end of a strike-slip fault: the Cerdanya basin (Eastern Pyrenees). Journal of the Geological Society, London, **145**, 261–268.

CALVET, M. 1992. Aplanissements sur calcaires et gîtes fossilifères karstiques. L'exemple des Corbières orientales (Pyrénées, France). In: Proceedings of the Karst-Symposium-Blaubeuren, CHARDON, M., SWEETING, M. & PFEFFER, K.-H. (eds) Tübinger Geographische Studien, **109**, 37–43.

CALVET, M. 1996. Morphogenèse d'une montagne méditerranéenne: les Pyrénées orientales. Mémoire BRGM, Orléans, 255.

CLARK, M. K., MAHEO, G., SALEEBY, J. & FARLEY, K. A. 2005. The non-equilibrium landscape of the southern Sierra Nevada, California. GSA Today, **15**, 4–10.

CONEY, P. J., MUÑOZ, A. J., MCCLAY, K. R. & EVENCHIK, C. A. 1996. Syntectonic burial and posttectonic exhumation of the southern Pyrenees foreland fold-thrust belt. Journal of the Geological Society, London, **153**, 9–16.

CRAVATTE, J., DUFAURE, J. F., PRIM, M. & ROUAIX, S. 1974. Les forages du Golfe du Lion. Stratigraphie, sédimentologie. Notes et Mémoires du C.F.P., **11**, 209–274.

CROCHET, B. 1991. Molasses syntectoniques du versant nord des Pyrénées: la série de Palassou. Document du BRGM, **199**, Orléans.

DAVIS, W. M. 1899. The geographical cycle. Geographical Journal, **14**, 481–504.

DAVIS, W. M. 1911. The Colorado Front range, a study on physiographic presentation. Annals of the Association of American Geographers, **1**, 21–83.

DE SITTER, L. U. 1952. Pliocene uplift of Tertiary mountain chains. American Journal of Science, **250**, 297–307.

EVANOFF, E. 1990. Early Oligocene paleovalleys in southern and central Wyoming: evidence of high local relief on the late Eocene unconformity. Geology, **18**, 443–446.

FAILLAT, J. P., AGUILAR, J. P., CALVET, M. & MICHAUX, J. 1990. Les fissures à remplissage fossilifère néogène du plateau de Baixas (Pyrénées-Orientales, France), témoins de la distension oligo-miocène. Comptes Rendus de l'Académie des Sciences Paris, série II, **311**, 205–212.

GARCIA-CASTELLANOS, D., VERGÉS, J., GASPAR-ESCRIBANO, J. & CLOETINGH, S. 2003. Interplay between tectonics, climate and fluvial transport during the Cenozoic evolution of the Ebro Basin (NE Iberia). Journal of Geophysical Research, **108**, 2347–2364.

GOTTIS, M. 1958. L'apport des travaux de la Compagnie d'Exploitation Pétrolière dans la connaissance du bassin tertiaire du Roussillon. *Bulletin de la Société Géologique de France*, **VIII**, 881–883.

GREGORY, K. M. & CHASE, C. G. 1992. Tectonic significance of paleobotanically estimated climated and altitude of the late Eocene erosion surface, Colorado. *Geology*, **20**, 581–585.

GUBBELS, T. L., ISACKS, B. L. & FARRAR, E. 1993. High-level surfaces, plateau uplift, and foreland development, Bolivian Central Andes. *Geology*, **21**, 695–698.

GUILCHER, A. 1994. Quaternary nivation, cryoplanation, and solifluction in western Brittany and the North Devonshire Hills. *In*: EVANS, D. J. A. (ed.) *Cold Climate Landforms*. Wiley, New York.

GUITARD, G. 1970. *Le métamorphisme hercynien mésozonal et les gneiss oeillés du massif du Canigou (Pyrénées orientales)*. Mémoire du BRGM, Orléans, 63.

GUITARD, G., LAUMONIER, B., AUTRAN, A., BANDET, Y. & BERGER, G. M. 1998. *Notice explicative, carte géologique de la France (1/50 000), feuille Prades (no. 1095)*. BRGM, Orléans (map by G. Guitard *et al.* 1992).

HORTON, B. K. & DECELLES, P. G. 2001. Modern and ancient fluvial megafans in the foreland basin system of the central Andes, southern Bolivia: implications for drainage network evolution in fold-thrust belts. *Basin Research*, **13**, 43–64.

KETCHAM, R. A. 2005. Forward and inverse modeling of low-temperature thermochronometry data. *Reviews in Mineralogy and Geochemistry*, **58**, 275–314.

LEONARD, E. 2002. Geomorphic and tectonic forcing of late Cenozoic warping of the Colorado piedmont. *Geology*, **30**, 595–598.

LEWIS, C. J., VERGÈS, J. & MARZO, M. (2000). High mountains in a zone of extended crust: insights into the Neogene–Quaternary topographic developement of northeastern Iberia. *Tectonics*, **19**, 86–102.

MATÓ, E., SAULA, E., MARTÍNEZ-RÍUS, A., MUÑOZ, J. A. & ESCUER, J. 1994. *Mapa geológico de España (1: 50, 000), hoja Berga, Memoria explicativa*. ITGM España.

MEIGS, A. J., VERGES, J. & BURBANK, W. D. 1996. Ten-million-year history of a thrust sheet. *Geological Society of America Bulletin*, **108**, 1608–1625.

MENENDEZ AMOR, J. 1955. *La depresión ceretana española y sus vegetales fósiles. Características fitopaleontológicas del Neógeno de la Cerdaña española*. Memorio Real Academia Sciencias Exactas, Fisicas y Naturales de Madrid, **XVIII**.

MÉRIAUX, A.-S., TAPPONNIER, P. *ET AL.* 2005. The Aksay segment of the northern Altyn Tagh fault: Tectonic geomorphology, landscape evolution, and Holocene slip rate. *Journal of Geophysical Research*, **110**, B04404; doi:10.1029/2004JB003210.

MEYER, B., TAPPONNIER, P., BOURJOT, L., MÉTIVIER, F., GAUDEMER, Y., PELTZER, G., GUO, S. & CHEN, Z. 1998. Crustal thickening in Gansu-Qinghai, lithospheric mantle subduction, and oblique, strike-slip controlled growth of the Tibet Plateau. *Geophysical Journal International*, **135**, 1–47.

MEYER, H. W. 2003. *The Fossils of Florissant*. Smithsonian Institution Press, Washington, DC.

MIALL, A. D. 1996. *The Geology of Fluvial Deposits*. Springer, Berlin.

MIGON, P. 2003. Cryoplanation – a unique Quaternary phenomenon? *XVI INQUA Congress* (Reno, NV), Programs with Abstracts, 64–65.

MORRIS, R. G., SINCLAIR, H. D. & YELLAND, A. J. 1998. Exhumation of the Pyrenean orogen: implications for sediment discharge. *Basin Research*, **10**, 69–85.

MUÑOZ, J. A., VERGÈS, J. *ET AL.* 1994. *Mapa geológico de España, (1:50,000), hoja Ripoll. Memoria explicativa*. ITGM España.

OLLIER, C. & PAIN, C. 2000. *The Origin of Mountains*. Routledge, London.

PHILLIPS, J. D. 2002. Erosion, isostatic response, and the missing peneplains. *Geomorphology*, **45**, 225–241.

REILLE, J.-L. 1971. Les relations entre tectorogenèse et sédimentation sur le versant sud des Pyrénées centrales d'après l'étude des formations tertiaires essentiellement continentales. Thèse doct. Sciences, USTL Montpellier.

RIBA, O., REGUANT, S. & VILLENA, J. 1983. Ensayo de síntesis estratigráfica y evolutiva de la cuenca terciaria del Ebro. Libro Jubilar J. M. Rios. IGME, Madrid, **II**, 131–159.

ROCA, 1986. Estudi geològic de la fossa de la Cerdanya. Tesis doctoral, Univ. de Barcelona.

SANZ DE CIRIA, A. 1981. Flora del Mioceno superior de la Bisbal (Baix Emporda). *Bulletin Informacion Institut Provincial de Paleontologia, Sabadell*, **XI**, 57–68.

SMITH, G. A. 2000. Recognition and significance of streamflow-dominated piedmont facies in extensional basins. *Basin Research*, **12**, 399–411.

SRTM. 2004, ftp://edcftp.cr.usgs.gov/pub/data/srtm

VERGÈS, J., MARTINEZ, A. *ET AL.* 1994. *Mapa geológico de España (1:50,000), hoja La Pobla de Lillet. Memoria explicativa*. ITGM España.

VINCENT, S. J. & ELLIOTT, T. 1997. Long-lived transfer zone paleovalleys in mountain belts: an example from the Tertiary of the Spanish Pyrenees. *Journal of Sedimentary Research*, **67**, 303–310.

South African pediments and interfluves

R. BRUCE KING

Kerry Stables, 4 Lammas Lane, Esher, Surrey KT10 8NY, UK (e-mail: rbkinguk@aol.com)

Abstract: From fifty 1:50,000 scale topographic maps of South Africa, the following attributes were recorded at every intersection of one minute longitude and latitude: basic land facet description, annual rainfall, rainfall concentration, rainfall seasonality, monthly rainfall, local relief, contour interval, rock type, vegetation, pediment length or interfluve width, stream order at valley bottom, local drainage pattern, altitude, terrain morphology, physiographic region and postulated age of planation surface as defined by King (*The Morphology of the Earth: a Study and Synthesis of World Scenery*, Oliver & Boyd, 1962) and Partridge & Maud (*South African Journal of Geology*, **90**, 179–208, 1987). A strong relationship was found between the propensity for concave slopes (or escarpments) where there is low annual rainfall and high local relief. The relationship with rock type was not so strong, but concave slopes are shown to be more likely on fine-textured rather than coarser-textured rock types. The relationship between drainage density and annual rainfall decreases in value from less than 200 mm annual rainfall up to 400–600 mm annual rainfall, and then increases above this value. Concave slopes are particularly prone in lightly vegetated areas, require some local relief, and are more common on easily eroded rocks and older land surfaces protected from recent drainage dissection. Sheetwash seems the most likely agent of erosion.

Semi-arid landscapes have characteristic landforms, significantly different from those produced under humid climates. In particular, they appear to be dominated by long concave slopes. South Africa is a particularly suitable country to investigate these landscapes because: (1) there is a gradual decrease in rainfall from east to west; (2) there has been little significant climatic change in the more arid western areas at least since the mid-Miocene (Ward *et al.* 1983; 1993; Cockburn *et al.* 1999; Tyson & Partridge 2000); (3) there is a lack of strong local tectonic influences over most of the country to complicate the understanding of the landscape evolution. Except for the upper slopes of some mountainous areas and some narrow coastal strips, the annual rainfall of the whole of South Africa is less than 1000 mm. As a broad generalization, the annual rainfall decreases westward from 600–1000 mm in the east to less than 100 mm in the extreme west.

Most of South Africa consists of a plateau, termed the Highveld (Fig. 1), which has been mainly stable since the last major uplift in the Late Pliocene (Partridge & Maud 1987). The plateau is bounded on its eastern, southern and western sides by a prominent escarpment. In the southeast, the Eastern Coastal Hills lead up to the escarpment. Much of the Highveld has little relief, but it does gradually increase from the southwest towards the northeast. To the north, the Bushveld Igneous Complex lies in a basin at a lower altitude. It is bounded in the north by the dissected Limpopo Valley, and to the east by the Lowveld below the bounding escarpment. In the west, the Great Escarpment separates the Western Coastal Plain from a zone of gentle dissection working back from the Orange River, mixed with deposits of Kalahari sand. Immediately to the south of this area is a landscape dominated by pans (the Panneveld) with mixed endoreic and exoreic drainage. A mainly densely dissected plain (the Karoo) separates the southern form of the Great Escarpment from the Cape mountain ranges, which themselves surround some intermontane plains. The Southern Coastal Plain lies to the south and west of the Cape mountain ranges.

Data collection

Fifty 1:50,000 scale topographic maps were randomly selected from those covering the whole of South Africa, excluding the topographic sheets dominated by Kalahari sand. Their locations are indicated in Figure 1. Twenty-seven attributes were recorded or measured at the intersection of every minute of longitude and latitude on all the randomly selected maps, except for where the intersection point occurred in an area of Kalahari or coastal sand. Thus except for those maps that include some areas of Kalahari or coastal sand or water, there were 196 recordings for each map, producing a total of 9280 sampled points. The following attributes were recorded at each sampled intersection point: basic landforms, rainfall, local relief (to the nearest 10 m), altitude (to the nearest 10 m), rock type, land cover, drainage and terrain morphology.

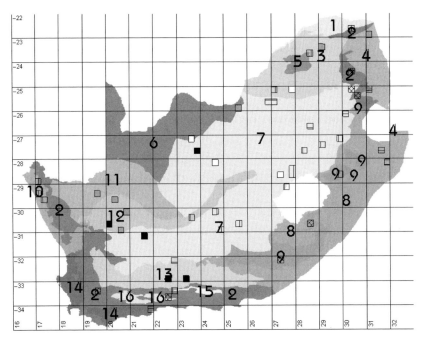

Fig. 1. Location of physiographic regions and randomly chosen topographic sheets. 1, Limpopo Valley; 2, mountains; 3, Bushveld Basin; 4, Lowveld; 5, Waterberg; 6, Kalahari; 7, Highveld; 8, Eastern Coastal Hills; 9, Eastern Middleveld; 10, Western Coastal Plain; 11, Kenhardt Plain; 12, Panneveld; 13, Great Karoo; 14, Southern Coastal Plain; 15, Willowmore Basin; 16, Little Karoo. Small black squares indicate tographic sheets displaying very low relief (see Table 3); grey squares, low relief; horizontal lines, moderate rolling relief; vertical lines, inselberg landscape; white squares, moderate flat relief; diagonal cross lines, high relief.

Landform

One of the following basic landforms (land facets) was recorded at each intersection point: convex slope, concave slope, valley-side slope, escarpment, floodplain, pan or badlands. The slope shape was determined from the 20 m contour spacing. Where the upslope contour spacing was greater than the downslope spacing, the slope was considered convex; where the upslope spacing was less than the downslope spacing, the slope was considered concave. (A truly rectilinear slope would require upslope and downslope contour spacing to be exactly equal.) On the extremely rare occasions where it was not easy to discern the difference in contour spacing, the slope was allocated the curvature of the neighbouring slopes. Crests were considered convex. However, where the spacing was less than 1 mm (i.e. indicating a slope greater than 22°), the landform was recorded as either valley-side or escarpment. (Such slopes are sometimes considered rectilinear.) A valley-side slope was distinguished if there was no gentler slope separating it from the valley bottom. Badlands were taken from the topographic map, where they are mapped as areas of 'erosion', but they only occupy 0.03% of the sampled area. A slope estimate was determined from the contour interval, although the interval was often too coarse to determine. In addition, for concave slopes, the length of slope was measured to the nearest 50 m (discussed below under *Pediment length*). The land facets upslope and downslope from the sampled intersection point were also recorded.

Local relief

Local relief was determined from the difference in contour interval between the highest contour value upslope from the minute intersection point to the lowest contour value downslope from the intersection point.

Rainfall

Rainfall values were taken from the *South African Atlas of Agrohydrology and Climatology* (Schulze et al. 1997). Their mean annual rainfall figures were derived from Dent et al. (1989). Median rainfall values were also recorded for the months of January, February, March, April, May,

September, October, November and December. (There is very little rain over most of the country in June, July and August.) Rainfall concentration and seasonality were also recorded. The concentration index (Markham 1970) indicates the number of months over which the annual rainfall occurs: the shorter the period, the higher the concentration index. Rainfall seasonality indicates the time of the year most of the annual rainfall takes place.

Rock type

The rock type was determined from the 1997 edition of the 1:1,000,000 geological map of South Africa produced by the South African Council for Geoscience. Where the sampled intersection point occurred on a concave slope, the rock type of the upslope escarpment or slope crest where there was no escarpment was also recorded.

Land cover

Land cover was taken from the topographic maps where indicated. Elsewhere the natural vegetation was taken from Tainton & Bosch (1999) and Acocks (1953).

Drainage

Drainage density was determined by measuring the interfluve width, defined as the distance from the valley bottom downslope from the intersection point to the valley bottom defined by projecting a line upslope from the intersection over the crest of the interfluve, downslope to the valley bottom of the neighbouring valley. Figure 2 represents drainage basin geometry. Then $\delta D = \delta l / \delta A$, where D is drainage density, and A is basin area. However, it can be seen from Figure 2 that $\delta A = \delta l * \delta I$. Thus $\delta D = \delta l / (\delta l * \delta I) = 1/\delta I$. The interfluve width for convex slopes was measured to the nearest 50 m. The local drainage pattern and stream order at the valley bottom downslope from the intersection point were also recorded.

Terrain morphology

The terrain morphology was recorded in various ways. *The South African Atlas of Agrohydrology and -Climatology* includes Kruger's (1983) terrain morphology map, which defines 30 different terrain types. The terrain type in which the intersection point occurred was recorded. Both the physiographic region defined by Wellington (1955) and a modified version (Fig. 1) derived from interpreting MODIS satellite imagery and a digital elevation model derived from the Shuttle Radar Topography Mission (SRTM), were also recorded. The postulated planation surface, or zone of dissection between planation surfaces, as defined by King (1962) and Partridge & Maud (1987), was recorded.

Primary analysis

Based on the 9280 sampled points, only 26% of the South African landscape consists of concave slopes, which somewhat contradicts the general impression of South African slopes being typically concave (e.g. Botha & Partridge 2000). Field observation demonstrates how this impression is created. From a distance many slopes look concave, but on closer inspection they often turn out to be dissected piedmonts (Fig. 3), i.e. concave longitudinal profile with convex transverse profile. (Only if the minute intersection point lies on the crest of the convex profile would the slope be recorded as concave.)

Most of the slopes are convex (65%), 4% valley, 4% escarpment and 1% floodplain and pan. Table 1 indicates that concave slopes are more common at lower rainfalls, although not at the lowest rainfall. Table 2 shows concave slopes are more likely as the local relief increases, and dominate convex slopes where the local relief is at least 60 m.

If the basic landform percentages are calculated for each local relief value for different annual rainfalls, the local relief value at which concave slopes dominate convex slopes increases as an apparently perfect ($R^2 = 1$) polynomial function over the annual rainfall range of 0–800 mm. The higher rainfalls occur in areas of higher relief, so that there are more valley-side and escarpment slopes than at lower rainfalls. Escarpments tend to be associated with concave slopes and these two landforms together are considered indicative of scarp retreat or 'pediplanation'. A curve was plotted of the local relief values at which concave slopes or concave slopes and escarpments $(C + E)$ together

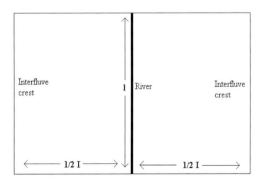

Fig. 2. Drainage basin geometry. l, river length; I, interfluve width.

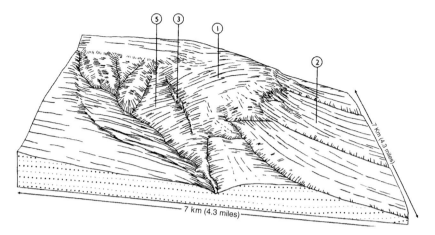

Fig. 3. Land facets, after King (1976). 1, interfluve; 2, dissected piedmont; 3, escarpment; 5, pediment.

are more likely than convex slopes (X) $[(C + E) > X]$ for different annual rainfalls, including the higher annual rainfalls (Fig. 4). The curve follows a second-order polynomial function with an R^2 of 0.9998:

$$y = 0.00007x^2 + 0.021x + 15.792 \qquad (1)$$

Analyses of rainfall concentration (as defined earlier), rainfall seasonality and the monthly rainfalls produced less significant correlations than the annual rainfall figures (more than 95% statistically significant).

Pediment length

It should be noted that, for the sake of objectivity, the analysis has not considered whether the concave slopes are erosional or depositional, although where alluvial fans were clearly recognizable on the topographic maps, they were recorded as such, but considered together with the other concave slopes in the analysis. Pediments are generally considered to be erosional features (e.g. Gilbert 1877; King 1963; Cooke & Warren 1973; Clark

Table 1. *Percentage concave slopes for different annual rainfalls*

Mean annual rainfall (mm)	Concave slopes (%)
100	25.9
300	38.9
500	24.0
700	22.8
900	14.8
1100	4.6

1998). Oberlander (1997), however, defined a pediment as 'a gently inclined slope of transportation and/or erosion that truncates rocks and connects eroding slopes or scarps to areas of sediment deposition at lower levels'. This concept has been used in this research with the extra provision that it should have a concave slope. The author particularly wished to consider why landscapes in arid areas appear to be characterized by concave slopes, but in order to be objective, avoids the pediment debate. For the sake of the analysis, concave slopes are here considered pediments.

The length of the pediment slope was found to be high in areas of low rainfall (annual rainfall less than 600 mm), decreasing asymptotically at higher rainfalls. As might be expected, it also increases with stream order from order 2 to 5 ($R^2 = 0.805$), but is lower at higher stream orders (although not statistically significantly) – possibly because of the greater dissection produced by higher-order rivers. Pediment slopes are generally higher at higher rainfalls, but not universally.

Rock type

Many authors have pointed out the significance of rock structure in arid and semi-arid landscapes (e.g. Twidale 1976), particularly pediments lying below resistant strata (e.g. Fair 1947; Koons 1955; Oberlander 1997). It was for this reason that the rock type above the pediment was recorded in addition to the rock type at the sampled intersection point. However, according to the analysis, only 22% of the concave slopes were produced in rock types different from the escarpment or slopes above the concave slope. Fieldwork did, however, demonstrate that concave slopes were often produced below small resistant bands that were not

Table 2. Land facet percentages for different local reliefs

Mean local relief to the nearest 10 m	Convex slopes	Concave slopes	Valleys	Escarpments	Floodplains	Pans	Badlands
0	92.5	1.5	0	0	2.3	3.4	0.1
10	89.5	10.0	0.1	0	0.4	0	0.1
20	79.0	20.4	0.4	0	0.1	0.2	0
30	66.3	32.2	1.2	0.1	0.2	0	0
40	61.1	34.7	3.3	0.6	0.3	0	0
50	58.5	36.6	2.3	2.6	0	0	0
60	41.3	49.9	6.0	2.6	0.3	0	0
75	31.3	53.5	10.4	4.1	0.8	0	0
95	28.5	49.0	14.9	7.3	0.3	0	0
115	18.8	58.0	11.4	11.8	0	0	0
135	27.2	42.9	19.0	9.2	1.6	0	0
155	18.4	44.0	20.8	18.1	0.7	0	0
175	20.8	43.6	16.8	17.8	1.0	0	0
200	20.1	46.9	13.4	19.6	0	0	0
245	14.5	37.6	24.8	23.1	0	0	0
360	12.9	34.7	27.9	22.4	2.0	0	0

mapped at the geological map scale used of 1:1,000,000.

The most common rock types that occur over a wide enough range of annual rainfalls to analyse are: shale, shale with sandstone horizons (Sh > SS) (e.g. the widespread Adelaide Subgroup of the Beaufort Group), sandstone with shale beds (SS > Sh) (e.g. the Vryheid Formation of the Ecca Group), basic rocks (mainly dolerite) and granite. Of these rock types, shale has the most concave slopes (31%) and the most pans (4%), and basic rocks the least (20%). Figure 5 demonstrates the considerable variability when percentage concave slopes are plotted against annual rainfall. Most notably, granite appears to have the most concave slopes in the most arid climate, but the least at higher rainfalls.

Figure 6 shows the $(C + E) > X$ values. While the plots do not follow as significant a trend as the overall one (Fig. 4), with the exception of basic rocks, they are similarly polynomially concave (second-order polynomials with positive coefficients). For annual rainfalls less than 400 mm, shale requires a lower local relief to produce concave slopes (greater than 98% statistically significant). Sh > SS requires a higher local relief, and SS > Sh requires a still higher local relief (greater than 99.9% statistically significant). Attempts were made to see if other more significant relationships could be obtained by:

1. choosing (randomly where possible) further 1:50,000 scale topographic maps to include more observations of specific rock types, and limiting the observations as much as possible to a particular formation, e.g. the Adelaide to represent Sh > SS;
2. limiting the analysis to planation surfaces;
3. adjusting the values according to the degree of dissection.

None of these attempts produced more significant trends than those shown in Figure 6. They did all indicate that shale generally required the least local relief with a progression from Sh > SS to SS > Sh. Various authors (e.g. King 1963; Tricart 1972; Gupta 2003) have described how shale is particularly devoid of vegetation in arid climates.

Where values for sandstone or quartzite alone were obtained, there was no discernible trend. Basic rocks generally required higher local relief for $(C + E) > X$ than other rock types; and it was evident from investigation of Landsat imagery that in mixed dolerite and shale arid to semi-arid areas, dissected dolerite zones (convex slopes) occurred among, generally above, long concave slopes. Granitic rocks need higher local relief at

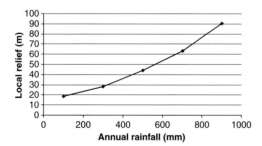

Fig. 4. Propensity for concave slopes or escarpments (values above the line).

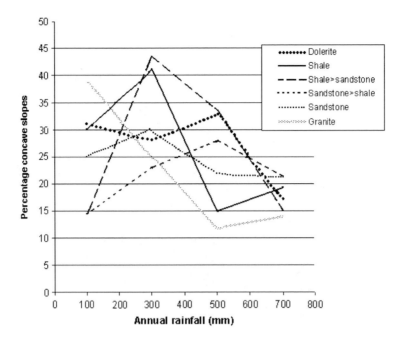

Fig. 5. Percentage concave slopes on different rock types at different annual rainfalls.

low rainfalls, but lower local relief at higher rainfalls compared with other rock types. Comparing Figure 6 with Figure 5 demonstrates the significance of local relief. The high number of concave slopes and the apparent need for higher local relief at low rainfalls are probably simply because most of the granite in the low rainfall zone is found in the Great Escarpment. Conversely, the low number of concave slopes and apparent need for lower local relief at higher rainfalls may simply demonstrate the lower local relief at higher rainfalls, which in turn may illustrate the greater resistance of granite in lower rainfall areas.

Drainage density

Since interfluve width is the inverse of drainage density, drainage density values can be plotted for different rainfalls (Fig. 7). The curve conforms well to that of Australian data as described by Abrahams (1972), while not conforming to the frequently-quoted Gregory's (1976) plot of mainly American data that indicate a maximal drainage density at 500–700 mm annual rainfall. Abrahams attributed the Australian drainage density minimum at 460 mm annual rainfall to the appearance of woodland and possibly 'prolonged high-intensity cyclonic storms'. For the South African data from

Fig. 6. Propensity for concave slopes or escarpments (values above the line) for different rock types.

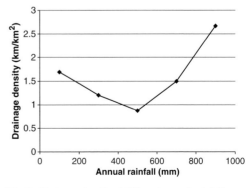

Fig. 7. Drainage density at different annual rainfalls.

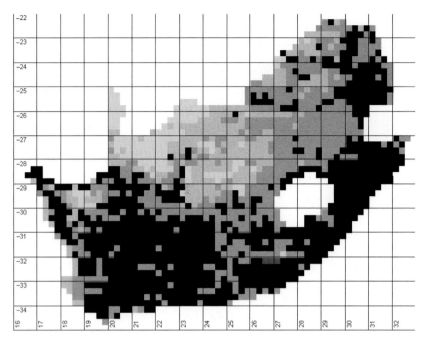

Fig. 8. Mean interfluve widths of areas occupied by 1:50,000 scale topographic maps. Less than 50 m (black), 50–200 m (dark grey), 200–1000 m (light grey), greater than 1000 m (pale grey).

the randomly selected topographic maps, most of the vegetation in the 400–600 mm annual rainfall range representing the South African drainage density minimum is 'bushveld' belonging to the savanna biome of grassland with trees (Tainton 1999). This is confirmed by examining the Landsat imagery analysis (Fig. 8), whereby the average interfluve width for each area covered by a 1:50,000 scale topographic sheet was determined from measuring the interfluve width subtended from randomly-selected points (4–10 for each topographic sheet). In this analysis, if the depositional areas are excluded, the greatest interfluve width is also found in the bushveld and climatic climax grassland (i.e. grassland not derived from forest).

Plotting local relief against interfluve width for different rainfalls produces different trends for low and high relief values (Fig. 9). The figure demonstrates that, where the local relief is greater than or equal to 50 m, it increases with interfluve width up to an interfluve width of 2 km. However, where the local relief is less than 50 m, mean local relief values remain constant as the interfluve width increases. Very low relief predominates at low rainfalls, and local relief values are consistently higher (greater than 90% statistically significant) for higher annual rainfalls, possibly due to increasing weathering with rainfall, allowing greater dissection. (Local relief does not correlate with altitude ($R^2 = 0.0165$); so that orographic precipitation control is not relevant, which has only a very slight correlation ($R^2 = 0.141$).)

Rock type. If the interfluve widths of landforms on different rock types are considered for different annual rainfalls (Fig. 10), we can see that the greatest variability is found where the annual rainfall is 400–600 mm (except for carbonate rocks), and least variability where the annual rainfall is 600–800 mm. Not surprisingly, carbonate rocks have the greatest interfluve width below 600 mm annual rainfall, followed by tillites, although there is less data for tillites. For the other rock types, for which there are more data, basic rocks have a consistently greater interfluve width, but which is not statistically significant. Surprisingly, arenaceous rock types appear to have a lower interfluve width than argillaceous rocks in the more arid climates (99.5% statistically significant at an annual rainfall of less than 200 mm). At higher rainfalls, argillaceous rock types have higher drainage densities (greater than 99.9% statistically significant at an annual rainfall of between 800 and 1000 mm), except for the tillites for which there were not much data. The argillaceous rocks would be more easily weathered and therefore more erosive in more humid climates.

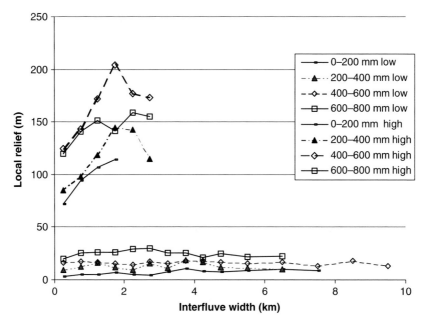

Fig. 9. Plot of mean local relief against interfluve width. 'Low' indicates local relief below 50 m; 'high', local relief above 50 m.

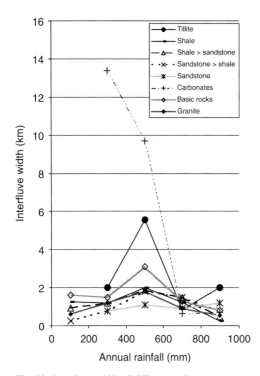

Fig. 10. Interfluve width of different rock types at different annual rainfalls.

Drainage pattern

Most of the drainage patterns (Fig. 11) are angulate (59%); 23% are angulate subdendritic, 5% dendritic, 4% karstic, 3% colinear, 1% pan (Fig. 12), 1% parallel, 1% radial, 1% floodplain and 1% rectangular. There is little correlation with rock type as conceived in traditional literature. The rock types that most display the angulate drainage pattern are Bushveld Complex granite and gabbro-norite, Goudplaats gneiss, Waterberg sandstone and mudstone and Volksrust shale, i.e. a significant variety of rock types. Unsurprisingly, karstic is the dominant drainage pattern on carbonates. Although the linear drainage pattern is not dominant for any rock type, 20% of tillites and Ecca shale are overlain by a linear drainage pattern. The pan drainage category was created to categorize the distinctive endoreic drainage centred on pans. A quarter of Price Albert shale, more than 10% of the rest of Ecca shale and 18% of the Ventersdorp Allonridge volcanics are overlain by the pan drainage pattern.

Drainage patterns appear to be due to a combination of climatic, landform and lithological factors. It would seem reasonable to relate the three dominant drainage patterns as indicative of structural control, viz. the angulate signifies the most structural control and dendritic the least. Most structurally controlled patterns in the most dissected zones ('2' in Fig. 1) lie on basic rocks, shales and carbonates, whereas tillite and sandstone

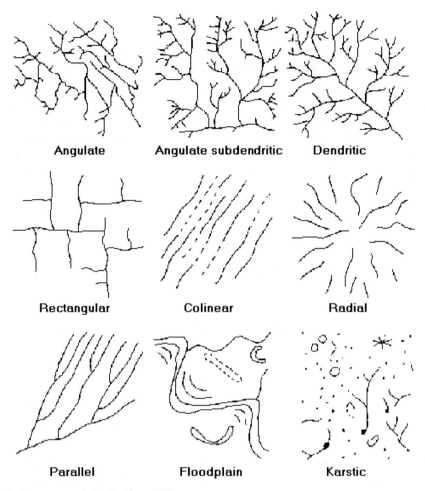

Fig. 11. Drainage patterns (after Rackham 1973).

exhibit the least structural control. In the highest rainfall areas (800–1000 mm/a), the dendritic pattern is the most common. On the Highveld, the greatest structural control appears somewhat surprisingly to be found on tillite and shale, although basic rocks have more structural control in the higher rainfall areas. In the Karoo, however, the drainage pattern on tillite is all dendritic or angulate subdendritic. The Panneveld has the most linear and, of course, pan drainage patterns.

Only linear, braided, distributary, pan and centripetal patterns are found more commonly on concave slopes than convex. In the most arid areas, there are more dendritic and angulate subdendritic than angulate drainage patterns on convex slopes. In flattish (<10 m local relief) areas, floodplain, pan, karstic and trellis patterns are the most common, whereas the angulate pattern is significantly more common in the most arid areas. For areas of higher relief at 600–800 mm annual rainfall, the angulate subdendritic pattern is more common and the angulate pattern is particularly common. Valley slopes are dominated by the angulate drainage pattern. However, they have significantly more angulate subdendritic and dendritic drainage patterns than other land facets.

Overall, there is considerable structural control, but there is slightly less structural control in areas of higher relief. Structural control is less predominant in arid areas, increases to the 400–800 mm annual rainfall range and then sharply decreases.

Secondary analysis

The primary analysis has demonstrated a strong relationship between concave slopes, annual

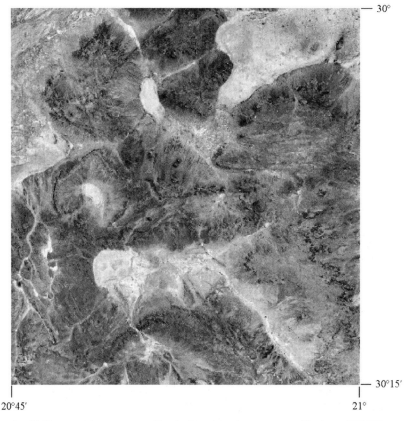

Fig. 12. Landsat TM image of the area covered by the Kareeboomleegte topographic sheet (3020BB).

rainfall and local relief. The drainage investigation indicated dense dissection at both low and high rainfalls with wide interfluves and strong structural control at annual rainfalls of 400–600 mm. The vegetation at this rainfall is dominated by grassland and bushveld, which is considerably more resistant to erosion than the dominant vegetation of the more arid areas. (Climatic climax) grassland (Tainton 1999) and bushveld are mostly found, respectively, in the Highveld and Bushveld physiographic regions, i.e. the older landscapes furthest from recent dissection. However, the Landsat imagery shows denser drainage in the grassland in the higher rainfall areas (still on the Highveld; Fig. 8), suggesting that rainfall is more important than landscape age for determining drainage density. The highest drainage densities are found in forest, grassland derived from forest, and valley bushveld in the higher rainfall areas and areas of recent dissection, particularly the Eastern Coastal Hills. The next densest drainage, both according to the topographic map and Landsat imagery analysis, is found in bushveld in the Lowveld, but which lies in the 400–600 mm annual rainfall zone, suggesting the importance of dissection. After that, the next highest drainage densities, according to the topographic map analysis, are found in the southern recently dissected areas and more arid landscapes under fynbos and succulent Karoo vegetation. The Landsat imagery analysis suggests the arid west has the same high drainage densities as the humid east, but the high drainage densities in the west are also found on the Highveld and Panneveld, and as indicated in Fig. 9, have a lower local relief than in the more humid east. In fact, the Highveld and Panneveld have significantly different landforms southwest of the Orange River in the 200–400 mm annual rainfall zone, compared with the landforms northeast of the river in the 400–600 mm annual rainfall zone. The pans in the southwest dominate the landscape (Fig. 12), often providing the local base-level, and are surrounded by concave slopes. Northeast of the river, the pans occur as indentations into interfluves (Fig. 13). All the topographic sheets analysed in the Panneveld display more concave slopes than elsewhere. The

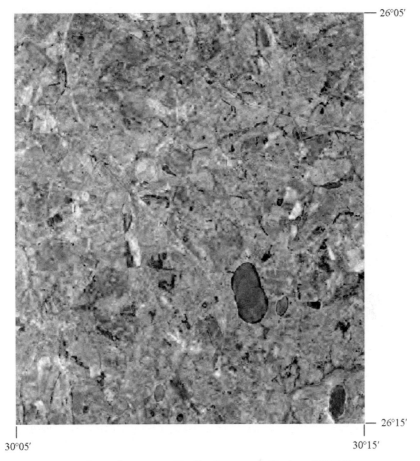

Fig. 13. Landsat TM image of part of area covered by Carolina topographic sheet (2630AA).

analysed topographic sheets of the Highveld in the arid west, with the exception of the extreme southwestern uplifted highland, also display more concave slopes than elsewhere; but the arid recently dissected Karoo has fewer concave slopes than elsewhere.

Figure 14 shows the 10 topographic sheets (from the 50 analysed) that indicate the greatest number of concave slopes. Four occur on the Highveld and two in the Panneveld. The other four are found in escarpment zones (two in the extreme arid western coastal area). It would appear that the propensity for concave slopes is not only expected in escarpment zones and arid areas, but also on the older land surfaces in arid areas. However there does not appear to be any relationship between concave slope propensity and the land surfaces described by King (1962, 1963) and Partridge & Maud (1987), indicating perhaps that the characteristics of the current landscape were carved out of the older landscapes defined by them, rather than being features of those landscapes. The more critical factor would appear to be the effect and extent of more recent dissection, or rather disjunction from it.

A field study of the landforms on the Darling Pluton, 50 km north of Cape Town, demonstrated the relative significance of fluvial dissection and apparent pediplanation. Landforms associated with both processes were displayed, with pediplanation more likely in the upper part of the catchment for catchments with high relief ratios (defined as the ratio of the total catchment relief to the catchment length), and in the lower part of the catchment for catchments with low relief ratios. The upper catchment pediments are below inselbergs, whereas the lower catchment pediments are on the outfacing hillslopes of the pluton. The strongest pediplanation correlation was a negative one with catchment penetration, defined as the number of other catchments

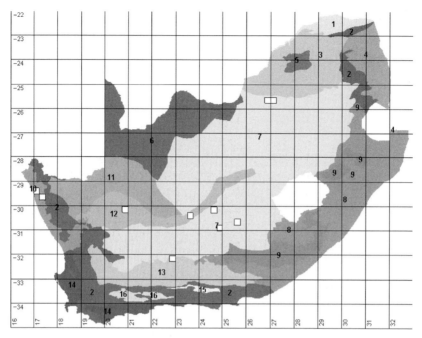

Fig. 14. Location of the 10 randomly selected topographic sheets indicating the most concave slopes (white squares). Numbers indicate physiographic regions as shown in Figure 1.

that border the catchment's perimeter: the greater the penetration the more likely it is that the landform will be dominated by fluvial dissection. Catchment penetration can be considered an indicator of the headward retreat of fluvial dissection. It is inversely related to age, as indicated by the catchments with the highest penetration values being where there is the most active stream rejuvenation.

General morphology

Figure 15 is a general morphology map based on visual interpretation of Landsat imagery. Slight and youthful dissection refer to the degree to which valleys dissect the interfluve. Mature dissection indicates complete dissection of the interfluve. A more objective derivation of general morphology was attempted with the topographic map data. Relative massiveness and roughness measures, as described by Mark (1980), did not describe the morphology sufficiently. Local relief quartiles seem to be a better measure. The general morphology definitions were defined as shown in Table 3, and tested on the 50 randomly chosen topographic maps. The definitions worked reasonably well, although a couple of the defined 'inselberg landscapes' would not normally be considered as such.

The definitions for each topographic sheet are shown in Figure 1.

Interpretation and conclusions

The foregoing analyses have indicated the following significant factors:

1. a very strong relationship between arid climates, local relief and the propensity for concave slopes, or landforms associated with pediplanation;
2. a weaker but similar relationship between pediplanation landforms and rock type, whereby pediplanation landforms are more likely on argillaceous rocks, bearing in mind that the dominant landform by area is the pediment rather than the escarpment;
3. the usual U-shaped curvilinear relationship between drainage density and annual rainfall, but whereby in South Africa the lowest drainage density is found in the bushveld/grassland zone;
4. the association between drainage density and rock type varies according to the annual rainfall – the difference between the drainage densities of rock types being most pronounced where the drainage density is lowest;

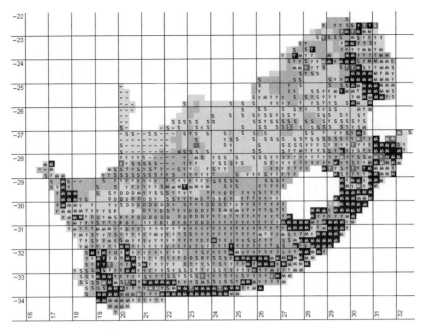

Fig. 15. General morphology of South Africa, based on visual interpretation of Landsat TM imagery. Mountains or hills (black), plain and hills or plain and mountains (dark grey), plain with inselbergs (light grey), plain (pale grey); slight dissection (S), youthful dissection (Y), mature dissection (M), trellised dissection (T); landscapes dominated by alluvial deposition (D), pans (P) and dunes (~).

5. although the greater factor affecting drainage density appears to be rainfall, there is also an association between drainage density and recent fluvial dissection;
6. drainage pattern seems to be strongly related to bedrock structural control;
7. Ecca shale and Ventersdorp volcanics have more than 10% of their area covered by pans. Where they occur in the arid west, they dominate the landscape, producing local endoreic base levels surrounded by pediplanation landforms. In the sub-humid east, they lie on interfluves and have little landform influence.

Thus there is a strong statistical relationship between climate, local relief and pediplanation, but there does also appear to be a weaker relationship with rock type, and probably more significantly with land surface age, as indicated by distance from recent dissection. The investigation of the Darling Pluton suggests fluvial dissection to be a stronger erosive process than pediplanation. Valley-side retreat would normally be expected when fluvial dissection has reached its base level, as revealed in the Darling Pluton landforms. However the process appears to be very slow, as suggested by Figure 16, where an indurated fossil pediment of possibly even early Cenozoic age (Partridge & Maud 1987) suggests very slow valley-side retreat since the time of induration at this locality.

Most landscapes are produced by their interaction with water. The form of the landscape will depend on:

1. how much water arrives as input (quantity and regularity of rainfall);
2. the underlying lithology and stratigraphy;
3. potential energy, indicated by the available relief above a local base level.

Table 3. *General morphology according to local relief quartiles*

Nomenclature	First quartile (m)	Third quartile (m)
High relief	>20	
Moderate rolling relief	20	<80
Inselberg plain	20	≥80
	<20	>30
Flat moderate relief	<20	30
Low relief	<20	20
Very low relief	<20	<20

Fig. 16. Three Sisters, Northern Cape. Indurated fossil pediment at 'A'.

Water removes soil initially by rainsplash (including falling from a tree canopy), but more significantly as wash flowing over the land surface. However rainfall increase, combined with rainfall regularity, encourages vegetation which not only protects the soil from erosion, but also concentrates wash into channels (Puigdefábregas & Sánchez 1996). Concentrated flow is more erosive than unconcentrated flow, and given sufficient potential energy, the land surface is rapidly downcut along the concentrated flow channel (river) to produce valleys between convex interfluves.

However, where the vegetation is not dense enough to protect the soil, episodic sheetwash is a more likely erosive agent. Tricart & Cailleux (1965) describe how, where the annual rainfall averages 240–300 mm, 'there is not enough rain to produce an integrated drainage net' and sheetwash occurs. The erosive power of sheetwash has been variously described (e.g. Bosworth 1922; King 1962), but its episodic nature, both in time and space, will ensure the overall erosive power over time will be slow. Partridge & Maud (1987) pointed out the episodic nature of landform development in southern Africa. While stressing the importance of pediplanation in carving the landscape, they pointed out that scarp retreat may not be continuous. The resulting landscape will be mainly governed by fluvial dissection, but escarpment zones and those areas that are sufficiently arid and old not to be affected by recent dissection will be undergoing pediplanation, or displaying landforms produced by pediplanation.

Although there is not universal agreement, most investigators consider pediments to be moulded by sheet wash (e.g. Bryan 1922; Yair & Lavee 1974; Dohrenwend 1994). It is quite possible that the longitudinal profile of a landform moulded by sheet wash will be similar to a thalweg, hence producing the concave slope typical of pediments, as described by Dohrenwend (1994). He wrote: 'the pediments of arid and semiarid regions are generally considered to be graded surfaces of fluvial transport where the surface slope is adjusted to the discharge and load of the upland/piedmont drainage, and the characteristic concave longitudinal profile is the result of this adjustment to fluvial processes'. In contrast, Fair (1947) considered the South African landforms under the more humid climate of Kwazulu–Natal would not be subjected to sheetwash due to a greater rate of weathering and thicker vegetation cover.

Thus the association between climate and pediments would appear to be explained. The importance of local relief could be explained in two ways:

1. higher relief produces more potential erosive energy (Taylor & Eggleton, 2001);
2. greater relief is more likely to produce harder resistant horizons.

The association of pediplanation with pans in the more arid climates has been noted. Short-distance flooding, allowing inundation of depressions, has been suggested as an initiation process of pan development (Bettenay 1962; Osterkamp & Wood 1987). Subsequent drying, salt accumulation and crystallization, followed by wind deflation have been suggested by Cooke et al. (1993) as a summary model of pan formation. The location of South African pans on fine-grained rocks would also support this model. Their lack of dominance in the more humid climate suggests that pan development there is no longer active, where the landscape is now dominated by fluvial dissection separating the interfluves on which the pans lie.

References

ABRAHAMS, A. D. 1972. Drainage densities and sediment yields in eastern Australia. *Australian Geographical Studies*, **10**, 19–41.

ACOCKS, J. P. H. 1953. *Veld Types of South Africa*. Botanical Survey Memoir **28**.

BETTENAY, E. 1962. The salt lake systems and their associated aeolian features in the semi-arid regions of western Australia. *Journal of Soil Science*, **13**, 10–18.

BOSWORTH, T. O. 1922. *Geology of the Tertiary and Quaternary Periods in the North-West Part of Peru*. Macmillan, London.

BOTHA, G. A. & PARTRIDGE, T. C. 2000. Colluvial deposits. *In:* PARTRIDGE, T. C. & MAUD, R. R. (eds) *The Cenozoic of Southern Africa*. Oxford University Press, Oxford, 88–99.

BRYAN, K. 1922. Erosion and sedimentation in the Papago country, Arizona. *United States Geological Survey Bulletin*, **730**, 765–775.

CLARK, A. N. 1998. *The Penguin Dictionary of Geography*. Penguin Books, London.

COCKBURN, H. A. P., SEIDL, M. A. & SUMMERFIELD, M. A. 1999. Quantifying denudation rates on inselbergs in the central Namib Desert using in situ-produced cosmogenic ^{10}Be and ^{26}Al. *Geology*, **27**, 399–402.

COOKE, R. U. & WARREN, A. 1973. *Geomorphology in Deserts*. Batsford, London.

COOKE, R. U., WARREN, A. & GOUDIE, A. S. 1993. *Desert Geomorphology*. UCL Press, London.

DENT, M. C., LYNCH, S. D. & SCHULTZE, R. E. 1989. *Mapping Mean Annual and Other Rainfall Statistics over Southern Africa*. Water Research Commission Report **109/1/89**.

DOHRENWEND, J. C. 1994. Pediments in arid environments. *In:* ABRAHAMS, A. D. & PARSONS, A. J. (eds) *Geomorphology of Desert Environments*. Chapman & Hall, London, 321–353.

FAIR, T. J. D. 1947. Slope form and development in the interior of Natal, South Africa. *Transactions of the Geological Society of South Africa*, **50**, 105–119.

GILBERT, G. K. 1877. *Report on the geology of the Henry Mountains*. Department of the Interior, Washington.

GREGORY, K. J. 1976. Drainage networks and climate. *In:* DERBYSHIRE, E. (ed.) *Geomorphology and Climate*. Wiley, London, 289–315.

GUPTA, R. P. 2003. *Remote Sensing Geology*, Springer, Heidelberg.

KING, L. C. 1962. *The Morphology of the Earth: A Study and Synthesis of World Scenery*. Oliver & Boyd, Edinburgh.

KING, L. C. 1963. *South African Scenery: A Textbook of Geomorphology*. Oliver & Boyd, Edinburgh.

KING, R. B. 1976. *Land Resources of the Northern and Luapula Provinces, Zambia. Volume 6: The Land Systems*. Land Resources Division, Surbiton.

KOONS, D. 1955. Cliff retreat in the southwestern United States. *American Journal of Science*, **253**, 44–52.

KRUGER, G. P. 1983. *Terrain Morphology Map of Southern Africa*. Soil and Irrigation Research Institute, Pretoria.

MARK, D. M. 1980. Geomorphometric parameters: a review and evaluation. *Geographiska Annaler*, **57A**, 1461–1467.

MARKHAM, C. G. 1970. Seasonality of precipitation in the United States. *Annals of the Association of American Geographers*, **60**, 593–597.

OBERLANDER, T. M. 1997. Slope and pediment systems. *In:* THOMAS, D. S. G. (ed.) *Arid Zone Geomorphology: Process, Form and Change in Drylands*. Wiley, Chichester, 135–163.

OSTERKAMP, W. R. & WOOD, W. W. 1987. Playa-lake basins on the southern High Plains of Texas and New Mexico, part I: hydrologic, geomorphic and geologic evidence for their development. *Geological Society of America Bulletin*, **99**, 215–213.

PARTRIDGE, T. C. 1993. The evidence for Cainozoic aridification in southern Africa. *Quaternary International*, **17**, 105–110.

PARTRIDGE, T. C. & MAUD, R. R. 1987. Geomorphic evolution of southern Africa since the Mesozoic. *South African Journal of Geology*, **90**, 179–208.

PUIGDEFÁBREGAS, J. & SÁNCHEZ, G. 1996. Geomorphological implications of vegetation patchiness on semi-arid slopes. *In:* ANDERSON, M. G. & BROOKS, S. M. (eds) *Advances in Hillslope Processes*. Wiley, Chichester, 1027–1060.

RACKHAM, L. J. 1973. *Notes to Accompany Land Unit Geomorphology Description Card*. Land Resources Division, Tolworth.

SCHULZE, R. E., MAHARAJ, M., LYNCH, S. D., HOWE, B. J. & MELVIL-THOMSON, B. 1997. *South African Atlas of Agrohydrology and -Climatology*. Water Research Commission Report **TT82/96**.

TAINTON, N. M. 1999. The grassland biome. *In:* TAINTON, N. M. (ed.) *Veld Management in South Africa*. University of Natal Press, Pietermaritzburg, 25–33.

TAINTON, N. M. & BOSCH, O. J. H. 1999. The ecology of the main grazing lands of South Africa. *In:* TAINTON, N. M. (ed.) *Veld Management in South Africa*. University of Natal Press, Pietermaritzburg, 23–53.

TAYLOR, G. & EGGLETON, R. A. 2001. *Regolith Geology and Geomorphology*. Wiley, Chichester.

TRICART, J. 1972. *The Landforms of the Humid Tropics, Forests and Savannas*. Longman, London.

TRICART, F. & CAILLEUX, A. 1965. *Introduction to Climatic Geomorphology*. Longman, London.

TWIDALE, C. R. 1976. *Analysis of Landforms*. Wiley Australasia, Sydney.

TYSON, P. D. & PARTRIDGE, T. C. 2000. Evolution of Cenozoic climates. *In:* PARTRIDGE, T. C. & MAUD, R. R. (eds) *The Cenozoic of Southern Africa*. Oxford University Press, Oxford, 371–387.

WARD, J. D., SEELY, M. K. & LANCASTER, N. 1983. On the antiquity of the Namib. *South African Journal of Science*, **79**, 175–183.

WELLINGTON, J. H. 1955. *Southern Africa: A Geographical Study, Volume 1*. Cambridge University Press, Cambridge.

YAIR, A. & LAVEE, H. 1974. Areal contribution to runoff on scree slopes in an extreme arid environment. *Zeitschrift für Geomorphologie, Suppl.* **21**, 106–121.

Summary of progress in geomorphologic modelling of continental slope canyons

NEIL C. MITCHELL

School of Earth, Ocean and Planetary Sciences, Cardiff University, Wales

Present address: School of Earth, Atmospheric and Environmental Sciences, The University of Manchester, Williamson Building, Oxford Road, Manchester M13 9PL, UK (e-mail: Neil.Mitchell@manchester.ac.uk)

Abstract: Far less is known of the processes involved in erosion of submarine channels compared with channels eroded subaerially by water runoff, but geometrical properties derived for canyons of the USA Atlantic continental slope reveal some intriguing similarities. Slope-confined canyons are concave-upwards, displaying decreasing channel gradient with increasing contributing area, as observed in many bedrock-eroding rivers. Tributaries join principal channels at the same elevation (without intervening waterfalls), in effect obeying Playfair's law, as do many river networks. Gradient and contributing area data for channels at confluences also reveal a tendency for tributaries to have steeper gradients than their associated principal channels, reflecting their smaller drainage areas. The concavities of bedrock-eroding rivers are often explained by a balance between river discharge, which increases with increasing rainfall catchment area, and gradient, which declines to offset the erosive effect of the discharge. It is unclear, however, if such a balance can be invoked for submarine canyons because erosion is probably caused when sedimentary flows are active only in individual canyon branches, originating from isolated slope failures. Instead, the frequency of sedimentary flows experienced by canyon floors may increase downstream simply because the area of unstable canyon walls available to source sedimentary flows increases, and this effect becomes compensated by declining gradient. Knickpoints created by faults in tectonically active slopes provide a further way to infer the form of erosion by sedimentary flows. Such knickpoints typically lie upstream of the faults that probably generate them, implying that detachment-limited erosion is enhanced where sedimentary flows become more vigorous on steep gradients, leading to knickpoint migration.

A major aim of subaerial geomorphology has been to understand quantitatively how runoff and other agents erode the Earth's surface, in order that climatic and tectonic effects can be interpreted from landscapes and from the sedimentary products of erosion (Burbank & Anderson 2001). The growing availability of high-quality bathymetry data from multibeam echo-sounders (Pratson & Haxby 1997) and associated geological datasets suggest that a similar more general and quantitative approach could also become possible in marine geology. In particular, continental slope canyons can appear remarkably similar to subaerial erosional systems (McGregor *et al.* 1982; Pratson & Ryan 1996) and, as shown below, they can be similar geomorphologically in a quantitative sense also. These observations may not necessarily imply equivalence of process in the two environments, but nevertheless they prompt the question of whether there could be parallels at least in the forms of the erosion equations. Furthermore, as the original sequence stratigraphy concept (Vail *et al.* 1977) implied that climatically driven changes in sea-level lead to predominantly deposition on continental slopes during high-stands and erosion by mobilized shelf sediment during low-stands, there is potentially a link between climate and erosion in the development of submarine canyons as is often discussed with subaerial erosion. The problem therefore has potential academic interest as well as more practical relevance to the petroleum industry in understanding slope architecture and transfer of sediment to the base of slope.

Submarine canyons are thought to be carved by turbidity currents (sedimentary flows in which suspended sediment is carried down-slope by flow turbulence driven by the fluid density excess caused by the sediment; Daly 1936), along with debris flows (denser flows in which particles are held in suspension by a viscous matrix), mass movements (landslides) and effects of oceanographic currents (Shepard 1981). Direct observations of erosion processes are not available as the flows destroy or displace instruments placed in their paths (Paull *et al.* 2003), although remote acoustic Doppler measurements of flow structure have recently been made

(Xu et al. 2004). Nevertheless, geological evidence suggests that erosion involves processes analogous to abrasion, plucking and quarrying in rivers (Hancock et al. 1998; Whipple et al. 2000). Smooth abraded surfaces and blocks quarried along joint planes have been observed in some submarine canyons (Shepard 1981; Robb et al. 1983; McHugh et al. 1993). Sidescan sonar images and more direct submersible observations have revealed large-scale erosive scours, flutes and channels (Robb et al. 1983; Farre & Ryan 1985; Malinverno et al. 1988; Hughes Clarke et al. 1990; Shor et al. 1990; Normark & Piper 1991; Piper et al. 1999; Klaucke & Cochonat 1999; Klaucke et al. 2000; Gee et al. 2001). Similar features have been observed at smaller scale on river delta fronts in fjords (Prior & Bornhold 1989, 1990; Bornhold & Prior 1990; Boe et al. 2004; Mitchell 2005a). Some of them might be explained by bed shear failure under the influence of flow stress (e.g. Klaucke & Cochonat 1999), a process that is not unlike that of river bedrock quarrying in which flow stress works against friction on joints (Hancock et al. 1998). Scours may be produced by concentrated abrasion by particles in the flows (Shor et al. 1990), in which case scour excavation rate should also be related to some measure of the vigour of the flow amongst other factors. Hughes Clarke et al. (1990) noted that scours were common in the lee of obstacles, presumably created from the kinetic energy of suspended particles where detached flowlines re-attached the bed, a spatial concentration of abrasion also observed in the lee of large boulders in rivers (Hancock et al. 1998). Erosional furrows are common (Farre & Ryan 1985; Piper et al. 1985, 1999), which Farre & Ryan (1985) likened to the effect of snow avalanche furrows, probably caused by relatively coherent flows (debris flows or slides). These observations of comparable erosive features suggest that describing submarine geomorphology in a similar quantitative manner to subaerial geomorphology could turn out to be extremely fruitful.

The purpose of this paper then is to summarize the previous work in quantifying and modelling submarine erosion and results by the author. This knowledge contributes towards the objective of ultimately being able to forward model numerically the evolution of continental slopes. Some likely difficulties that may be encountered and gaps of knowledge or techniques are outlined.

Earlier work on submarine erosion

Quantifying slope and canyon geomorphology and models with sedimentary flow erosion

Although there has been extensive work on the geology of submarine canyons in steep continental slopes, some of which is cited above, and efforts to model and image acoustically the passage of turbidity currents (Komar 1969, 1977; Bowen et al. 1984; Dengler & Wilde 1987; Johnson et al. 2001; Paull et al. 2003; Xu et al. 2004), there has been relatively little quantitative work addressing the development of submarine canyons by erosion that might be considered equivalent to that of river valley development in subaerial geomorphology. Some work has addressed the relief of canyons and empirical relationships to slope form. For example, Goff (2001) quantified an average measure of canyon relief (root-mean-square relief derived from along-slope profiles) for parts of the USA Atlantic continental slope that are linear (New Jersey), exponential (Virginia) and Gaussian (Maryland) in across-slope profile. Relief was found to decrease down-slope for the linear profile, increase for the exponential and was approximately constant for the Gaussian profile. Goff suggested that varied resistance to erosion could potentially explain some of these differences, for example hard rock strata outcrop in the lower New Jersey slope which probably inhibit erosion. It also seems likely that the plan-view geometry of the canyon channels, in particular their spacing, also dictates relief as the canyons shown have different tributary networks and spacing. Pratson & Haxby (1996) characterized variations in gradient derived from multibeam echo-sounder data, which they interpreted generally in terms of slope erosion and deposition. O'Grady et al. (2000) carried that work further with a global bathymetry dataset. The presence of deeply incised slopes was shown to be associated with steep margins where there is relatively little modern river sediment input. Pratson & Ryan (1996) adapted drainage extraction software to map channels of the Monterey Bay canyon system. Although they emphasized the difficulties with applying such software to submarine canyons (channels are not necessarily as widespread as in fluvial systems because channelled erosion only occurs where there are erosive sedimentary flows), geomorphologic measures of channel branching were roughly similar to those of river systems.

A few researchers have attempted forward modelling of erosion by sedimentary flows. Pratson & Coakley (1996) developed a simple numerical model for the development of canyons by landsliding and erosion of channels by sedimentary flows created by the failed material. In the model, the process ultimately causing landslides is the accumulation of hemipelagic sediments around canyon walls, which fail where deposits oversteepen and then erode the canyon floors downstream. Incorporating a stochastic function to represent the spatial variability of deposition of hemipelagic sediment led to spatial variability in landsliding and canyon enlargement. The model predicted canyon morphologies that appear remarkably similar to those in multibeam sonar data. The depth of erosion caused by an individual flow was represented by $(SV)^{0.5}$, where S is the local gradient and V is the volume of failed material. Interestingly, their erosion scheme is not greatly dissimilar from that of 'stream power' erosion laws for river bedrock erosion, where a slope exponent of $2/3$ to 1 has been suggested for plucking and quarrying (Whipple et al. 2000).

Cao & Lerche (1994) developed a model for transport, deposition and erosion by turbidity currents using depth-averaged flow properties. Erosion rate was represented by the excess bed shear stress produced by the flows above a critical threshold. The model appears to have produced realistic deposits. A further model by Tetzlaff & Harbaugh (1989) showed incision of canyon systems by a similar erosion scheme and then deposition of turbidity currents to form sedimentary fans. Mulder et al. (1998) and Skene et al. (1997) modelled bed erosion by a turbidity current created by sediment-laden river water (a hyperpycnal flow) in the Saguenay Fjord, Canada. They assumed that the eroded material was cohesive with a linear shear strength depth profile. Assuming that erosion was limited to where the sediment's shear strength was overcome by the flow's shear stress, erosion depth was varied with u^2/a, where 'u' is flow velocity and 'a' is the rate of shear strength variation with depth (normally compacting cohesive sediments typically increase linearly in shear strength with burial depth; Skempton 1970). The model illustrates that erosion of cohesive muds is likely to be limited if erosion occurs by bed shear failure alone and there is no other (e.g. biological) modification of the sediment. Some two-dimensional stratigraphic models have also incorporated bed erosion by sedimentary flows (e.g. Syvitski & Hutton 2001).

Erosion of muds by oceanographic currents

Part of the relief of continental slope canyons can arise because hemipelagic fallout causes interfluves to aggrade with time while hemipelagic sediment is prevented from accumulating along canyon floors by occasional gravity flows or internal wave currents (Durrieu de Madron et al. 1999; Faugères et al. 1999; Cacchione et al. 2002). Currents within canyons become strongly enhanced if the bed gradient is close to the gradient at which internal waves break (Cacchione & Drake 1986), a possible mechanism for causing non-deposition along canyon floors. Growth of canyon relief by this kind of mechanism is suggested by stratigraphy in some high-resolution seismic reflection images which reveal interfluve topography evolving from smooth and low relief to tall and sharp while adjacent canyon floors have eroded or been maintained sites of non-deposition (Robb et al. 1981; Alonso et al. 1985; Farre 1987; Pratson et al. 1994).

Erosion may also occur where strong boundary currents impinge on the slope (Knebel 1984; Land et al. 1999). Some areas of the continental slope have become smoothed by bottom current interactions (e.g. Smooth Ridge of Monterey Bay; Jordahl et al. 2004), suggesting that they can sometimes lead to diffusive-like tendencies in slope morphology (Mitchell & Huthnance 2007, 2008). As fine-grained sediment covers much of continental slopes in areas such as the Atlantic USA margin (Doyle et al. 1979), it would be useful to know quantitatively how erosion rate varies with the current flow and other properties in order to predict its effect on morphology and stratigraphy, but there has not been much work so far in this area.

In numerical models of shallow erosion by oceanographic currents, erosion rate is typically represented by a linear or power law function of the bed shear stress in excess of a critical value (Gross & Nowell 1990; Fohrmann et al. 1998; Kampf et al. 1999; Davies & Xing 2001, 2002). Such functions represent erosion of very shallow deposits of fine-grained sediment, as some erosion experiments of unconsolidated muds show that erosion rate is initially linear with shear stress (McCave 1984). Erosion of other muds, however, has been found to follow power law (e.g. Johansen et al. 1997) and exponential (e.g. Parchure & Mehta 1985; Andersen et al. 2002) dependencies on an excess shear stress. These experiments represent very shallow buried muds that often still retain aggregates produced by flocculation in the water column. In contrast, McCave (1984) interpreted experiments of Krone (1978) as showing that an overburden equivalent to only 25 mm of sediment can destroy aggregate structure and lead to bed hardening.

Furrows tens of metres deep incising abyssal muddy sediments (McCave & Tucholke 1986) demonstrate that oceanographic currents can remove consolidated deeper sequences. Some Atlantic slope sediments have larger shear strengths than expected from normal consolidation (Keller et al. 1979; Silva & Booth 1986), possibly because of this erosion. Flood (1983) suggested that furrows are probably excavated by abrasion by sand-grade particles carried by bottom currents. Abraded surfaces have been observed in other areas of seabed erosion (Carter & Carter 1985; Land et al. 1999). For modelling erosion of slope cohesive sediments, therefore, an approach of representing erosion rate by functions of solely the excess bed shear stress imposed by currents is probably too simplistic. As suggested for river bedrock abrasion (Hancock et al. 1998; Sklar & Dietrich 2001, 2004), erosion rate of consolidated slope sediments may turn out to be a function of the kinetic energy flux of particles impacting the bed. If so, the model described by Hancock et al. (1998) suggests that erosion rate should vary with the particle mass concentration in the water as well as with the square of current flow velocity. Predicting the abundance of material in saltating and suspension modes would be challenging, as would be the erosional resistance of mud with varying degrees of consolidation, mineralogy and biological influence.

Quantifying exhumation

At present fewer tools are available to measure exhumation of submarine strata than are available in subaerial geomorphology, where researchers can call upon cosmogenic radionuclide methods and dating from fission-tracks and radiogenic particles trapped in minerals (Burbank & Anderson 2001). The issue could be important in some erosion problems because, as outlined earlier, canyon relief can potentially arise by non-deposition along canyon floors while interfluves aggrade (Faugères et al. 1999), implying negligible exhumation, while others may arise by a more simple progressive history of erosion. Strata dipping towards canyon floors are evidence of a non-deposition (interfluve growth) origin rather than entrenchment, but they can easily be removed by landsliding as the canyons widen, so seismic stratigraphy may not unequivocally resolve the alternative origins. Martin et al. (2004) estimated erosion depths from the temperature and pressure imbalances implied by variations in depths of gas hydrate bottom simulating reflectors (BSRs). It is unclear, however, if heat flow and erosion age (as the BSR displacement is a transient effect) can be known with sufficient accuracy for this to be used in geomorphological modelling and the BSR is not easily and consistently followed beneath canyon areas. Furthermore, canyons can distort the pattern of fluid escape (Orange et al. 1994) and therefore heat flow also.

A geotechnical study of samples from ODP Site 902 within Middle Berkeley Canyon of the New Jersey slope found remarkably little evidence for compaction (Blum et al. 1996) and hence showed that the canyon relief arose from the non-deposition mechanism. Consolidation trends in sediments recovered from beneath that canyon's floor were shown to be consistent with almost negligible exhumation ($\sim$10 m) despite up to 100 m of canyon relief at the site. Tests on the sediment samples also showed that they had been variably and strongly affected by pore fluid overpressures. Porosity–depth data for the New Jersey slope drill sites (Mountain et al. 1994) therefore do not show simple compaction trends, so porosity unfortunately cannot be used to estimate exhumation depth reliably. Measures of overburden stress such as the over-consolidation ratio (OCR) have also been used elsewhere in slope geology, with near-surface OCR reaching >100 in some areas of erosion (Booth 1979; Silva & Booth 1986). Some combination of geotechnical assessments (in areas where overpressure is limited) with other more traditional dating methods (hiatuses in $\delta^{18}O$ trends in cores; e.g. McHugh & Olson 2002) may help to resolve exhumation depth, although we still lack the possibility at present to quantify exhumation history as is possible in subaerial geomorphology by combining radiometric techniques with different decay constants (Burbank & Anderson 2001).

Geomorphological characteristics of USA Atlantic slope canyons

The following summarizes reports by the author mostly on a passive margin setting, which are broadly similar to results of Ramsey et al. (2006) from the tectonically active slope of Taiwan. Figure 1 shows bathymetry data collected off the central USA Atlantic coast using multibeam echo-sounders. The results of a study of these and other data along this margin (Mitchell 2004, 2005b) are summarized by the graphs in Figures 2–4, which show various data on canyon floor gradient S and contributing area A. Figure 2 shows longitudinal profiles of two canyons located in Figure 1. Canyon L3 is concave-upwards so that gradient S declines almost systematically with distance from the shelf edge (the small irregularity at 4–5 km distance could be a result of slump deposits from a failure of the canyon wall). The large shelf-incising Norfolk Canyon, in contrast, has moderately declining gradient both shorewards and seawards of the shelf break, and appears more nearly linear or slightly upwards convex in profile. Gradient–area graphs for the slope canyons (Fig. 3) are not exactly inverse power-law but they nevertheless show an inverse relationship that is not too different to those of bedrock eroding rivers – the dashed line shown is such a relationship for Taiwanese rivers (Whipple & Tucker 1999). Other data (Mitchell 2005b) show that shelf-incising canyons and the slope-confined canyons form end-members of an apparent evolutionary series in which slope canyons have the greatest concavity while the deep shelf-incising canyons have negligible average concavity. Other canyons that incised the shelf to a lesser degree than Norfolk and Washington

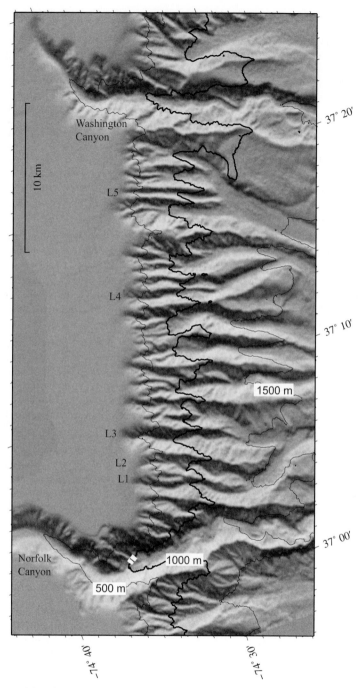

Fig. 1. Bathymetry of the USA Atlantic continental slope off Virginia collected with multibeam sonars and made available by the National Geophysical Data Center (Coastal Relief Model). The canyons marked include the major shelf-incising canyons Norfolk and Washington and linear continental slope canyons L1–L5.

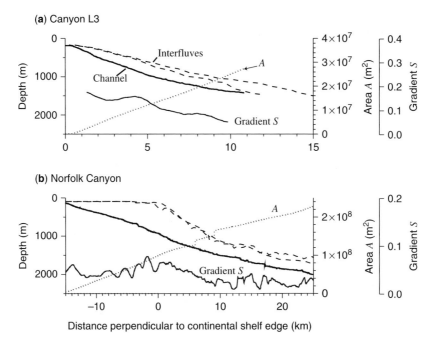

Fig. 2. Longitudinal profiles of two canyons marked in Figure 1 (note differing horizontal scales). Notice that the slope canyon (L3) is concave-upwards, whereas the shelf-incising Norfolk Canyon has non-systematic gradient shorewards of the shelf break (0 km) and is slightly upwards-convex about the break. Whereas (**a**) has similar concavity to rivers, (**b**) appears to have evolved more towards a graded profile.

canyons have intermediate concavity. Canyons seem to evolve towards a state that is similar to graded rivers when the sediment input to the head of the canyons is enhanced because of capture of shelf sediment or from direct river supply during sea level low-stands (Twichell *et al.* 1977).

In common explanations for the concavities of bedrock eroding rivers, the erosive effect of increasing discharge that occurs down the network and with increasing contributions from tributaries becomes balanced by declining gradient to

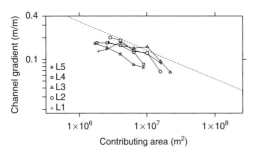

Fig. 3. Logarithmic graph of channel gradient S v. contributing area A for the linear slope canyons shown in Figure 1. The oblique dotted line is an inverse power-law relationship derived for rivers in Taiwan (Whipple & Tucker 1999).

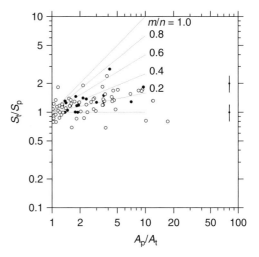

Fig. 4. Logarithmic graph of the ratio of channel gradients and contributing areas for principal and tributary channels measured at confluences. Open symbols represent confluences with at least one contributing canyon heading at the shelf edge. Solid symbols represent confluences where both contributing canyons head within the slope.

achieve a state of spatially equilibrated erosion rate, or at least a state in which erosion rate does not increase towards base-level. It is not clear that this mechanism applies in submarine canyons because tributaries are probably not simultaneously active. Rather they are active when an individual slope failure occurs, sourcing a sedimentary flow (Malinverno et al. 1988; Pratson & Coakley 1996), so discharge and flow power do not increase downstream in the same way. It has been suggested instead (Mitchell 2004, 2005b) that an increasing downstream erosive effect occurs because the frequency and size of sedimentary flows experienced by the canyon floor will increase down-canyon, reflecting the upstream area of potentially unstable canyon walls available to source such flows. The upwards concavities of slope canyons then partly reflect this effect and other possible changes in flow behaviour downstream.

The topography at confluences is also a clue to the style of erosion. In these slope canyons, tributaries join principal channels smoothly towards the same elevation, in effect obeying Playfair's law (Playfair 1802). This implies that channels here are eroded by many small flows rather than isolated large landslides or flows, which might be expected to leave tributaries forming hanging valleys (which are observed in some other submarine canyons; Mulder et al. 2004). Figure 4 shows the ratios of gradients and areas derived from the data at confluences, where the subscripts 'p' and 't' represent the principal and tributary channels, respectively. Such a graph was developed by Seidl & Dietrich (1992) to quantify the tendency for a tributary stream draining small area to have a steeper gradient reflecting its smaller discharge. As the lowering on two branches should be nearly equal at a confluence, the data should produce a linear trend if erosion follows a stream power-type law in which erosion rate $E = kA^m S^n$ with constant k, m and n. These data clearly do not form a simple linear array (the dispersion of the data is larger than the estimated uncertainties shown). This, however, could largely be caused by the many knickpoints seen in detailed contour maps of the data, and the average tendency nevertheless is for tributaries with small contributing area to have steeper channels, as seen in rivers.

The evolution of knickpoints (reaches of anomalously steep gradient) can potentially reveal the style of erosion because detachment-limited erosion schemes (in which erosion rate is related to bed gradient because of its influence on flow vigour) are predicted to cause knickpoints to migrate upstream whereas transport-limited erosion schemes (in which bed material is easily detached and erosion or deposition is dictated by how transport flux varies down-stream) cause knickpoints to smooth out (e.g. Weissel & Seidl 1998; Tucker & Whipple 2002). Many river knickpoints in uniform lithologies (i.e. knickpoints are not caused by variations in resistance to erosion) appear to have migrated upstream (e.g. Seidl & Dietrich 1992; Seidl et al. 1994; Rosenbloom & Anderson 1994), indicating that erosion is enhanced on steep gradients where the flow speeds up, a feature of stream-power erosion models. However, knickpoints typically also decline in relief, which is not predicted by stream-power erosion models and indicates that other processes may be operating in detail (Seidl et al. 1994).

Figure 5 shows a map and relief profiles for channels of the Alaskan continental slope, where knickpoints (highlighted by circles) have been created by movement on thrust faults. Near the range front, knickpoints 1a, 1b and that along profile 2 lie upstream of the steep sections of fault-generated relief. Knickpoints have the greatest relief near the range front, where they have hundreds of metres in relief (1a and 1b). They are much smaller farther shorewards (only tens of metres of relief) where knickpoints may instead be caused by erosionally resistant strata in the uplifted cores of anticlines. Erosion might therefore be partly modelled by stream-bed erosion laws (enhanced erosion rate where the flows accelerate on steep gradients), but erosional resistance probably also plays a strong role.

Discussion: erosion laws for modelling continental slope development

As we know far less about submarine erosive process than fluvial erosion, any attempt at numerical modelling will be more speculative, but it is nevertheless useful to highlight some possible differences with fluvial erosion. Some similarities were outlined in the introduction: erosion is probably in both cases caused by abrasion, plucking and quarrying by a flow of excess density relative to its surroundings. A 'stream power' law for erosion by turbidity currents (Mitchell 2004, 2005b) can be derived as for subaerial streams by starting with an assumption that erosion rate is related to the shear stress imposed by the flow on its bed (Howard 1994) and using the Chezy formula to represent how flow velocity depends on depth-averaged flow properties. Incorporating an area term A for the reasons outlined in the previous section, the model time-averaged erosion rate was suggested to follow:

$$E = KA^m G^n S^n \qquad (1)$$

where the term K represents the erodibility of the substrate as well as some other rate-setting

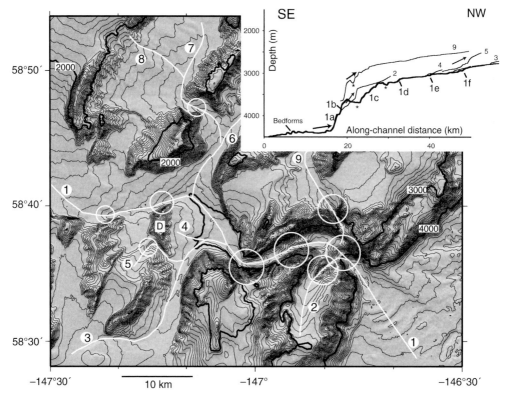

Fig. 5. Bathymetry data and (inset) channel profiles for the tectonically active continental slope in the Gulf of Alaska. The data were collected with multibeam sonars of National Ocean Service survey ships and made available by the National Geophysical Data Center. Circles on the map highlight knickpoints and arrows on channel profiles suggest their migrations. Asterisks mark depressions along channels which indicate on-going tectonic activity not removed by recent sedimentary flows.

factors, such as those representing how frequently flows occur, which likely depends on how rapidly hemipelagic sediments accumulate in canyon walls. An exponent m is applied to the area term to allow for downstream changes in channel width (unfortunately not resolved in these Atlantic USA multibeam sonar data) and to allow for the fact that some flows may not fully run out to the continental rise, as suggested by deposits found in drill cores recovered from many slope areas (e.g. McHugh et al. 2002) or that may intensify (Parker et al. 1986). The term $G = g'h/(f_o + f_i)$ where h is the flow thickness (m) and g' represents the depth-averaged specific weight of the flow equal to $g\Delta\rho/\rho$ (g is the acceleration due to gravity (m/s^2), ρ is the flow's density (kg/m^3), and $\Delta\rho$ is the flow excess density relative to the ambient water (kg/m^3)). The symbols f_o and f_i are the Darcy–Weisbach friction factors of the flow's base and overlying water column, respectively. Variations in G could create more significant changes in erosive behaviour compared with those predicted for fluvial systems with the stream power law. For example, a flow may rapidly become more vigorous and erosive if it incorporates significant material from its base affecting g' (Parker et al. 1986). Furthermore, many of the criticisms of the stream power approach applied to river bed erosion also apply here, for example that it ignores the importance of availability of tools to cause abrasion (Sklar & Dietrich 2004).

If erosion in submarine channels occurs by shear failure of the bed (Mulder et al. 1998), erosion will be limited to the level at which the depth-increasing shear strength of the sediments is smaller than the sum of shear stresses due to the flow and sediment weight. This is a different situation to quarrying and plucking along bedrock channels if progressive unloading of deeply buried rock by the erosion creates new joints by stress-release fracture and weathering, allowing bedrock erosion to continue indefinitely (Whipple et al. 2000). Bio-erosion (Dillon & Zimmerman 1970; Warme et al. 1978; Paull et al. 2005) might provide one possible mechanism by preparing substrates for shear erosion by sedimentary flows, allowing exhumation to continue, although the rates of such processes remain to be quantified. In the study of Atlantic USA canyons (Mitchell 2004, 2005b), channels of canyons with many tributaries were found to decline to similar gradients at a given distance from the shelf edge to channels of linear canyons without tributaries (hence with much smaller contributing area). This is possible evidence that erosion was increasingly inhibited with burial depth of the sediment, which implies that biological conditioning cannot be strongly efficient.

Abrasion is also potentially different. Hancock et al. (1998) developed a scaling argument for how abrasion varies with flow velocity by assuming that erosion rate relates to the kinetic energy of individual particles impacting the bed. They suggested that increases in flow velocity both increase the kinetic energy of individual particles and promote more bed material into suspension and saltation mode. It is unclear if the latter effect will be exactly the same in submarine channels.

Conclusions

Submarine erosion by sedimentary flows is likely to be different in a number of respects from erosion by rivers. Flow vigour is more greatly affected by changes in solid load, erosion by shear failure in cohesive muds is limited to the depth at which shear strength can be overcome and the availability of tools for abrasion does not clearly follow similar rules. Despite the likely differences and the relatively sparse observations of the processes involved in erosion, morphological characteristics of submarine canyons can show striking similarities to those of subaerial drainage systems, which suggests that the basic processes in both cases are abrasion, plucking and quarrying by flows of excess density relative to their surroundings. Similarities in the data presented here include tributary confluences obeying Playfair's law and tributaries showing a tendency to have steeper gradients reflecting their smaller contributing areas. Slope canyons are concave-upwards with similar concavity index (inverse power-law exponent between gradient and catchment area) to some bedrock-eroding rivers. Furthermore, knickpoints in tectonically active slopes lie upstream of their generating topography, implying that erosion is enhanced where flows are more vigorous on steep gradients, as seen in river knickpoint evolution. Significant difficulties remain in predicting erosion of variably consolidated mud by oceanographic currents, because of varied biological effects, substrate resistance and the potential role of suspended particles in abrasion. Nevertheless, these results imply some scope for submarine slope geomorphological modelling by adapting some of the methodologies of subaerial geomorphology to describe how submarine landscapes evolve by erosion.

Brian Dade is thanked for thought-provoking discussion and guidance in the initial stages of this work. The NGDC are thanked for supplying the multibeam sonar data used in this study. Figures in this paper were produced with the help of the GMT software system (Wessel & Smith 1991). Thanks are due to two anonymous reviewers for their helpful suggestions for improving the paper. Much of this work was supported by a research fellowship from the Royal Society.

References

ALONSO, B., KASTENS, K. A. ET AL. 1985. Morphology of the Ebro Fan valleys from SeaMARC and Sea Beam profiles. *Geo-Marine Letters*, **5**, 141–148.

ANDERSEN, T. J., HOUWING, E. J. & PEJRUP, M. 2002. On the erodibility of fine-grained sediments in an infilling freshwater system. *In*: WINTERWERP, J. C. & KRANENBURG, C. (eds) *Fine Sediment Dynamics in the Marine Environment*. Elsevier Science, Amsterdam, 315–328.

BLUM, P., XU, J. & DONTHIREDDY, S. 1996. Geotechnical properties of Pleistocene sediments from the New Jersey upper continental slope. *In*: MOUNTAIN, G. S., MILLER, K. G., BLUM, P., POAG, C. W. & TWICHELL, D. C. (eds) *Proceedings of Ocean Drilling Progress, Scientific Research*, **150**. Ocean Drilling Program, Austin, TX, 377–384.

BOE, R., BUGGE, T., RISE, L., EIDNES, G., EIDE, A. & MAURING, E. 2004. Erosional channel incision and the origin of large sediment waves in Trondheimsfjorden, central Norway. *Geo-Marine Letters*, **24**, 225–240.

BOOTH, J. S. 1979. Recent history of mass-wasting on the upper continental slope, northern Gulf of Mexico, as interpreted from the consolidation states of the sediment. *In*: DOYLE, L. J. & PILKEY, O. H. (eds) *Geology of Continental Slopes*, Special Publication **27**. The Society of Economic Paleontologists and Mineralogists, Tulsa, OK, 153–164.

BORNHOLD, B. D. & PRIOR, D. B. 1990. Morphology and sedimentary processes on the subaqueous Noeick River delta, British Columbia. *In*: COLELLA, A. & PRIOR, D. B. (eds) *Coarse-grained Deltas*, Special Publication **10**. International Association of Sedimentologists, 169–181.

BOWEN, A. J., NORMARK, W. R. & PIPER, D. J. W. 1984. Modelling of turbidity currents on Navy Submarine Fan, California Continental Borderland. *Sedimentology*, **31**, 169–185.

BURBANK, D. W. & ANDERSON, R. S. 2001. *Tectonic Geomorphology*. Blackwell, Malden, MA.

CACCHIONE, D. A. & DRAKE, D. E. 1986. Nepheloid layers and internal waves over continental shelves and slopes. *Geo-Marine Letters*, **6**, 147–152.

CACCHIONE, D. A., PRATSON, L. F. & OGSTON, A. S. 2002. The shaping of continental slopes by internal tides. *Science*, **296**, 724–727.

CAO, S. & LERCHE, I. 1994. A quantitative models of dynamic sediment deposition and erosion in three dimensions. *Computers and Geosciences*, **20**, 635–663.

CARTER, L. & CARTER, R. M. 1985. Current modification of a mass failure deposit on the continental shelf, north Canterbury, New Zealand. *Marine Geology*, **62**, 193–211.

DALY, R. A. 1936. Origin of submarine 'canyons'. *American Journal of Science*, **231**, 401–420.

DAVIES, A. M. & XING, J. 2001. Modelling processes influencing shelf edge currents, mixing, across shelf exchange, and sediment movement at the shelf edge. *Dynamics of Atmospheres and Oceans*, **34**, 291–326.

DAVIES, A. M. & XING, J. 2002. Processes influencing suspended sediment movement on the Malin-Hebrides shelf. *Continental Shelf Research*, **22**, 2081–2113.

DENGLER, A. T. & WILDE, P. 1987. Turbidity currents on steep slopes: application of an avalanche-type numeric model for ocean thermal energy conversion design. *Ocean Engineering*, **14**, 409–433.

DILLON, W. P. & ZIMMERMAN, H. B. 1970. Erosion by biological activity in two New England submarine canyons. *Journal of Sedimentary Petrology*, **40**, 542–547.

DOYLE, L. J., PILKEY, O. H. & WOO, C. C. 1979. Sedimentation on the eastern United States continental slope. *In*: DOYLE, L. J. & PILKEY, O. H. (eds) *Geology of Continental Slopes*, Special Publicaton **27**. The Society of Economic Paleontologists and Mineralogists, 119–129.

DURRIEU DE MADRON, X., CASTAING, P., NYFFELER, F. & COURP, T. 1999. Slope transport of suspended particulate matter on the Aquitanian margin of the Bay of Biscay. *Deep-Sea Research II*, **46**, 2003–2027.

FARRE, J. A. 1987. Surficial geology of the continental margin offshore New Jersey in the vicinity of Deep Sea Drilling Project Sites 612 and 613. *In*: POAG, C. W. & WATTS, A. B. (eds) *Initial Report on Deep Sea Drilling Project* **95**. United States Government Printing Office, Washington, DC, 725–759.

FARRE, J. A. & RYAN, W. B. F. 1985. 3-D view of erosional scars on U.S. mid-Atlantic continental margin. *American Association of Petroleum Geology Bulletin*, **69**, 923–932.

FAUGÈRES, J.-C., STOW, D. A. V., IMBERT, P. & VIANA, A. 1999. Seismic features diagnostic of contourite drifts. *Marine Geology*, **162**, 1–38.

FLOOD, R. D. 1983. Classification of sedimentary furrows and a model for furrow initiation and evolution. *Geological Society of America Bulletin*, **94**, 630–639.

FOHRMANN, H., BACKHAUS, J. O., BLAUME, F. & RAMOHR, J. 1998. Sediments in bottom arrested gravity plumes: numerical case studies. *Journal of the Physics of Oceanography*, **28**, 2250–2274.

GEE, M. J. R., MASSON, D. G., WATTS, A. B. & MITCHELL, N. C. 2001. Passage of debris flows and turbidity currents through a topographic constriction: seafloor erosion and deflection of flow pathways. *Sedimentology*, **48**, 1389–1409.

GOFF, J. A. 2001. Quantitative classification of canyon systems on continental slopes and a possible relationship to slope curvature. *Geophysical Research Letters*, **28**, 4359–4362.

GROSS, T. F. & NOWELL, A. R. M. 1990. Turbulent suspension of sediments in the deep sea. *Philosophical Transactions of the Royal Society, London*, **A331**, 167–181.

HANCOCK, G. S., ANDERSON, R. S. & WHIPPLE, K. X. 1998. Beyond power: bedrock river incision process and form. *In*: TINKLER, K. J. & WOHL, E. E. (eds) *Rivers Over Rock: Fluvial Processes in Bedrock Channels*. Geophysical Monographs **107**. American Geophysical Union, Washington, DC, 35–60.

HOWARD, A. D. 1994. A detachment-limited model of drainage basin evolution. *Water Resources Research*, **30**, 2261–2285.

HUGHES CLARKE, J. E., SHOR, A. N., PIPER, D. J. W. & MAYER, L. A. 1990. Large-scale current-induced erosion and deposition in the path of the 1929 Grand Banks turbidity current. *Sedimentology*, **37**, 613–629.

JOHANSEN, C., LARSEN, T. & PETERSEN, O. 1997. Experiments on erosion of mud from the Danish Wadden Sea. *In*: BURT, N., PARKER, R. & WATTS, J. (eds) *Cohesive Sediments*. Wiley, New York, 305–314.

JOHNSON, K. S., PAULL, C. K., BARRY, J. P. & CHAVEZ, F. P. 2001. A decadal record of underflows from a coastal river into the deep sea. *Geology*, **29**, 1019–1022.

JORDAHL, K. A., PAULL, C. K., USSLER, W., AIELLO, I. W., MITTS, P., GREENE, H. G. & GIBBS, S. 2004. Geology of Smooth Ridge: MARS-IODP cabled observatory site. *EOS Transactions of the AGU*, **85**, Fall Meeting Suppl., Abstract OS43B-0574.

KAMPF, J., BACHAUS, J. O. & FOHRMANN, H. 1999. Sediment-induced slope convection: Two-dimensional numerical case studies. *Journal of Geophysical Research*, **104**, 20509–20522.

KELLER, G. H., LAMBERT, D. N. & BENNETT, R. H. 1979. *Geotechnical Properties of Continental Slope Deposits – Cape Hatteras to Hydrographer Canyon*. SEPM Special Publication **27**. The Society of Economic Paleontologists and Mineralogists, 131–153.

KLAUCKE, I. & COCHONAT, P. 1999. Analysis of past seafloor failures on the continental slope off Nice (SE France). *Geo-Marine Letters*, **19**, 245–253.

KLAUCKE, I., SAVOYE, B. & COCHONAT, P. 2000. Patterns and processes of sediment dispersal on the continental slope off Nice, SE France. *Marine Geology*, **162**, 405–422.

KNEBEL, H. J. 1984. Sedimentary processes on the Atlantic continental slope of the United States. *Marine Geology*, **61**, 43–74.

KOMAR, P. D. 1969. The channelized flow of turbidity currents with applications to Monterey deep-sea fan channel. *Journal of Geophysics Research*, **74**, 4544–4558.

KOMAR, P. D. 1977. Computer simulation of turbidity current flow and the study of deep-sea channels and fan sedimentation. *In*: GOLDBERG, E. D., MCCAVE, I. N., O'BRIEN, J. J. & STEELE, J. H. (eds) *The Sea: Ideas and Observations on Progress in the Study of* the Seas, Volume 6. Marine Modeling. Wiley, New York, 603–621.

KRONE, R. V. 1978. Aggregation of suspended particles in estuaries. In: KJERFVE, B. (ed.) Estuarine Transport Processes. University of South Carolina Press, Columbia, SC, 177–190.

LAND, L. A., PAULL, C. K. & SPIESS, F. N. 1999. Abyssal erosion and scarp retreat: Deep Tow observations of the Blake Escarpment and Blake Spur. Marine Geology, 160, 63–83.

MALINVERNO, A., RYAN, W. B. F., AUFFRET, G. A. & PAUTOT, G. 1988. Sonar images of recent failure events on the continental margin off Nice, France. In: CLIFTON, H. E. (ed.) Sedimentological Consequences of Convulsive Geologic Events. Geological Society of America Special Paper 229. Geological Society of America, Boulder, CO, 59–75.

MARTIN, V., HENRY, P., NOUZÉ, H., NOBLE, M., ASHI, J. & PASCAL, G. 2004. Erosion and sedimentation as processes controlling the BSR-derived heat flow on the Eastern Nankai margin. Earth & Planetary Science Letters, 222, 131–144.

MCCAVE, I. N. 1984. Erosion, transport and deposition of fine-grained marine sediments. In: STOW, D. A. V. & PIPER, D. J. W. (eds) Fine-grained Sediments: Deep Water Processes and Facies. Geological Society of London Special Publication 15. Blackwell Scientific, Oxford, 35–69.

MCCAVE, I. N. & TUCHOLKE, B. E. 1986. Oceanic particles and pelagic sedimentation in the western North Atlantic Ocean. In: VOGT, P. R. & TUCHOLKE, B. E. (eds) The Geology of North America: Vol. M, The Western North Atlantic Region. The Geological Society of America, Boulder, CO, 451–468.

MCGREGOR, B. A., STUBBLEFIELD, W. L., RYAN, W. B. F. & TWICHELL, D. C. 1982. Wilmington submarine canyon: a marine fluvial-like system. Geology, 10, 27–30.

MCHUGH, C. M. G. & OLSON, H. C. 2002. Pleistocene chronology of continental margin sedimentation: new insights into traditional models, New Jersey. Marine Geology, 186, 389–411.

MCHUGH, C. M., RYAN, W. B. F. & SCHREIBER, B. C. 1993. The role of diagenesis in exfoliation of submarine canyons. American Association of Petroleum Geologists Bulletin, 77, 145–172.

MCHUGH, C. M. G., DAMUTH, J. E. & MOUNTAIN, G. S. 2002. Cenozoic mass-transport facies and their correlation with relative sea-level change, New Jersey continental margin. Marine Geology, 184, 295–334.

MITCHELL, N. C. 2004. Form of submarine erosion from confluences in Atlantic USA continental slope canyons. American Journal of Science, 304, 590–611.

MITCHELL, N. C. 2005a. Channelled erosion through a marine dump site of dredge spoils at the mouth of the Puyallup River, Washington State. Marine Geology, 220, 131–151.

MITCHELL, N. C. 2005b. Interpreting long-profiles of canyons in the USA Atlantic continental slope. Marine Geology, 214, 75–99.

MITCHELL, N. C. & HUTHNANCE, J. M. 2007. Comparing the smooth, parabolic shapes of interfluves in continental slopes to predictions of diffusion transport models. Marine Geology, 236, 189–208.

MITCHELL, N. C. & HUTHNANCE, J. M. 2008. Oceanographic currents and the convexity of the uppermost continental slope. Journal of Sedimentary Research, (in press).

MOUNTAIN, G. S., MILLER, K. G. & BLUM, P. 1994. Proceedings of Ocean Drilling Progress, Initial Reports, 150. Ocean Drilling Program, Austin, TX.

MULDER, T., SYVITSKI, J. & SKENE, K. 1998. Modeling of erosion and deposition by turbidity currents generated at river mouths. Journal of Sedimentary Research, 68, 124–137.

MULDER, T., CIRAC, P., GAUDIN, M. ET AL. 2004. Understanding continent-ocean sediment transfer. EOS Transactions of the AGU, 85, 257–262.

NORMARK, W. R. & PIPER, D. J. W. 1991. Initiation processes and flow evolution of turbidity currents: implications for the depositional record. In: OSBORNE, R. H. (ed.) From Shoreline to Abyss. SEPM Special Publication 46. Society of Economic Paleontologists and Mineralogists, Tulsa, OK, 207–230.

O'GRADY, D. B., SYVITSKI, J. P. M., PRATSON, L. F. & SARG, J. F. 2000. Categorizing the morphologic variability of siliciclastic passive continental margins. Geology, 28, 207–210.

ORANGE, D. L., ANDERSON, R. S. & BREEN, N. A. 1994. Regular canyon spacing in the submarine environment: the link between hydrology and geomorphology. Geological Society of America Today, 4, 29–39.

PARCHURE, T. M. & MEHTA, A. J. 1985. Erosion of soft cohesive sediment deposits. Journal of Hydraulic Engineering, ASCE, 111, 1308–1326.

PARKER, G., FUKUSHIMA, Y. & PANTIN, H. M. 1986. Self-accelerating turbidity currents. Journal of Fluid Mechanics, 171, 145–181.

PAULL, C. K., USSLER, W., GREENE, H. G., KEATEN, R., MITTS, P. & BARRY, J. 2003. Caught in the act: the 20 December 2001 gravity flow event in Monterey Canyon. Geo-Marine Letters, 22, 227–232.

PAULL, C. K., USSLER, W., GREENE, H. G., BARRY, J. & KEATEN, R. 2005. Bioerosion by chemosynthetic biological communities on Holocene submarine slide scars. Geo-Marine Letters, 25, 11–19.

PIPER, D. J. W., SHOR, A. N., FARRE, J. A., O'CONNELL, S. & JACOBI, R. 1985. Sediment slides and turbidity currents on the Laurentian Fan: sidescan sonar investigations near the epicenter of the 1929 Grand Banks earthquake. Geology, 13, 538–541.

PIPER, D. J. W., COCHIONAT, P. & MORRISON, M. L. 1999. The sequence of events around the epicentre of the 1929 Grand Banks earthquake: initiation of debris flows and turbidity current inferred from sidescan sonar. Sedimentology, 46, 79–97.

PLAYFAIR, J. 1802. Illustrations of the Huttonian Theory of the Earth. Dover, London.

PRATSON, L. F. & COAKLEY, B. J. 1996. A model for the headward erosion of submarine canyons induced by downslope-eroding sediment flows. Geological Society of America Bulletin, 108, 225–234.

PRATSON, L. F. & HAXBY, W. F. 1996. What is the slope of the U. S. continental slope? Geology, 24, 3–6.

PRATSON, L. F. & HAXBY, W. F. 1997, Panoramas of the seafloor. Scientific American, 276, 66–71.

PRATSON, L. F. & RYAN, W. B. F. 1996. Automated drainage extraction for mapping the Monterey submarine drainage system, California margin. *Marine Geophysics Research*, **18**, 757–777.

PRATSON, L. F., RYAN, W. B. F., MOUNTAIN, G. S. & TWICHELL, D. C. 1994. Submarine canyon initiation by downslope-eroding sediment flows: evidence in late Cenozoic strata on the New Jersey continental slope. *Geol. Soc. Am. Bull.*, **106**, 395–412.

PRIOR, D. B. & BORNHOLD, B. D. 1989. Submarine sedimentation on a developing Holocene fan delta. *Sedimentology*, **36**, 1053–1076.

PRIOR, D. B. & BORNHOLD, B. D. 1990. The underwater development of Holocene fan deltas. *In*: COLELLA, A. & PRIOR, D. B. (eds) *Coarse-grained Deltas*. Special Publications of the International Association of Sedimentologists **10**, 75–90.

RAMSEY, L. A., HOVIUS, N., LAGUE, D. & LIU, C.-S. 2006. Topographic characteristics of the Taiwan orogen. *Journal of Geophysics Research*, **111**, Paper F02009, doi:10.1029/2005JF000314.

ROBB, J. M., HAMPSON, J. C., KIRBY, J. R. & TWICHELL, D. C. 1981. Geology and potential hazards of the continental slope between Lindenkohl and South Toms Canyons, offshore Mid-Atlantic United States. United States Geological Survey Open-File Report, 81–600.

ROBB, J. M., KIRBY, J. R., HAMPSON, J. C., GIBSON, P. R. & HECKER, B. 1983. Furrowed outcrops of Eocene chalk on the lower continental slope offshore New Jersey. *Geology*, **11**, 182–186.

ROSENBLOOM, N. A. & ANDERSON, R. S. 1994. Evolution of the marine terraced landscape, Santa Cruz, California. *Journal of Geophysics Research*, **99**, 14013–14030.

SEIDL, M. & DIETRICH, W. E. 1992. The problem of channel erosion into bedrock. *In*: SCHMIDT, K. H. & DE PLOEY, J. (eds) *Functional Geomorphology: Landform Analysis and Models*, Catena Suppl. **23**, 101–124.

SEIDL, M. A., DIETRICH, W. E. & KIRCHNER, J. W. 1994. Longitudinal profile development into bedrock: An analysis of Hawaiian channels. *Journal of Geology*, **102**, 457–474.

SHEPARD, F. P. 1981. Submarine canyons: multiple causes and long-time persistence. *Bulletin of the American Association of Petroleum Geologists*, **65**, 1062–1077.

SHOR, A. N., PIPER, D. J. W., HUGHES CLARKE, J. E. & MAYER, L. A. 1990. Giant flute-like scour and other erosional features formed by the 1929 Grand Banks turbidity current. *Sedimentology*, **37**, 631–645.

SILVA, A. J. & BOOTH, J. S. 1986. Seabed geotechnical properties and seafloor utilization. *In*: VOGT, P. R. & TUCHOLKE, B. E. (eds) *The Geology of North America, Volume M, The Western North Atlantic Region*. Geological Society of America, Boulder, CO, 491–506.

SKEMPTON, A. W. 1970. The consolidation of clays by gravitational compaction. *Quarterly Journal of the Geological Society, London*, **125**, 373–411.

SKENE, K., MULDER, T. & SYVITSKI, J. P. M. 1997. INFLO1: a model predicting the behaviour of turbidity currents generated at a river mouth. *Computers and Geosciences*, **23**, 975–991.

SKLAR, L. S. & DIETRICH, W. E. 2001. Sediment and rock strength controls on river incision into bedrock. *Geology*, **29**, 1087–1090.

SKLAR, L. S. & DIETRICH, W. E. 2004. A mechanistic model for river incision into bedrock by saltating bedload. *Water Resources Research*, **40**, doi: 10.1029/2003WR002496.

SYVITSKI, J. P. M. & HUTTON, E. W. H. 2001. 2D SEDFLUX 1.0C: an advanced process–response numerical model for the fill of marine sedimentary basins. *Computers and Geosciences*, **27**, 731–753.

TETZLAFF, D. A. & HARBAUGH, J. W. 1989. *Simulating Clastic Sedimentation*. van Nostrand Reinhold, New York.

TUCKER, G. E. & WHIPPLE, K. X. 2002. Topographic outcomes predicted by stream erosion models: sensitivity analysis and intermodel comparison. *Journal of Geophysics Research*, **107**, doi:10.1029/2001JB000162.

TWICHELL, D. C., KNEBEL, H. J. & FOLGER, D. W. 1977. Delaware River: evidence for its former extension to Wilmington submarine canyon. *Science*, **195**, 483–485.

VAIL, P. R., MITCHUM, R. M. & THOMPSON, S. 1977. Seismic stratigraphy and global changes of sea level; Part 4, Global cycles of relative changes of sea level. *In*: PAYTON, C. E. (ed.) *Seismic Stratigraphy – Applications to Hydrocarbon Exploration*. American Association of Petroleum Geologists Memoir **26**, American Association of Petroleum Geologists, Tulsa, OK, 83–97.

WARME, J. E., SLATER, R. A. & COOPER, R. A. 1978. Bioerosion in submarine canyons. *In*: STANLEY, D. J. & KELLING, G. (eds) *Sedimentation in Submarine Canyons, Fans, and Trenches*. Dowden, Hutchinson and Ross, Stroudsburg, 65–70.

WEISSEL, J. K. & SEIDL, M. A. 1998. Inland propagation of erosional escarpments and river profile evolution across the southeast Australian passive continental margin. *In*: TINKLER, K. J. & WOHL, E. E. (eds) *Rivers over Rock: Fluvial Processes in Bedrock Channels*, Geophysical Monographs **107**. American Geophysical Union, Washington, DC, 189–206.

WESSEL, P. & SMITH, W. H. F. 1991. Free software helps map and display data. *EOS, Transactions, American Geophysical Union*, **72**, 441.

WHIPPLE, K. X. & TUCKER, G. E. 1999. Dynamics of the stream-power river incision model: Implications of height limits of mountain ranges, landscape response timescales, and research needs. *Journal of Geophysics Research*, **104**, 17661–17674.

WHIPPLE, K. X., HANCOCK, G. S. & ANDERSON, R. S. 2000. River incision into bedrock: Mechanics and relative efficacy of plucking, abrasion, and cavitation. *Geological Society of America Bulletin*, **112**, 490–503.

XU, J. P., NOBLE, M. A. & ROSENFELD, L. K. 2004. *In-situ* measurements of velocity structure within turbidity currents. *Geophysics Research Letters*, **31**, doi:10.1029/2004GL019718.

Index

Note: Page numbers in *italics* denote Figures, those in **bold** refer to Tables.

abrasion 184, 191
alluvial systems 18–20, *24*
   see also fluvial systems; sediments
Alpine debris flows 3, 63–78
   case studies 73–6
   conceptual modelling 68–70, *71*
   event volumes 64–6
   sediment availability 72–3
   volume thresholds 72
anisotropic dissolution 50–3, 57–60
aquatic molluscs **90**

bedload 3, 129–45
   bed form development 125
   Cenozoic and Pleistocene rivers 130–4
   entrainment 120–4
   erosion 117–27
   experiments 118–23
   gravels and sand–gravel 119, 123–5
   minimum estimates **140**
   palaeohydraulic data 134–40
   Rio Cinca 131–2
   spatial distribution 117–18
   vertical structures 142
bedrock river erosion 12–14
bimodal gravel experiments 121–3
biochronology 154
bivergent progenic wedge *10*
braided river systems 79, 80–5
buffered landscape 22

canyons, continental slope 183–94
carbon dioxide dissolution
   anisotropic 50–3, 57–60
   isotropic 49–50, 54–7
   simulations 53–60
catastrophic events 1–5
   debris flows 65–6, *67*, 71–3, 76
   experiments 118–23
   Quaternary 79–104
   recognition 99–101
   semi-arid catchment *39*
catchment–fan systems *11*, 15–18, *19*
catchments
   Alpine 3, 63–78
   equilibrium *34*
   event volumes 64–6
   evolution 12–14
Cenozoic rivers
   bedload transport 129–45
   palaeohydraulic data 134–42, 143
Cordage intermundane basin *159*
channel erosion 4, 12–14, 183–94
channel flow velocities **138**
channel form
   cross-section 31
   dry lands 34–6

   semi-arid areas 32
   variation 31
CHILD landscape evolution model 48–9
   parameters **57**
   surface/subsurface flows 51–2, 58, 60
clays 85
climate
   erosion relationship 183
   fluvial response 79, 97–9
   pediments relationship 180–1
   relief relationship 178–81
   South Africa 167, 168–9, 172–3
cluster analysis 67–8
cockpit karts landforms 2–3, 47–62
   apiarist concept 50–1, 60
   hydraulic conductivity 52–3, 58
   simulation comparisons 53–60
   surface/subsurface flows 51–2, 58, 60
Coleopteran 91–5
concave slopes 169, 170–2, 176–7, *178*
Confluent basin 155, *156*
conglomerates 131–4, *135*
continental slope
   canyons 183–94
   development model 189–91
continuous processes 1–5, 63–78
coupled tectonic–surface process systems
   7–9, *10*

Davisian theory 148, 163, 164
debris flows 3, 63–78
   case studies 73–6
   catastrophic events *67*, 71–3, 76
   event volumes 64–6
   modelling 68–70, *71*
   sediment availability 72–3
   statistical analysis 66–8
denudation
   chronology, planar landforms 147–66
   cockpit karts landscapes 50
deposition model 48–9
Devonian 86, 87–9
dissolution of carbon dioxide
   CHILD model 49–50
   fractures in karts 53
   time–space simulations 50–60
drainage density
   rock types 172–4
   South Africa 169, 176
drainage patterns 174–5
dry lands 29–46
   channel morphology 34–6
   equilibrium 2, 32–42, *33*
   flood event *39*
   process and form 33–7
   spatial variability 37–8
   time 38–40

Eastern Pyrenees planar landforms 147–66
   age 153–6
   formation/palaeoelevation 158–64
   two generations 150–2
   uplift 157–8, 163–4
efficiency of rivers 138–40, 142–3
elevation
   digital data base 148–9, *150*
   erosion surfaces 148
   palaeoelevation 158–64
entrainment of bed materials 120–4
epikarst concept 50–1, 60
equilibrium 2, 29–46
   definitions 29, **30**
   drylands 32–42, *33*
   see also geomorphological equilibrium
erosion 4, 8–9
   Alpine catchments 63–4
   bed materials 117–27
   bedrock channels 12–14
   CHILD model 48–9, 60
   climate relationship 183
   extreme events 65–6, **66**, *67*, 71–3, 76
   gradual *67*, 71–3, 76
   hillslopes 14–15
   intermontane basins 154–7
   muds 185–6
   oceanographic currents 185–6
   orogenic scale 9
   rivers 12–14
   sheetwash 180
   submarine channels 183–94
   turbidity currents 183–5, 191
   water 179–80
erosion surfaces 4, 148, 150–1, 153–6
   see also pediplanation
events
   extreme *33*, *35–6*, *41*, 63, 65–6, *67*, 76
   gradual *67*, 71–3, 76
   single annual floods 80, 82, 99–101
   volume/catchment area 64–6
   see also catastrophic events
exhumation of submarine sediments 186
experiments
   bed load 117–27
   gravels 120–1
   sand–gravel 121–3
extensional faults 9–12, *24*
extreme events *33*, *35–6*, *41*, 63, 65–6, 76

fans
   slope 16–17
   see also catchment–fan systems
fault displacements 9–12
feedbacks 31
floods
   effectiveness 63–78
   experiments 118–23
   single annual events 80, 82, 99–101
   see also catastrophic events; erosion
flora 86–9
fluvial solar signals 105–15
   heliohydrology 107
   palaeometeorology 107–12

   solar variability 105–7
fluvial systems
   climatic phases 97–9
   mixed-gravel systems 20
   Pyrenees 129–45
   response times 18–20
   sedimentology **84**
   see also rivers
fold belt planar surfaces 153–6
footwall topography 9–12
forcing, tectonic 21–3
foreland basins *160*
form and process concept 33–7
fractures in karst 50–3
France, storm events *35–6*

geochronology, Nene Valley 95–7, 100–1
geology
   Pyrenees 129–30
   South Africa 169, *170–2*, 173
   see also rock types
geomorphological equilibrium 29–46
   constant environmental factors 40
   drylands 32–42
   measurement 29–32
   nonlinearities 40–2
   process and form 33–7
   spatial variability 37–8
   time 38–40
geomorphology, slope canyons 186–9
global atmospheric circulation 107–12
grade concept 31
graded pediment model 158–64
gradual erosion 71–3, 76
grain size distribution 118, 123–5
grass pollen 86, **88**
gravels
   bedload transport rate *139*
   bimodal experiments 121–3
   laboratory study 117–27
   Pyrenees *135*
   UK 80–5
   unimodal experiments 120–1
gullies *34*

hazard assessment 76
heliohydrology 107
herb pollen 86, **87–8**, **89**
Highveld, South Africa 167, *168*, 177
hillslopes
   development 33–4
   erosion 14–15
   sediment transport *42*
hydraulic conductivity 52–3, 58

ice-wedge casts 82, 85
Illgraben catchment 73, *74*
image analysis techniques 119
inter-flood low flows 118–23
interfluves
   South Africa 167–81
   width 172–3, *174*
intermontane basins *152*, 154–7, *159*
isotropic dissolution 49–50, 54–7

INDEX

Jamaica, cockpit karst simulations 53–60

karst landforms 2–3, 47–62
   fissure fillings 154
   karstification 49–53
   see also cockpit karst landforms

laboratory study of gravels 117–27
landscape analysis 168, 169–75, 178–81
landslides *69*, 184
large flood experiments 118–23
longitudinal profiles
   canyons 186, *188*, *190*
   Rio Cinca *137*
   river systems 12–14

mass balance numerical model 16–17
mass transfer model 48
mechanical erosion and deposition model 48–9, 60
Mediterranean valley fill 109, **111–12**
Métin catchment 73–6
Miocene
   fluvial systems **140**, 141–3
   palaeoelevation 161–2, 164
mixed-gravel fluvial systems 20
modelling
   CHILD model 48–9, 51–2, **57**, 58, 60
   cockpit karst landforms 47–62
   continental slope canyons 183–94
   coupled tectonic–surface processes 8–9
   debris flows 68–70, *71*
   deposition 48–9
   drylands 43
   graded pediments 158–64
   mass balance 16–17
   see also laboratory study
Mollusca 89–91
morphometric analysis 4
mountain belt erosion surfaces 148, 153–6
Mt Canigou, Eastern Pyrenees 155, *157*
mud erosion 185–6

negative feedbacks 31
Nene Valley, UK 79–104
   depositional succession 80–5
   fluvial response to climate 97–9, *100*
   geochronology 95–7, 100–1
   palaeontology 85–95, 99–100
Neogene *152*, 154–7
New Zealand 22, 23
nonlinearities 40–2
North Atlantic depressions *108*

oceanographic currents 185–6
Oligo–Miocene fluvial systems **140**, 141–3
optically stimulated luminescence 95, **96**, 97, **98**, 100, *101*
orogenic scale 9

Pacific anticyclone 107–8, *109*
palaeodepth estimation 135
palaeoflow 137–8

palaeometeorology 107–12
palaeontology 85–95, 99–100
palaeoslope 136–7
palaeohydraulic data
   error analysis 140–1
   Pleistocene rivers 134–40, **136**, 143
palynology 86–9, 108–9
pediments
   length 170
   rock type 170–2
   South Africa 167–81
pediplanation 177, 179–80
permafrost 82
perturbation 21–3
piedmont backfilling 158–61
planar landforms 147–66
   age 153–6
   formation/palaeoelevation 158–64
   two generations 150–2
   uplift 157–8, 163–4
plant macrofossils 85–6, **87–8**
Pleistocene
   bedload transport 129–45
   deposits *83*
   palaeohydraulic data 134–42, 143
Polio–Pleistocene rivers **140**, 141–3
post-orogenic phase
   erosion surfaces 153
   uplift 163–4
precipitation
   cockpit karst landforms 47
   debris flows *69*, 76
   drainage density 172–3, 176
   South Africa 167, 168–9
process and form concept 33–7
process rates 1
pulse events 63–78
Pyrenees
   bedload transport 129–45
   geology 129–30
   intermontane basins *152*, 154–7
   planar landforms 147–66
   structure 149, *152*

Quaternary
   catastrophic events 79–104
   flood event recognition 99–101
   fluvial sequence sedimentology **84**

radiocarbon
   Mediterranean valley fill **111–12**
   Nene Valley, UK 95–7, 100
   tree rings 106, 109–10
rainfall *see* precipitation
reactive landscape 22
red beds 154, *155*
regional uplift 157–8, *159*
relief
   climate relationship 178–81
   land facets **171**
   South Africa 168–9
   submarine canyons 184
residual summit surface 150–1

response times
    catchment–fan systems 17
    river systems 13, *14*, 18–20
    sediment routing systems 22–5, *24*
Rio Cinca, Spain 130–1, 134–40
river systems
    erosion 12–14
    longitudinal profile 12–14
    Pyrenees 129–45
    response times 18–20
    terraces 131–2, *134*, 142
    see also fluvial systems
rock types
    drainage density 173, *174*
    slope analysis 170–2
    South Africa 169

sand–gravel mixtures 117–27
sandstones 132–4, *135*
Saxe catchment 73–6
sedimentary flows *see* turbidity currents
sedimentology 98
sediment routing systems 2
    catchment–fan systems 15–18, *19*
    response times 22–5, *24*
sediments
    debris flows 72–3
    fluxes 124–5, 129
    submarine 186
    suspended 141
    transport 3–4, 18–20, *42*
    see also bedload
semi-arid landscapes 167–81
    see also drylands
shear failure 184, 190
sheetwash erosion 180
short-term events 3
silts 85
siltstones 132, *133*
simulations
    karst landforms 53–60
    see also laboratory study; modelling
single annual flood events 80, 82, 99–101
slip rate 16–17
slopes
    analysis 169–72
    canyons 186–9
    concave 169, 170–2, 176–7, *178*
    continental 183–94
    fans 16–17
    hillslopes 14–15, 33–4, *42*
    palaeoslope 136–7
small flood experiments 118–23
snowmelt flooding 80, 82
solar signals 3, 105–15
South Africa
    Land sat imagery analysis 175–8
    landscape analysis 169–75, 178–81
    landscape and rainfall data 167–9
    pediments and interfluves 167–81
south central Pyrenees
    bedload transport 129–45
    palaeorivers comparison 142–3
Southern Alps, New Zealand 22, 23

Spain
    bedload transport 129–45
    catchment equilibrium *34*
    major storm events 33, *41*
    Rio Cinca 130–1, 134–40
spatial patterns
    anisotropy/isotropy of karst 49–53, 54–60
    bed load transport 117–18
    entrainment 120–3
    equilibrium 37–8, 42
    steady-state landscape 1, 8, *22*, 23, 63, 76, 147
stratigraphic successions 129–45
stream power
    proto-Rio Cinca 134–5, 140, 141
    rule 12–13, *14*
STRM digital elevation data base 148–9, *150*
subarctic nival regime 80, 85
submarine canyons 4, 183–94
submarine sediments exhumation 186
surface runoff
    CHILD model 51–2, 58, 60
    simulation *55*, 58, 60
surfaces *see* erosion surfaces
suspended sediment 141
Switzerland, debris flows 73–6
system variables 31

tectonics
    active environments 1, 7–28
    displacement 8–9
    forcing 21–3
    inversion 145–66
    orogenic scale 9
    perturbation 21–3
    uplift 147, 157–8, 163–4
terrestrial molluscs **90**
timescales
    anisotropic/isotropic dissolution 50–3, 54–60
    entrainment 120, 121–2
    equilibrium 38–40
    tectonic landscapes 2, 7–28
topography
    footwalls 9–12
    planar landforms 147–66
    slope canyons 186–7, *190*
    South Africa 178, *179*
transient landscape *22*, 23
tree pollen 86, **87, 89**
turbidity currents 183–5, 191

UK, Nene Valley catastrophic events 79–104
unimodal gravel experiments 120–1, *123*
uplift 147, 157–8, 163–4
USA, Atlantic slope canyons 186–9

valley fill 109, **111–12**
vegetation 169, 176
velocity, proto-Rio Cinca 137–8

water erosion 179–80

Younger Dryas 91, 97, 99
younger pediments 151–2